Understanding Electric Utilities and De-Regulation

Second Edition

POWER ENGINEERING

Series Editor
H. Lee Willis
KEMA T&D Consulting
Raleigh, North Carolina

Advisory Editor
Muhammad H. Rashid
University of West Florida
Pensacola, Florida

Understanding Electric Utilities and De-Regulation

Second Edition

Lorrin Philipson
H. Lee Willis

CRC Press
Taylor & Francis Group
Boca Raton London New York

CRC Press is an imprint of the
Taylor & Francis Group, an **informa** business

CRC Press
Taylor & Francis Group
6000 Broken Sound Parkway NW, Suite 300
Boca Raton, FL 33487-2742

First issued in paperback 2019

ISBN-13: 978-0-8247-2773-4 (hbk)
ISBN-13: 978-0-367-39204-8 (pbk)
Library of Congress Card Number 2005050630

Library of Congress Cataloging-in-Publication Data

Philipson, Lorrin, 1945-
    Understanding electric utilities and de-regulation / Lorrin Philipson, H. Lee Willis.-- 2nd ed.
        p. cm. -- (Power engineering ; 27)
    Includes bibliographical references and index.
    ISBN 0-8247-2773-8 (alk. paper)
    1. Electric utilities. 2. Electric utilities--Deregulation. 3. Deregulation. I. Title: Understanding electric utilities and deregulation. II. Willis, H. Lee, 1949- III. Title. IV. Series.

HD9685.A2P48 2005
333.793'2--dc22   2005050630

Visit the Taylor & Francis Web site at
http://www.taylorandfrancis.com

and the CRC Press Web site at
http://www.crcpress.com

Series Introduction

Power engineering is the oldest and most traditional of the various areas within electrical engineering, yet no other facet of modern technology continues to undergo a more significant evolution in technology or industry structure. In the 120 years since its first commercial use, electricity has grown to be a cornerstone of our civilization for three reasons. First, it is incredibly flexible in application, capable of making motors turn, lights glow, or boom boxes play hip-hop. Further, it is very controllable in any of these and thousands of other applications. Well-engineered equipment can dole out amounts as small as one microwatt or as great as one billion watts, and control flow at any of these levels to within a few hundredths of a percent. Finally, electric power is quite inexpensive, a fact lost on many people spoiled by the success of an industry that is often taken for granted.

Understanding Electric Utilities and De-regulation – Second Edition presents a broad, non-technical look at the electric power industry, its technology, structure and organization, as it makes its transition from the regulated framework within which it functioned for over a century to the partly de-regulated structure that is its future. De-regulation has driven many changes in the power industry, among them an influx of executives, managers, and skilled professionals from other industries – de-regulated industries – to help drive its success within that new structure. In addition, despite dire predictions of disaster, the industry is in many ways thriving under de-regulation, even as it wrestles with many new challenges. As a result it is adding new engineers, computer scientists, accountants, and business managers, many only recently out of college. Seasoned veterans from other industries and recent graduates alike will find this book a practical, accessible explanation of their new industry, its technologies, operation, history, habits (good and bad), and future.

Like all the books in the Taylor & Francis (formerly Marcel Dekker) Power Engineering Series, *Understanding Electric Utilities and De-regulation – Second Edition* puts modern technology in a context of practical application, useful as a reference book as well as for self-study and advanced classroom use. The Power Engineering Series includes books covering the entire field of power engineering, in all of its specialties and sub-genres, all aimed at providing practicing power engineers with the knowledge and techniques they need to meet the electric industry's challenges in the 21st century.

H. Lee Willis

Preface

Understanding Electric Utilities and De-Regulation – Second Edition is, like the first edition, a non-technical description of the electric power industry, what it does, how it works, its history, and its future, along with an examination of the major issues revolving around its transition from a regulated to a de-regulated industry. Both the electric power industry and the character of its de-regulation have changed rapidly since the first edition was published, in ways not completely anticipated by the industry, or the authors. Some aspects of de-regulation worked as planned. But many did not, and as a result industry structure, government policy, and grid operating rules have evolved in ways not originally foreseen. Other factors, including massive blackouts and new technologies, have also had their impact. This second edition brings these and other aspects of the industry up to date, and discusses some of the shortcomings of early de-regulation and how they were addressed. It also covers three topics new to the second edition, each of great growing importance to the industry: aging electric infrastructures, service reliability, and blackouts.

This book is intended as a reference book and a tutorial guide for those many non-engineering professionals who find themselves part of an industry dominated at every turn by esoteric engineering concepts and technical terminology. It will also serve engineers, economists, and utility managers by providing a non-technical overview of how their particular fields interact with the whole. Electric power has become such a wide landscape that many of the most experienced experts have little opportunity to see how their contributions fit into the big picture, particularly with respect to how all its interconnected facets are evolving under de-regulation. The authors have endeavored to make the entire discussion as understandable as possible, but nonetheless complete.

This book was organized and written with an expectation that most people who use it will not read the book through from beginning to end, that busy professionals want to read selectively about only the matter of immediate interest. Therefore, the authors have included what they believe is a particularly detailed index and a lengthy glossary of terms. Most important, this book is organized into chapters and sections by topical areas (e.g., Retail Sales, Blackouts, Distributed Generation, Fuel Cells) that have been written so that while each fits into the whole, each is also a stand-alone tutorial on its particular topic. This makes the book, if read through from start to finish, somewhat redundant, in that the same issue or consideration might be presented and discussed several times in various places, if it bears on several different areas within the power field.

The first three chapters cover the power industry. Chapter 1 discusses electric utilities and its traditional regulated structure, under which the industry operated for over a century and within which many of the industry's most haloed and institutionalized habits and expectations were forged.

Chapter 2 summarizes the larger issues of de-regulation – its what, why, when, where and how in overview. It is impossible, even in an entire book, to delineate every detail or every concept or every structure that has been proposed, debated, or even actually tried in the power industry. Instead, this chapter sticks to just the major concepts and how they interact with one another, and discusses how and why some issues are the subject of intense debate and concern.

Chapter 3 gives a history of the power industry from three perspectives: electricity and electric power itself – how the use of electric power developed and how that fueled the industry; technical – how the equipment, systems, and technology developed and evolved; and business – how and why people invested in and built an industry that is today a cornerstone of our technology and culture.

Of course, the power industry is built upon a base of fundamental electric engineering concepts and principles and depends wholly on the performance of complicated electrical equipment whose designs have been honed to near perfection over the past 120 years. The next six chapters discuss electric power, power systems, and the various equipments and functions involved. Chapter 4 covers basic electric power and electrical engineering concepts: voltage, current, and power, and the basics of power systems and their operation. Chapter 5 looks at the myriad ways that electric power is used. Chapters 6, 7, and 9 discuss how electric power is manufactured, or generated. Traditional central station generating plants, typically the size of large office buildings and capable of producing power for an entire town, are covered in Chapter 6, which also presents a basic overview of how and why a generator works. Renewable energy – solar, wind, and other natural sources – are discussed in Chapter 7, along with the pros and cons of renewable versus fueled types of generation.

Small "single household size" generators of various types, along with the energy storage systems needed so that they can dependably meet peak demand levels, are covered in Chapter 8. Such distributed resources (DR) were once thought to be the future of the entire industry, but they have succumbed, at least for now, to a combination of disappointing real-world performance coupled with higher than hoped for fuel costs. Regardless, they fill a number of specialized energy needs quite well and are a permanent part of the industry.

Chapter 9 looks at power systems, the amalgamations of thousands, perhaps even hundreds of thousands, of generators, transmission lines, buses, breakers, switches, reclosers, control relays, distribution circuits and other equipment that forms an "electric utility system." The chapter reviews the basic types of equipment used in a power system and the role of each, and summarizes some of the key challenges that owners and operators of large power grids and distribution systems face.

Chapters 10 through 14 examine various aspects of regulation and de-regulation. Chapter 10 focuses on the basic concepts: What is regulation and why was it preferred when the industry was in its infancy and for nearly a century thereafter? What does de-regulation really mean? Why was it suddenly preferred over the traditional regulated approach? How do the various approaches to de-regulation (there are several) differ, and why would a government or a nation pick one over the others?

Chapters 11 - 13 then examine de-regulation as it applies to three distinct levels of the power industry: wholesale generation and transmission, local power distribution, and retail sales. Chapter 11 discusses the main "why" of de-regulation: a desire to create competition at the power generation level, and shows why and how rules and regulations aimed at that goal are immutably tied and constrained by how the transmission grid behaves. Chapter 12 discusses de-regulation from the perspective of the part of the industry that is, functionally, least affected by de-regulation: local distribution is and always will be regulated, but de-regulation has nonetheless had a noticeable impact, among other things forcing it to grow from an ancillary operation within traditional utilities to a business that stands on its own feet. Chapter 14 discusses competition at the retail level, including competitive power vendors, customer choice and the issues that work at the small-purchaser level under de-regulation.

Chapter 15 is new to the second edition, presenting two interconnected issues that are of growing concern to the power industry. The first is aging infrastructures. Electric utility equipment such as wooden poles, power transformers, and high voltage circuit breakers can last for fifty years or more if well cared for, but many utilities own large amounts of equipment at or past that age. Simply put, many of the physical assets in electric utility systems throughout North America and around the world are nearly worn out. This is creating maintenance, reliability, and financial problems that are quite challenging, at a time when the industry would rather focus on making a clean transition to its new de-regulated structure. Older power systems prove to be less inherently reliable, which means they make keeping the customers' lights on even more of a challenge. Reliability of service is discussed in the second half of Chapter 15, including basic concepts and definitions concerning customer service, along with a summary of the key issues utilities must wrestle with in owning and operating a reliable power delivery system.

Chapter 16 discusses blackouts, the rare but widespread interruptions of power typically affecting tens of millions of people. Blackouts and their causes, which are deeply rooted in some of the most complicated electrical engineering phenomena known, are explained through the use of non-technical descriptions and analogies. The chapter also summarizes how blackouts can be prevented, and presents the authors' somewhat pessimistic view: blackouts will probably be with the industry forever, because while they can be prevented by sound technical means, their root cause is political and economic, not technical.

The glossary following Chapter 16 provides basic definitions of industry acronyms and abbreviations, and many concepts and topics which the reader will encounter in using the book or working within the power industry.

We wish to thank our many colleagues and friends who have provided so much assistance and advice on this book, in particular, Drs. Richard Brown, Gerard Cliteur, John Finney, Ralph Masiello, and Damir Novosel at KEMA, Randy Schrieber and Gary Rackliffe at ABB, Mike Engel at Midwest Energy, Jim Bouford at National Grid USA, Jim Sanborn at PG&E, and Terry Henry at OG&E. Their frank and insightful discussions of the industry's critical issues, as well as their comments on the layout and topics covered in this book, are deeply appreciated, as is their continuing dedication to the power industry.

Lorrin Philipson Willis
H. Lee Willis

Contents

Contents

1

The Electric Industry and Its Traditional Regulated Structure

1.1 INTRODUCTION

Electricity is a very effective form of energy. It can be produced by a variety of methods, moved quite efficiently and safely, and fashioned into light, heat, power, or electronic activity with ease. Without it, few of the industrial, technical, or cultural levels achieved by the human race would be possible. Over eighty percent of the people on this planet have some access to the use of electric power on a daily basis. It is provided to each of them by their local electric utility, the company or governmental department that produces and delivers electric power to them. A massive industry and infrastructure has developed worldwide to support the production, transportation, use, and business of electric energy. The electric power industry is, depending on how one measures it, somewhere between the second and the fourth largest industry in the world. And, along with food and water, health, housing, transportation, and communication/computing, one of the core infrastructure areas without which our culture and society could not exist as it does.

Electric utility is a term that is actually difficult to define in today's power industry, for de-regulation has fragmented the traditional industry structure so that very often, no one company or organization has total responsibility for electric power in a region. For this second edition, the authors have decided upon a rather simple set of rules. First, the electric power industry refers to everyone and everything involved in the production, delivery, sales, control, and business side of electric power. Second, electric utility refers to the company or organization that energy consumers in an area think of as the business source of their electric power: the company responsible for the quality of the power they buy and to whom they send payment. In a modern power industry there are many, many companies and players who never interact with the final, retail consumer of power. Many are quite important, in fact vital, to the industry's viability. But as far as "electric utility" is concerned, the term will be used throughout this book for the entity that electric power consumers in any area think of as their supplier of electric power.

This chapter provides an overview of electric utilities: what they do, who they are, and how the industry was structured during its development and prior to de-regulation. Although the industry is now de-regulated, electric utility needs and functions – what has to be accomplished, how that is done, and what equipment and resources are used to do that – are best covered by looking at the traditional vertically-integrated electric utility. Largely a thing of the past, that one-utility-who-is-responsible-for-everything is, nonetheless, the best starting place for developing an understanding of the electric power industry and what is required to manufacture and deliver electric power to consumers. Therefore, Section 1.2 begins the quest to "understand" the electric utility industry by looking at the functions that an electric utility needs to perform and the system of equipment it must own and operate. Section 1.3 discusses the people and organization needed to operate the system and serve the utility's customers.

Sections 1.4 and 1.5 discuss, respectively, the functional and business structures that categorize utilities and their different ownership and operating implications. Government regulatory agencies and commissions, the groups that make the policy and regulations and implement their enforcement, are covered in section 1.6. Section 1.7 briefly discusses the companies that research, develop, and manufacture the equipment utilities use in their power systems and control buildings. Section 1.8 summarizes key points about utilities and the industry.

1.2 ELECTRIC UTILITY FUNCTIONS AND SYSTEMS

Traditionally, the term "electric utility" denoted a vertically-integrated company operating under a monopoly franchise. *Vertically integrated* meant it performed *all* of the functions involved to produce and sell electric power for an area of the country – be that only a small town, or a region consisting of several states. Under regulation, this local electric utility held a government-granted *monopoly franchise*, giving it the exclusive right to provide electric service in its territory – that small town or those states. In return for that lack of competition, the utility had to agree that it would serve all customers in the region, not just those it saw as advantageous to its business, and to limit its prices to a level deemed reasonable, by the government, based on a review of its costs and spending.

Today, in the United States and much of the world, few utilities with this traditional structure exist. The industry is de-regulated, which among other things means that there is no longer a strict monopoly in electric power business and that many of the functions performed by that single vertical utility are fragmented and shared among a number of separate companies. However, the best way to understand the power industry and how it works is to look at this traditional, vertically integrated utility, as it existed until the mid 1990s. Here, a person found in one organization, in correct proportion and with all its gears meshing properly, the entire mechanism needed to produce, deliver, and sell power to home and industry, and to do so as a viable business. The discussion of the traditional, vertically-integrated utility is a sound starting point for this chapter's subsequent look at de-regulation, since it was that utility and its industry that de-regulation sought to change.

Therefore, this section will look at Big State Power and Light Company, a hypothetical vertically-integrated electric utility of the traditional type, serving a customer base of 1,000,000 connected meters (individual homes and businesses) in a service territory that contains a total population of about two million. Big State serves a large city and several nearby towns in a territory of about 10,000 square miles. In every respect, Big State Power and Light is a typical "large utility" as it existed prior to de-regulation.

Four Key Functions

In order to do its job, Big State Power and Light Company has to perform four major functions. It must manufacture, or *generate,* the electric power it will sell. It must *transmit* that power – move it – often over long distances from where it is produced or available to where it is needed. It must *distribute* it by routing it to the thousands, in this case one million, homes and businesses where it will be consumed. And finally, it must *sell* that power, which means it must perform a number of tasks, some large and many small, needed in order to count its sales, bill its customers, handle and resolve questions and complaints about service, in order to run a business supporting all of the aforementioned functions.

Table 1.1 lists these four functions and the types and numbers of power system equipments that are required by an electric utility the size of Big State. Electric utilities are *very capital intensive* businesses – in order to perform their work they need *a lot* of durable machinery and equipment and they spend more capital, proportionate to their revenues, than just about any other type of business. For this reason their identity is often very much tied to their system (the aggregated sum of all the equipment they own) which is depicted in Figure 1.1. But it is people and resources who run a utility system and who make the utility what it is. Table 1.2 lists all of the resources and people Big State would have, circa 1985 (a decade before de-regulation), in order to do its business, and explains a bit about their functions and locations throughout Big State's system.

Table 1.1 The Basic Electric Utility Functions and the Equipment That Might Be Used to Accomplish Them in a Utility Power System Serving a Typical Large American City

Function	Composed of	Number	Description
Generation	Stations	8	The actual manufacture of electric
	Generators	35	power, by converting some other form of energy, be it coal, nuclear fission, falling water, wind, or sunlight, into electricity.
Transmission	Trans. lines	180	Transportation of bulk quantities of
	Switch stations	50	power long distances, as from hydro-electric power plants deep in the mountains to large cities on the coast.
Distribution	Substations	165	Local delivery of power to consumers
	Feeders	850	involves breaking up bulk quantities of
	Service transf.	125,000	power into "household" size amounts, and routing it to homes and businesses.
Retail Sales	Meters	1,000,000	Measuring and billing consumers for the power delivered, and perhaps providing other services such as energy efficiency or power quality automation. Operation of the system and business.
Service to	Customers	1,000,000	
	Peak demand	5,000 MW	
	Revenue (yr)	$2.6 billion	

Chapters 6-9 describe electric power equipment such as generators, transmission lines, and distribution transformers, as well as power systems (the assemblages of them), in considerable detail. A lengthy description will not be given here. But it order to appreciate the structure of these power systems and the constraints that utilities faced, the reader needs to appreciate one critical fact about power engineering: There was and still is a tremendous economy of scale in nearly every aspect of power systems equipment. If a particular type and size of generator is efficient (i.e., produces power at an economical cost), a larger one of the same type will be still more efficient – a giant, "economy size" generator. Similarly, larger, high-voltage transmission lines cost much less per unit and carry power more effectively than low-voltage lines. A large transformer costs less per unit of capacity than a small one. This qualitative rule applies to nearly all types of equipment required to move and control power, at all levels of a utility system.

Thus, Big State wants to own and operate *big* equipment in order to take

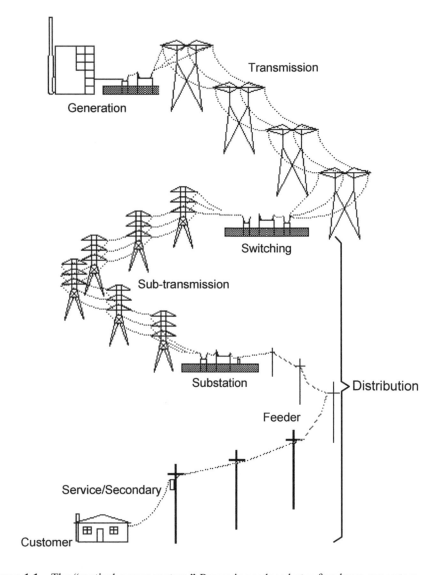

Figure 1.1 The "vertical power system." Power is produced at a few large generators (only one is shown) and moved over a transmission system consisting of dozens, even hundreds of regional power lines (only one path is shown). Once brought to the local community, it is reduced in voltage and shipped to neighborhoods, and to the individual consumer, on a distribution system (only one of thousands of lines and customers is shown). Some utilities perform all the functions shown, others only a portion.

advantage of those economies of scale to reduce its costs. This preference for size is limited by only two factors. First, Big State cannot have "too many eggs in one basket" for both reliability and business risk reasons. Second, ultimately it must deliver its product (electric power) in very small, household-size units. The amount of power that a home or even a large office building uses is miniscule compared to the overall total handled on Big State's system, and is tiny compared to the power produced by even a small generator, or that carried on the smallest power line.

In looking at Figure 1.1, the reader must keep in mind that only one of every type of major equipment unit is shown – one each of all the essential elements in the connected "chain" of electric flow from generator to customer. The drawing shows the progression of power flow from manufacture at a generating plant (top) to consumption by a customer (bottom). Electric utility systems like Big State's consist of a few very large units of equipment used to generate and move power in bulk quantities to cities and towns, and smaller, local equipment that divides and sub-divides the power as it is distributed down streets and to individual homes. In general, equipment like that shown at the top of Figure 1.1 is individually large but of small numbers; that at the bottom, small in physical size and electric capacity, but quite numerous.

Thus, although Big State ultimately delivers power to 1,000,000 separate locations (1 million metered houses and businesses) it owns only 8 sites where generators are operated: one for every 125,000 homes and businesses. Each plant has about four large generators (machines that actually make the power) at it – one for every 37,500 sites Big State serves. It owns several hundred transmission lines and substations, but thousands of distribution feeders (it needs one for every "neighborhood" it serves), and over a hundred thousand service transformers (each serves only a handful of its customers). Finally, it maintains one million service lines (one leading to each customer) and the same number of meters to measure usage by each. Basically, the power splits and re-splits its pathway onto smaller but more numerous pieces of equipment, distributed more widely throughout the system, as it makes its way from generation to customer. In total, Big States' investment in its system, if it had to buy all this equipment at today's prices, is on the order of six *billion* dollars, or about six thousand dollars per connected meter.

Chapters 6-9 provide a more lengthy and comprehensive discussion of electric power systems, the equipment in them. Tables 1.1 and 1.2 and Figure 1.1 will provide a useful basis for the rest of this chapter's discussion.

Generation

Electric power does not exist naturally. It must be produced by machinery that turns some other form of energy into electricity. That other form of energy can be heat from burning coal, oil, natural gas, bio-waste, or nuclear fission;

sunlight; wind or water currents; or falling water. Machinery – a power generator – converts this energy into electric power. Electric generators vary in size from very small (about the size of a clothes washer and capable of providing power to only a single home) to very large (as large as an office building and capable of powering 250,000 homes). Generally, larger and newer ones are more efficient and cost less to run per unit of electricity produced.

A *generating station*, or power plant, includes one or more generators along with all the ancillary equipment needed to provide operation and control the generators. Most utility generating sites have several large generators at them, and thus the entire site produces a very large amount of "bulk power." Most large utilities own or draw power from many generators located at several strategic sites (stations) scattered throughout their service territory. The city of Houston, for example, is at any one time drawing power from about one hundred large generators located at about two-dozen generating plant sites.

Transmission

The bulk power produced at a generating plant is moved to where it is needed over bulk power transmission lines. Each transmission line operates at relatively high voltages, somewhere between 35,000 volts (35 kV) and 750,000 volts (750 kV) depending on design. Higher voltage lines cost more, require bigger towers and equipment and thus have a greater negative esthetic impact, but carry *much* more power: A line with twice the voltage carries four times as much power. Thus, utilities prefer to use high voltage when they can: It costs less and avoids the need for a greater number of lines.

A large electric utility will own many transmission lines – perhaps several thousand if it serves a very populated or multi-state region. These are linked together in a transmission *power grid* or what is often called a *transmission network* that crisscrosses its service territory. This transmission network permits the utility to route power from its many generating locations to the many locations (cities and towns) where it is needed, and to re-route power instantly if a particular generator breaks down or a transmission line has to be withdrawn from service for maintenance, etc.

Large utility power grids usually consist of two distinct levels, or sets of lines, as depicted in Figure 1.1. A set of very high voltage lines, each rated somewhere between 750 kV and 230 kV, simply referred to as transmission, criss-crosses the service territory and connects all the major generating plants together. At certain key locations, two to four of these lines intersect at a *switching substation.* Equipment at this substation controls power flow on the lines and reduces the voltage of the incoming power to route it out onto a number of lower-voltage *sub-transmission* lines.

These sub-transmission lines operate at what are still quite high voltages – anywhere from 161 kV to 35 kV. A typical switching substation might have two

230 kV transmission lines and four 138 kV sub-transmission lines. Power is routed into it on the transmission lines, lowered in voltage, and channeled out on the four sub-transmission lines. In many cases, some of the power coming in on the transmission will pass through the substation: perhaps 500 MW flows in on one of the 230 kV lines, with 150 MW passing on through, and the remaining 350 MW being lowered in voltage and routed onto the four 138 kV lines.

The sub-transmission lines lead to different *distribution substations*, each one routing through about two to six. A large utility will have several hundred or more distribution substations scattered throughout its system, typically two to ten miles apart. At these, power is taken off the sub-transmission line, further lowered in voltage, to primary distribution voltage (somewhere between 2,100 to 25,000 volts) and routed onto the distribution system.

Distribution

Distribution lines, called *feeders,* take power from each substation and route it to every neighborhood. Feeders are most often built on wooden poles, as shown in Figure 1.1 and carry from one to four wires depending on the voltage, type, and amount of power they are designed to carry. At periodic locations along each feeder, service transformers further reduce power to the voltage level actually used in offices and homes, and service lines from that transformer route it into those office buildings, stores, and houses.

Retail Sales

The set of generation-transmission-substation-distribution equipment comprises the electric utilities system. In addition, it owns metering equipment located on every house and building and at every industrial plant that it serves. These meters measure the power consumed. The utility uses that information to prepare, mail, and process bills and payments of its customers. It maintains people and facilities to answer phones for "lights-out," service request, and billing questions and other inquiries, and maintains resources to repair and operate its system. These are its retail sales functions.

1.3 ELECTRIC UTILITY RESOURCES AND ORGANIZATION

The identity of an electric utility is often completely intertwined with its *power system,* the large, geographically distributed system discussed above. Taken as a whole, such power systems are quite expensive, making electric utilities among the most capital intensive of businesses because of their need to buy equipment for this massive yet often very intricate infrastructure. The huge cost, and the fact that the system must literally extend everywhere that the utility does business, means that many people view the power system as the electric utility itself.

But in addition to that power system, electric utilities also employ numerous very skilled workers and specialists without whom that power system could not function. These include power engineers, line workers, power plant operators, regulatory practices attorneys, customer service representatives, equipment troubleshooters, power line surveyors, maintenance technicians, and many others. Traditional regulated electric utilities had about one employee for every 200 customers. Downsizing and productivity improvements due to de-regulation and modern business pressures have improved that ratio to about one employee for every 400 customers. To a great extent these people *are* the electric utility, for it could not function without their skills and dedication.

These people work within each utility's *operating infrastructure,* consisting of control centers, engineering departments, meter reading systems and meter departments, equipment and repair offices, line trucks and repair crews, billing systems and information technology (data processing) systems, and a host of other resources required to keep the power system in prime shape to deliver the power required by its customers.

Although there was a great deal of variation in organization, the traditional vertically integrated electric utility was typically organized into divisions along the lines described below. Although utilities were greatly re-structured and therefore re-organized as the industry de-regulated, as will be discussed in Chapter 2, the functions discussed below are all performed in one manner or another. The descriptions shown in Table 1.2 provide a perspective on what is accomplished in "providing electric power."

1.4 VERTICAL INTEGRATION AND MONOPOLY REGULATION

Electric utilities can be classified in two broad ways. First, they can be categorized by the part of the electric supply chain that they manage. Traditionally, until de-regulation, larger electric utilities performed all of the four functions listed in Table 1.1 – generation, transmission, distribution, and sale of power – as one business. They owned all of the equipment and system infrastructure depicted in Figure 1.1. Some of smaller utilities, however, performed only a portion (generation-transmission, or distribution and sales).

A second way to characterize utilities is by their ownership or business type. Some electric utilities are investor-owned companies, others are government agencies or departments, while still others are "cooperatives" owned by their customers. This will be discussed more in section 1.5. But regardless, every electric utility is a *business* of some kind. It cares about revenue, making profit and/or holding to budgets, and about customer service and satisfaction. Whether a profit-making investor-owned utility, or a government department, it exists to sell electric power, with revenues from sales going toward covering the cost of producing and delivering electric power and services. In the United States alone, that business amounts to over $300 billion dollars annually.

Table 1.2 People and Resources Needed to Operate the Electric Utility

Function	Resources	Number	Description
Generation or Energy Division	Managers	20	This division is responsible for keeping the generation plants up and running, and in good condition. They "operate" them in the sense of keeping them running and fixing problems, but system operations controls how much power each generator provides and when and how much power it makes. New generation plants are usually planned according to type, timing by personnel in Engineering and Planning, but the utility usually contracts with outside construction companies to design and build new power plants it needs
	Engineers/CompSci	80	
	Operators	200	
	Maintenance	500	
	Support & Admin.	100	
	Total Energy Division	900	
System Operations	Managers	20	These people "run the system" – they actually control the power plants and the transmission system as one "system." The core of this group resides at a heavily computerized Operations Center that controls generation plants and key transmission sites through a series of high-speed data communication lines and computers. This is a relatively small but vital function for a utility. It is sometimes put within either the Energy or the Transmission division.
	Engineers/CompSci	50	
	Operators	100	
	Maintenance	50	
	Support & Admin.	30	
	Total System Operation	250	
Transmission Operations	Managers	25	These people keep the transmission system in good repair and condition. They do not control its operation – System Operations does that. There is a very small core located at headquarters but the majority are scattered at service centers, which have offices, garages, parts storage, spare parts, etc. This group also has about 100 field vehicles including special trucks and cranes, considerable special test equipment and tools, and a central warehouse with spare parts, etc.
	Engineers/CompSci	75	
	Skilled O&M Techs	400	
	Less skilled helpers	200	
	Support & Admin.	100	
	Total Trans. Operations	800	

Table 1.2 cont.

Function	Resources	Number	Description
Distribution Operations	Managers	25	These people keep the distribution system in repair and "up and running." Unlike transmission, they both maintain and operate the system. A very small core of management and support is located at the headquarters building. Roughly 100 operators and technicians work at one or two "Operations Centers" monitoring outages and dispatching resources to repair them on a 24/7 basis. The people who actually do repairs and maintenance are distributed at about 12 "service centers" scattered around the system. This group has a large number of vehicles, some of them quite specialized, And many tools, jigs, and other special equipment for repair and maintenance.
	Engineers/CompSci	100	
	Skilled O&M/Techs	525	
	Less skilled helpers	250	
	Support & Admin.	150	
	Total Distr. Ops.	1050	
Engineering & Planning	Managers	25	This group is located mostly at the central office or at one "Engineering" building, although about 250 engineering personnel are distributed at the various T&D district operations service centers, a few at each. They are in charge of planning, designing authorizing, and checking all equipment specifications, changes, and settings for everything in the system.
	Engineers	200	
	Technicians	300	
	Support and Admin	200	
	Total Engineering	725	
Management and Services	Managers	200	This includes "everything" else needed to run the business and includes all the functions needed by any business: information systems, public relations, accounting, legal, human resources, mail room, payroll, billing, marketing, etc.
	Specialized (legal, etc.)	100	
	Skilled/degreed	300	
	Computer sci./IT	200	
	Support and Other	400	
	Total Overhead	1200	
Big State Elec.	**Total employees**	**4,975**	
	Total offices, etc.	40	Including a headquarters building located "downtown," and district/division offices, warehouses, repair facilities, etc.

Vertically Integrated Utilities

Traditionally, most electric utilities were vertically integrated as described earlier. A single company owned all the equipment shown in Figure 1.1 and Table 1.1, employed all the people and did all the work summarized in Table 1.2, and had total responsibility for all aspects of electric service in its exclusive service territory. Figures 1.2 and 1.3 show this with a diagram that will be used and varied through the rest of this chapter to discuss organization, operation, and de-regulation. The four functions – denoted as G, T, D, and S for generation, transmission, distribution, and service – are shown, vertically, inside one oval to indicate that they are all part of one business.

These functions were not just integrated in the sense of ownership but also in terms of business and operation. For example, although the utility might have separate "Transmission" and "Distribution" departments to engineer and operate each of those levels of its system (see Table 1.2), it had only one Accounting department to track and distribute costs of both, and for all other departments, too. Also, the utility integrated all costs and revenues: the costs of running "T" and "D" were lumped together (in fact, they were not really tracked separately) and they and all other costs were covered by the revenues made from only that last link in the functional chain – sales done at the service level.

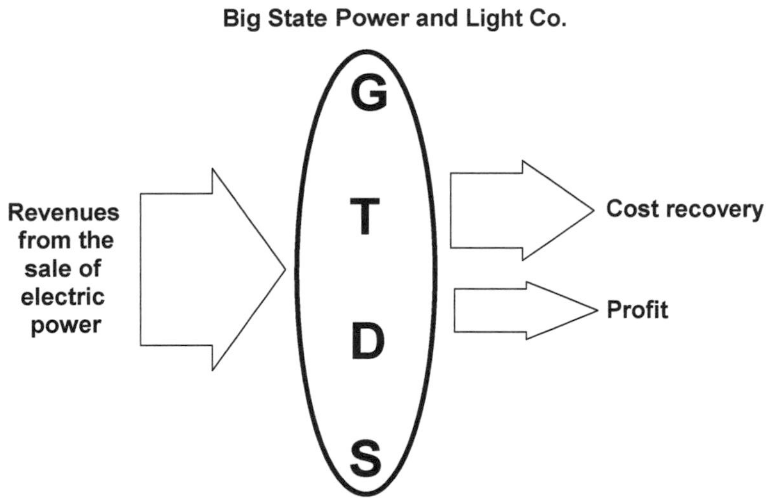

Big State Power and Light Co.

Figure 1.2 The traditional vertically integrated utility – a thing of the past in a de-regulated utility industry – performed four vertical functions: Generation, Transmission, Distribution, and retail Sales of electric power as one owner-operator company (oval).

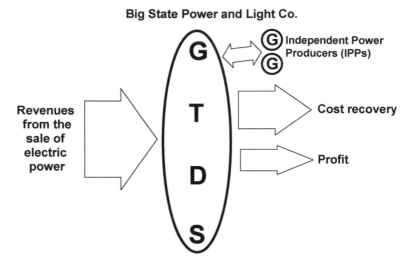

Figure 1.3 After 1978, in the United States, the PURPA Act required that the vertically integrated utility buy power from any independent company that would sell power to it at less cost than it would spend to produce it with its own generators. Some independent power produces (IPPs) were able to sustain a viable business in this manner.

Generation and Transmission versus Local Distribution Utilities

There was one popular alternative to the vertically integrated utility in the traditional, regulated power industry, particularly when there were a number of smaller utilities – municipal power departments of small towns and electric cooperatives – in a region. This is depicted in Figure 1.4. Utilities in small towns and rural areas did not own generating plants or transmission lines (their total demand was often far less than the power produced by even one medium-sized generator). They handled only local distribution of power along with all retail sales and services, and were referred to as local distribution companies (LDCs).

The LDCs bought power from their regional supplier. A large generation and transmission (G&T) company would operate generators and bulk transmission over a region and sell bulk power to these LDCs. In some cases this G&T company was owned by the many small utilities it served. In other cases it was a government agency or authority. Sometimes a utility like Big State would sell a few municipal LDCs in its area power, using its generators and transmission. Regardless, the G&T's transmission lines would route power to one or two incoming switching substations at each LDC, which would then take ownership of the power and route it through its sub-transmission lines to its substations and

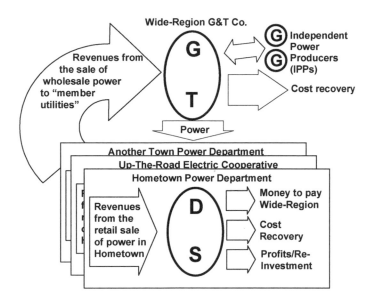

Figure 1.4 An alternative regulated structure involved a number of small utilities (three in this case) jointly forming and owning a generation transmission company that met their needs.

distribution system. Generally, G&Ts owned by the local LDCs they served made no profit, or returned all profits to their share-owning LDC members.

Make Up of the Power Industry Prior to De-regulation

Prior to de-regulation, there were about 250 large vertically integrated utilities in the US, which together served about 85% of all electric demand. There were roughly 4,000 small municipal electric departments and rural electric LDCs – served by about 50 G&Ts of one type of another, which together served the remaining 15% of the nation's electric demand. Table 1.3 outlines the major types of utilities in this traditional industry by function.

Regulators

Who regulated electric utilities? Usually there were several organizations which exercised something between influence and outright control over aspects of the utility's operations. First, the franchise-granting authority had some authority over the utility. This was most often the city government in a municipal area,

Table 1.3   Major Types Electric Utility Companies Prior to De-Regulation

Vertically integrated electric utility. Owns facilities and manages all four functions for producing, delivering, and selling electric power to the end users (Table 1.1, Figure 1.2). *Vertically integrated* means that all of the functions needed are intertwined into one system, company business, with the costs of all covered by one revenue stream from the final product (retail sales). An example is Houston Lighting and Power Company (circa 1995) which provided generation through retail sales in the area in and around Houston, Texas. It owned generation plants sufficient to meet the demand of all its customers, operated transmission and distribution facilities to move the power from those plants to its customers, and performed all retail services required. It, like many large vertically integrated utilities, performed all the functions associated with electric service in its franchise area, and was the only seller of electric power there.

Generation and transmission (G&T) utilities. G&Ts produce electricity, move it and sell it in bulk (wholesale) to local distribution companies (see below). They do not distribute or sell it at the retail level (to individual homeowners and businesses). Often, a G&T is a power supplier for a group of local distribution and municipal utilities (LDCs) in its region and actually owned, on a share basis, by those companies. For example, Tri-States G&T Association, Inc., is the G&T for 44 rural electric and public power districts in Colorado, Nebraska, and Wyoming. Those utilities each have no or only limited generation and transmission, for the most part performing just the distribution and retail sales functions in each of their territories. Tri-States runs generation and transmission over the entire 44-utility region.

Local distribution companies (LDCs). These are local electric utilities that own and operate only a distribution system, which they use to move power produced elsewhere to the local consumers. They also provide retail sales and services. Traditionally, LDCs have distributed and sold electric power to all the customers in their service territory and acted as "the local electric utility." Examples of such companies are the many municipal electric departments in smaller communities, which own no generation equipment, and many rural electric cooperative utilities which own little or no generation. Usually, smaller LDCs group together and form a G&T co-operative (see section 1.5 on business/ownership models) to make and transport bulk power for them.

Independent power producers (IPPs) and non-utility generators (NUGs). In the United States, under the traditional utility industry structure that existed prior to 1996, there were two types of companies, really not electric utilities, which nonetheless owned generators and produced and sold electric power. IPPs and NUGs are both private companies that own generators and produce electric power. Some NUGs, e.g., a large factory that has its own generator, consume the power they produce, owning and running their own generation to avoid the cost of buying power from the local electric utility. Many IPPs sell the power they make to the local utility. Under the Public Utility Regulatory Practices Act of 1978, an electric utility is required by law to buy power from an IPP if that IPP will sell power for less than it would cost the utility to produce it itself.

but perhaps no one at all in a county or the country (a utility doing business under a franchise within a city might extend its lines outside the city limits in order to obtain additional business).

By the mid 1970s most states had a public utility commission (PUC) which had regulatory authority over most utilities in the state. It standardized and applied on a uniform basis regulations and laws pertaining to how utilities would operate, compute prices and bill customers, access public facilities and exercise eminent domain, and resolve disputes with cities, landowners, or customers. Certain policies, rules, and laws affected by the PUC would apply to all electric utilities equally, but typically, pricing and performance were *not* among them. Each individual utility entered into its own negotiations with the state PUC regarding its rates. Such "rate cases" often led to considerable differences in electric rates between neighboring utilities. For example, in the early 1980s, the authors lived east of Pittsburgh, PA, within a block of the service territory boundary between the then West Penn Power Company and Duquesne Light. There was nearly a 35% difference in the price that neighbors across a street paid for power. This was one of the more dramatic local differences, but there were many such situations. Traditionally, PUCs regulated the price (rates) permitted for each utility based on the "local costs" the utilities paid and documented. Two neighboring utilities could have different labor, construction, and operating costs, or one might simply make a better case for higher rates. Beginning in the mid 1990s, PUCs, and changes in business conditions for utilities, forced regulated utilities to "compete" on the basis of business efficiency. Chapter 10 will discuss this in more detail.

Beginning in 1977, the United States government formed the Federal Energy Regulatory Commission (FERC), which regulates all aspects of electric power that involves interstate trade. Utilities that operated wholly within only one state often did not fall under regulation by FERC. But many larger ones that had lines that crossed state boundaries, and many others that operated by exchanging energy at the transmission level across state lines in a power pool arrangement, did. Initially, FERC set rules and policy on bulk power operation and exercised only mild control over utilities – most of their regulation came from state PUCS. Eventually, it ordered de-regulation: without FERC, de-regulation would not have happened in the US.

1.5 ELECTRIC UTILITY BUSINESS FRAMEWORKS

Regardless of whether the electric industry is regulated or de-regulated, a number of different *types* of ownership/business structures can act as electric utilities. Some are governmental departments – many countries and cities provide their own electricity through utilities they own and control. Others are essentially corporations owned by investors. Still others are owned by the customers directly or indirectly. When combined with the categorization by function, discussed early (e.g., vertically integrated utilities, G&Ts, LCDs, etc.), this results in a considerable number of types of utilities, as shown in Table 1.4, which gives overall statistics comparing the typical of electric utilities as discussed here prior to de-regulation.

Investor-Owned Utilities

Investor-owned utilities (IOUs) are companies owned by stockholders. Although technically an "investor-owned company" could be owned by one or a small group of individuals (investors) who do not trade the stock publicly, regulated utilities are required to be owned via publicly traded stock and to have no single investor dominate ownership.

Investor-owned utilities have a business focus similar to profit-motivated companies in other industries, such as ABB, General Motors, FIAT, American Airlines, or Microsoft. They try to make a profit and they are interested in both profit and stock appreciation. However, the business practices, investment, and prices (electric rates) are subject to government regulation, usually on several levels, e.g., federal, regional, and local. On the other hand, as explained earlier, they face no competition within their service territory in the sale of electric power (they do face competing energy companies, such as gas, oil, and propane).

Roughly half the electric power on earth is sold by investor-owned utilities. Generally, IOUs tend to be very large utility companies, exceeded only in size by national utilities, such as *Electricité de France.* For example, of the 4,200 electric retail utilities (sellers of power to consumers) in the United States, only a few more than 200 (5%) are investor-owned, yet they sell more than 85% of the electric power consumed in the United States. Traditionally (prior to de-regulation), IOUs tended to have at least half a million customers and sell thousands of GWh (gigawatt hours) of power per year. Today, due to mergers and growth that in some cases was spurred directly or indirectly by de-regulation, a "small" IOU has a million customers or more. It takes a minimum of four million or more customers and an annual revenue in excess of 10 billion dollars to make it into the top half dozen with respect to business size.

Economy of scale is the major reason that IOUs are large utilities, and that many of the largest IOUs continue to grow through mergers, e.g., in the late 1990s Union Electric and Central Illinois Power became Ameren; National Grid is the union of the former Niagara Mohawk and New England Electric systems, both formed by earlier mergers of utilities. Despite the large managerial hierarchy that size and geographic diversity inevitably create, many IOUs have found that size creates better organizational and business efficiency. For example, a utility will need only one standards department, regardless of its system's size, and an operations department spread over several states can respond to bad storms by sending repair crews from other states to the site of the disaster, increasing service quality and response while keeping costs lower. A study done in the United Kingdom in the 1990s estimated that the fixed cost of running a large electric utility was about $40 million/year regardless of size: mergers generally target to produce this or larger savings.

Municipal Utilities

Within the United States, about 800 communities own and operate their own electric utility system. Cities as large as Los Angeles (with about 1,400,000 connected meters[1]) and as small as Hobgood, North Carolina (less than 400 connected meters) own and operate their own electric system as part of their municipal public works departments. Generally, municipal utilities provide service only within their city limits, and do not have rural systems or extensions outside their jurisdictional boundaries, although there are some exceptions.

Many large municipal utilities, like those in Austin or Los Angeles, are vertical utilities. They own generation and transmission as well as distribution and retail sales resources, so that they have all the facilities they need to produce power and sell it within their municipal boundaries. However, the vast majority of municipal utilities, particularly those in smaller communities, own no generation facilities and have very limited or no transmission, but instead buy power wholesale and resell it within their service territory, through a distribution system they own and operate.

Some municipal utilities are profit-making in practice if not in mandate, with the electric rates set by the city fathers determined so that revenues exceed costs. A few communities obtain a noticeable portion of total municipal revenues through profits obtained from sales of electric power and energy services. However, in other municipalities, rates for some or all customers are subsidized. In particular, a few cities use their electric department and the rates it charges as a tool for economic expansion, offering particularly attractive price and service packages to attract large employers to their area, in order to stimulate a healthy local economy.

Within the United States municipal utilities are the least regulated of all electric utilities. Most are subject to no oversight or to less regulation than investor owned utilities, by state or federal utility regulatory agencies. As a result, the quality of their electric systems and their operating practices and performance vary widely, and municipal utilities represent both the best and the worst performance in the electric power industry. Larger municipal utilities, particularly vertically-integrated ones who were involved in power pools prior

---

[1] Electric utilities count their customers in terms of "connected meters," or points where the transaction of electric power sales takes place. A home that buys power from the local utility is counted as one "connected meter," even though there may be several people living in the home, all of whom use electricity.

to de-regulation, generally adhere to the standards of their neighbors, many of which are IOUs, and they also make good use of the economy of scale their size allows. A few are industry leaders. The city of Colorado Springs has an outstanding distribution system that excels at both efficiency and reliability. The city of Austin (Texas) has for many years been considered a world leader in effective planning of its electric system expansion and utilizes some of the most advanced substation and transmission technologies. However, many smaller electric municipal utilities lag far behind the industry, and are likely to find themselves uncompetitive in the 21st century and their customers somewhat unhappy.

Public Utility Districts (PUDs)

These are essentially "county-owned" municipal utilities. Sometimes a political entity other than a county can also "own" them (an irrigation district, for example). Quite common in the Pacific-Northwest of the United States, they vary from small to medium in size and function. Unlike municipal utilities, most serve rural or suburban areas rather than urban centers. They are often subject to some amount of regulation and oversight at the state level, at least on a voluntary basis.

Electric Membership Cooperatives (EMCs)

Cooperative utilities are owned by their customers, at least in theory. During the period when the electric industry was forming, local farmers and businesses in many rural areas of the United States would pool their resources to build a jointly-owned rural electric cooperative system serving their community. Most cooperatives obtained financial support from the United States government, under the Rural Electrification Act of 1936. If the cooperative met certain requirements, and operated within guidelines established by the United States federal government, the Rural Electric Authority (REA) would provide financing. As a result, electric membership cooperative utilities are sometimes referred to as "REA utilities" (particularly those originally set up by the REA sponsorship). Most people in the power industry make little distinction between an EMC and an REA utility, for the simple reason that in almost all cases they are the same. Almost all electric cooperatives use REA funding, and with only very rare exceptions, all utilities that seek REA funding are electric cooperatives.

Almost all electric cooperatives are quite small compared to IOUs. They are community-sized utilities that own little or no generation but instead buy power from a local G&T and provide electric sales to only a limited area of their state or county. Electric cooperatives constitute over 50% of the utility companies in the United States, but distribute less than 12% of the power consumed. At the time of this writing, six decades since the Rural Electrification Act, these "co-ops" are for the most part stable, complete businesses. Their corporate values and manner of doing business are broadly similar to those of IOUs, in that they must juggle investment and finances in the manner of a private company. Like IOUs, they seldom have their policy and prices defined by political objectives, as do some municipal utilities and public utility districts (PUDs). But unlike all other types of electric utilities, they must adhere to REA rules on how to build and operate their system, and how to price their product.

Many electric cooperatives serve rural and sparsely populated areas: farms, scattered businesses, and very small communities in agricultural areas of the United States. All served only such areas when founded decades ago. A few larger EMCs, such as Cobb EMC, in Georgia, have service territories that caught a good deal of the expansion from a metropolitan area. Cobb EMC is northwest of Atlanta, and that city's growth has caused suburban and urban growth, making about 94% of Cobb's customers residential.

Looking to the future, many electric cooperatives seem unprepared for competition and de-regulation. Often, inflexible governmental rules require REAs to build their electric system to partly outdated, non-optimal standards that do not take advantage of modern technological advances. Lacking the engineering and business clout that accrues from large size, as with IOUs, and enjoying a cultural comfort factor from decades of government financial support, they are not highly aggressive or competitive. There are exceptions, like Cobb EMC, whose rapid urban growth has forced the utility to become a vital organization, and Midwest Energy, a cooperative in Kansas, which has uniquely financed itself and has an aggressive attitude that has made it an industry leader.

National Utilities

A number of nations own and operate their electric utility on a national basis, either as a governmental department or as a single, government-owned company, which, while legally separate, has a symbiotic relationship with the government and is closely constrained by its policies. An example is Electricité de France, which serves France. Such a utility will be organized into regions, and districts within regions, each responsible for local operations and sales, and with some degree of autonomy over local policy and procedure. But, the utility sets major policy, and makes decisions, based on a national perspective.

State-owned utilities most often operate their electric system according to national policy. In developing nations, the national utility is very much involved in economic policy decisions, its *purpose* being both to provide available power to stimulate industrial growth, and to support public infrastructure development. In other countries, the national utility provides more than just electric services. For example, Electricité de France provides research and technological services and has a *cabinet level* position in the nation's culture and economy.

State-owned utilities

In many nations, the federal government owns the electric industry, but has organized it into several separate companies, or owns only part of the utility system. Most often this separation is done by function, with the most frequent national utility being a generation company that produces all electricity for the country, regardless of whether it is distributed throughout. For example, government-owned EGAT produces all the electric power in Thailand and transports it in bulk throughout the country. The power is distributed within the city of Bangkok by the Metropolitan Electric Authority of Bangkok, while the Provincial Electric Authority (PEA) distributes power throughout the rest of the country. All are government-owned, but separately organized and run, each with a different focus and priorities.

Administrations, Authorities, Agencies, and Government Utilities

Collectively, through a number of its agencies and component parts, the U.S. government operates the largest G&T in North America, and, in fact, one of the largest electric utility companies in the world. The U.S. Department of the Interior, the U.S. Army Corps of Engineers, and the Tennessee Valley Authority own dozens of hydro-electric dams throughout the United States. Collectively, they provide a peak output of more than 50,000 MW.

Two of the largest utilities in the United States are the Tennessee Valley Authority (TVA) and the Bonneville Power Administration (BPA). TVA is essentially a large government-owned G&T, created by an act of Congress during the 1930s in order to harvest the considerable hydro-electric potential within the Appalachian Mountains in the eastern United States, and to provide electric power to the many electric cooperatives and municipal utilities throughout that region. It provides electric power for the entire area, and also has responsibility for some resource management, e.g., water and river flow, etc. With about 24,000 MW of generation capability, much of it hydro, but including some modern nuclear, coal, and gas generators, it maintains a large transmission grid, which it uses to move power to cities like Knoxville and Nashville, and to the many small communities and rural electric utilities throughout its region. In general, the name *authority* denotes a government-owned generation and transmission utility. TVA is the largest but there are others, e.g., the Lower Colorado River Authority in central Texas.

The Bonneville Power Administration (BPA) is a part of the United States Department of Energy, which distributes wholesale power in the northwestern United States. It sells power produced by generating plants owned and run by the U.S. Army Corps of Engineers and the U.S. Department of the Interior, which operate 21 power plants in the northwest United States, which cumulatively produce up to 20,000 MW of power. BPA operates an extensive

Table 1.4 Examples of Various Types of Electric Utility Companies - 1994

Type	Name	Functions*	Customers	Peak Demand	Employees
Large IOU	Pacific Gas and Electric	G, T, D, S	4,000,000	15 GW	10,000
Small IOU	Maine Public Service Co	G, T, D, S	33,000	134 MW	150
Large Muni	City of Austin Electric	G, T, D, S	284,000	1,450 MW	1000
Small Muni	Hobgood (North Carolina)	D, S	350	350 kW	3
Large EMC	Midwest Energy	G, T, D, S	44,000	250 MW	175
Small EMC	Patuala EMC	D, S	4,100	14 MW	23
National Utility	Electricité de France	G, T, D, S	19,000,000	64 GW	70,000
State Utility	MEA (Thailand)	T, D, S	1,000,000	1300 MW	8,000
Power Authority	Lower Colorado Riv. Auth.	G, T	82 wholesale	1,900 MW	900
Power Admin.	Western Area Power Adm.	T	637 wholesale	6 GW	900
Power Agency	Florida Muni. Power Ag.	G	19 wholesale	450 MW	100

* Generation, Transmission, Distribution, and retail energy Services.

transmission grid spanning seven states, and delivers this power to over 900 sites, including cities, towns, and rural electric utilities in its region. In general, a power *administration* is a government agency that does not own generation but sells or manages it when produced by other governmental resources. For example, the U.S. Army Corps of Engineers operates nine hydropower plants in the Appalachians producing up to 900 MW of power, which is marketed in the eastern central United States through the Southeastern Power Administration.

Often, a group of municipal utilities will form a Generation and Transmission agency, a *power agency* or *generating district* of their own, so that they can jointly own and operate generation plants to provide power to themselves. This was explained earlier (Figure 1.4). The Florida Municipal Power Agency in Orlando, Florida, is owned by 19 municipal local distribution utilities and operates five generating plants, whose power output is shared by its owner municipalities.

Table 1.4 gives information on a number of typical utilities in the United States prior to de-regulation, along with pertinent facts about their size and operation.

1.6 GOVERNMENT REGULATORY AGENCIES AND COMMISSIONS

Since its inception, the electric utility industry everywhere in the world has been closely regulated by a combination of local and national government agencies. And they will continue to be regulated, for "de-regulation" is more properly termed "re-regulation," because governments are not giving up their regulatory authority over the electric industry. They are simply changing the rules to permit competition in some areas of it. In many ways, the level of regulatory scrutiny will increase. Important regulatory authorities for the electric power industry in the United States include:

Department of Energy (DOE). A major branch (cabinet position) of the U.S. government, the Department of Energy, oversees all federal policy on energy, one of the most important segments of which is electric power. DOE finances significant research into new electric technology, and operates both the Bonneville Power Administration and many other entities dealing with national electrification.

Federal Energy Regulatory Commission (FERC). FERC is a federal agency that regulates interstate trade in electrical energy. Basically, this means the wholesale electricity market – power and transmission sales and service between utilities and between utilities and non-utility generators. An independent agency of the Department of Energy, FERC was established in 1977, and is composed of five commission members appointed by the President of the United States and confirmed by the Senate, all supported by an extensive technical and legal staff. Commissioners, who serve staggered five-year terms, each have an equal vote on all regulatory matters.

Nuclear Regulatory Commission (NRC). This agency oversees the licensing and operation of nuclear power plants, regardless of ownership (IOU or municipal). It approves and constantly watches the operation of all commercial nuclear power plants.

State Public Utility Commissions (PUCs). Every state has established utility commissions that oversee the operation of investor owned utilities within their borders. Called variously the Public Utilities Commission, Public Service Commission, or another similar name, they regulate the rates, planning and spending practices, customer service, and operating policies of the utilities in their jurisdiction. PUCs do not regulate municipal utilities per se, but this is changing in a de-regulated industry. First, state PUCs, by federal mandate (FERC order 888), control all distribution (as opposed to transmission) policy and procedures in their state. Second, due to various reciprocity requirements, this implies that many municipal utilities fall at least partly under explicit or implied PUC regulation. Paradoxically, as a result of de-regulation, more of the distribution of power is being regulated.

Laws and Major Rulings Governing De-regulation

De-regulation is usually a political process driven from the top down by the federal government. Often, it is intertwined with privatization efforts. It usually is based upon a major law defining the changes and setting broad guidelines and regulations established by the appropriate agency of the government, interpreting these policies in detail. Among the more important laws and regulations are:

Public Utility Regulatory Policies Act (PURPA). One of the five bills signed into law on November 8, 1978 as the National Energy Act, PURPA was a broad statute aimed at expanding the use of co-generation and renewable energy resources in the United States. It created a new class of power producers called Qualifying Facilities (QFs), which are basically independent (non-utility-owned) power generators who meet certain stipulations. The PURPA *requires* utilities to buy power from these non-utility generators at each utility's avoided cost – a price equal to the incremental cost that particular utility would incur to produce the power itself (i.e., what the utility saves (avoids spending) by not generating that same amount of power with its own generators).

PURPA left some details of pricing and interpretation to the individual state regulatory commissions, since avoided cost definitions and pricing fell within the venue of state regulators. Interpretations varied, but QFs sold power to utilities in nearly every state. It has been estimated that between 1994 and 2005, electric consumers will pay about $38 billion above utilities' current avoided costs for power purchased under PURPA's requirements.

Energy Policy Act of 1992 (EPAct) was signed into law in the United States on October 24, 1992. This comprehensive bill had over 30 titles (sections) covering more than just electric power, but in that sector it addressed many important issues, including nuclear plant licensing, environmental impacts, energy efficiency and electric vehicle technology applications, and more. By far its most sweeping impact on the electric industry was the fundamental changes it mandated by creating open access for transmission. Section 211 of Title VII provides that any wholesale generator or buyer can petition FERC to mandate wheeling over any electric utility transmission facilities.

While there were limitations placed on open access, and numerous details left for the FERC to work out, the EPAct essentially opened the floodgates of competition in the power industry. The EPAct also authorized FERC to order utilities to provide access to their transmission lines to other utilities, non-utility producers, and other participants in the wholesale electricity market.

FERC Orders 888 and 889. While the EPAct introduced competition, it was up to the FERC to define the way it would be implemented. In March 1995, after two years of study following the EPAct, FERC issued what was called the "mega-NOPR" (Notice of Pending Regulation), which described its intended direction: toward full wholesale competition and open transmission access. During the next year, FERC received over 20,000 pages of commentary on its rulings. In April 24, 1996, it issued order 888, which orders all public utilities with transmission assets to offer non-discriminatory, open access and ancillary services (see Chapter 9) to wholesale sellers and buyers of power. Order 889 specified details of how open access was to be implemented.

These are rules to expand competition in the wholesale electric industry. They require utilities under FERC jurisdiction to file non-discriminatory open access transmission tariffs and other comparable transmission services to eligible third parties. They also allow these utilities to recover stranded costs from departing customers for whom the costs were incurred. In addition, FERC required utilities to develop same-time information systems to make simultaneous transmission information data available to those selling power.

FERC "Millineum" Order (Order 2000). This FERC order encouraged "voluntary" membership by electric utilities in Regional Transmission Organizations (RTOs). FERC did not go so far as ordering utilities to join groups with regard to transmission, but it did strongly recommend that every transmission-owning utility join one group or another, and also recommended that groups of utilities form RTOs.

FERC Standard Market Design (July 2003). This FERC NOPR outlined its thinking on the operation of a regional transmission system market. It proposed a single, flexible pricing policy by uniformly applied, along with the use of locationally-based marginal pricing (rather than, for example, zonal pricing) for congestion-management of situations where there is not enough transmission capability (see Chapter 2). FERC proposed that RTOs have procedures to plan for needed expansion and for investment in new lines, equipment, and facilities in the region. FERCS "SMD" (Standard Market Design, or if one is a proponent, "Successful Market Design) is a mostly workable approach that was immediately supported by some utilities and regional organizations (PJM) but opposed by others (California's Gray Davis, whose lamentable and in the author's opinion misguided stewardship of electric policy for his state led to widespread outages, price volatility and chaos among utilities and consumers, was culminating in bitter opposition). At the time of this writing, SMD, in some evolved form, appears likely to become the common basis for all wholesale electricity trading nationwide in the US.

Energy Bill (1994 and 2005). *Not* passed by Congress in 1994, the proposed Energy bill then contained laws and wording that would have simplified and clarified the situation for many utilities. Most important, it would have "covered" investment in transmission lines needed to support workable regional grids. As approved in 2005, it rescinded PUHCA and brought about various changes in cost recovery that improved the utility landscape, but it failed to make many big transmission-level changes that the industry wanted. First, there is no eminent domain for inter-state transmission lines and it is not clear who regulates many aspects of electric transmission. Second, regulated utilities still cannot always recover investment they must make in "inter-state" lines whose purpose is to support regional grid strength in order to make de-regulated wholesale markets work well. This will be described in Chapters 2 and 12. The 1994 Energy bill would have made such recovery not just legal, but required, and set in motion a process to institutionalize a mechanism whereby utilities could do that. The 2005 bill is less effective there.

1.7 ELECTRICAL EQUIPMENT SUPPLIERS AND DEVELOPERS

Electric utility companies – whether traditionally regulated and vertically integrated, or the Gencos, Transcos, and Discos of de-regulation – do not manufacture the generators, transformers, towers, poles, control systems, and other equipment that they use in their power systems. Nor, for the most part, do they assemble and program the vast computer and data communications systems that control their geographically distributed systems. Instead, they buy these types of equipment, and services associated with them, from *electric equipment manufacturers*. Among other things, the equipment manufacturers perform the vast majority of research and development of new electrical technologies. They sponsor a great deal of the international industrial activities needed to put uniform definitions and criteria in place throughout the world, and determine workable and safe standards for electric operations. The equipment manufacturing element of the power industry has always been de-regulated and is intensely competitive.

The manufacturing and service segment of the power industry includes several global giants, companies like ABB, General Electric, Mitsubishi, Areva, and Siemens, most of which can trace their heritage to the dawn of the electric power era (see Chapter 3). These very large "full service" electrical suppliers consist of dozens of divisions, each focusing on the manufacture of specific product ranges, including generating plants, transformers, high voltage lines, circuit breakers, low voltage power lines, phase shifters, capacitors, relays, meters, substations, control software, metering systems, cable and transmission towers, and all the other equipment required for electric systems.

Typically, these global companies have hundreds of manufacturing plants scattered all over the world, so that they are close to the markets, and because equipment needs and standards are often best addressed locally. For example, since electric metering needs in Africa are somewhat different from those in North America, many companies manufacture meters on both continents aimed at their respective markets.

In addition, there are hundreds of "niche" suppliers and manufacturers, small companies that manufacture and supply just one, or only a few, specialized products or product lines, often restricting their sales to or focusing on marketing exclusively in one country or continent. Examples are Howard Transformers, which produces only certain types of transformers needed by utilities, and Milsoft, a company that produces software for the design and operation of distribution systems. Usually, these companies have two things in common. They are small, usually with less than a thousand employees, and they are recent start-ups, founded on the basis of a better idea or new invention.

1.8 A COMPLEX INDUSTRY UNDERGOING MAJOR CHANGES

The electric utility industry is a complicated interaction of electric utilities, government, equipment suppliers, and researchers. More than a century old, the electric utility industry currently provides power to nearly three-quarters of the world's population and is expanding daily. Electricity is perhaps the single most useful energy form applied by mankind, infinitely controllable, and applicable to everything from transportation to entertainment.

Worldwide, the electric industry is undergoing major changes as it shifts to a "de-regulated" structure. Hopefully, this transition will improve both how it operates, and its economic and financial benefits to mankind. But it will little alter the fundamentals of what electricity is, what it can do, how it does it, and the ways mankind uses it. Chapter 2 will discuss de-regulated structures and their rationale.

2

The Electric Industry Under De-Regulation – An Overview

2.1 INTRODUCTION

This chapter provides an overview of the most common type of operational and regulatory structure used for de-regulation, particularly as it applies in the United States, and discusses in somewhat simplified terms why, what, and how de-regulation was implemented. Later chapters in this book will delve into more details; here, only the big picture and the most salient aspects are covered, often without the many ifs, ands, or buts or variations in hue and texture that complicate de-regulation in some instances.

De-regulation did not occur everywhere at the same. Timing varied worldwide by well over a decade. Some countries and some states in the US wanted to move forward quickly. Others waited until they had a clear indication of what worked well and what didn't, and could see that there really were benefits to electric de-regulation. Neither did de-regulation occur instantly where it was implemented. Everywhere, it was done in stages. And in fact it is far from complete at the time of this writing. In many parts of the world, particularly throughout the US, all stages of electric utility de-regulation have not been implemented – although it is questionable if they ever will be.

Globally several forces pushed governments to de-regulate the power industry in their countries. For many second- and third-world nations, part of the motivation was to garner foreign investment and cash flow through privatization of state-owned utilities, as will be discussed in Chapter 10. But in the US, UK, and other first-world countries, the goal was to create competition at the top, and at the bottom, of the traditional industry "vertical integration" chain discussed in Chapter 1 and shown again in Figure 2.1. There, the four functions – denoted as G, T, D, and S for generation, transmission, distribution, and service – are shown vertically inside one oval to indicate that they are all part of one business. That business had traditionally been a regulated monopoly. Monopoly operation had been good for the electric industry in its infancy, no one disputed that (see Chapter 10), but after nearly a century, in a mature industry, it was perceived as stifling cost reduction and service innovation. Competition would benefit society and the consumer as a whole, proponents maintained.

Regardless, de-regulation proceeded, at first slowly and in only a few nations, then somewhat quickly, spreading around the world. Section 2.1 will discuss the original motives and concept of de-regulation, and Section 2.2, how it evolved and what it became, particularly in the US. Section 2.3 to 2.5 will discuss de-regulation as it affected the retail, generation, and transmission levels, respectively. Section 2.6 will explore the difference between the real electrical flows in a power system, and the "fictitious" flow patterns used in business dealings. How utilities and business re-organized to accommodate the new industry structure is discussed in Section 2.7. Section 2.8 concludes with a summary of what still needs improvement if de-regulation is to work well.

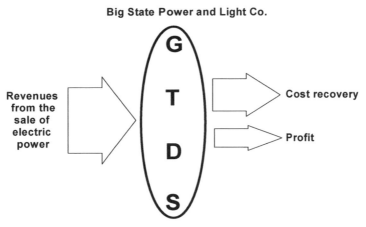

Figure 2.1 The traditional vertically-integrated, regulated electric utility as described in Chapter 1. It performed generation, transmission, distribution, and customer sales of electric power as one owner-operator-seller (oval), under prices and following practices laid out by government regulators. Although its regulated monopoly approach was good for industry during its development and maturation, policy makers decided it did not best fit modern needs as they looked to the 21st century.

2.2 DE-REGULATION: CONCEPT AND EVOLUTION

The Original Double-Ended Competitive Concept

Figure 2.2 depicts the basic concept that drove de-regulation, showing a version of the traditional industry structure as de-regulation proponents saw it at the beginning. Comparison to Figure 2.1 will show the effect it would have had on the industry. Here, a number of unregulated competitive generation companies (Gencos – the circled Gs at the top) compete to sell power. Their prices and practices are not regulated any more than those of companies in other "unregulated" industries (all businesses are regulated somewhat, in accounting practices, employee rights, safety, environment impacts, etc., by government bodies like the Securities and Exchange Commission, OSHA, EPA).

At the other end of the G-T-D-C vertical chain, the retail level, there would also be competition among "electric service companies" (ESCos) who vied for retail sales much like competitive long-distance telephone companies vie for consumers' business. Regulated utilities would still exist, as the owner-operators

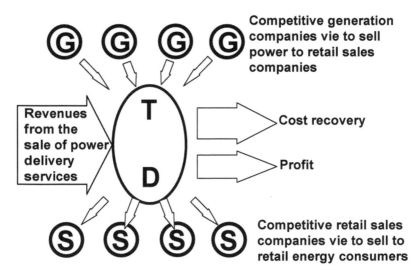

Figure 2.2 The basic idea that drove most de-regulation in first-world countries was to create competition at the generation and retail levels. It was hoped that this would lower the cost of power, and provide a wider range of services to consumers. Transmission and distribution would continue to be a local monopoly run by a regulated utility whose job would be to maintain an efficient electrical delivery system, billing those who use it.

of the various electric grids in each region: no one saw the need for more than one transmission and distribution system in any locality. Thus, in every city, town, or rural area the T&D system ("set of electric highways and roads") would be operated as a regulated monopoly franchise by what had once been the vertically integrated utility there, now devoid of its generation assets, and without any mandate or responsibility to sell power to end consumers.

The regulated utility running the local T&D system would earn money to cover its costs and provide a reasonable profit margin by billing generators and/or retail sales companies for their use of the transmission and distribution system. T&D companies would still be subject to the same type of regulation as vertical utilities traditionally had had. The prices they charged would have to be approved by the government. They would have to follow certain business practices deemed appropriate by the government. Their books would be subject to review by the government. They could show no favoritism toward or against any particular generation or retail company in either pricing or performance of delivery services.

The system would work like this. Competitive ESCos would vie for the attention and business of retail energy consumers (businesses, homeowners). They would sign up households and businesses to be their customers, and determine how much power they would need. Each of the ESCos would then shop around among the different generation companies, which were all vying for their business. Demand and supply ("market forces") would determine price. Competition at the generation and at the retail level would reward innovation, better customer service, and cost reduction with larger market share and higher profit. Everyone would benefit.

That was the basic concept of de-regulation. Chapter 10 will discuss it in much greater detail along with variations on its themes and explain why and how it evolved into de-regulation as actually implemented. Regardless of his and her own opinions or feelings about the matter, the reader can no doubt see that de-regulation had an appeal. Competition worked well in just about every other human endeavor – why not electric generation?

At the generation level, proponents of de-regulation expected it to drive down costs. A considerable number of studies predicted reductions of up to 25%. Benefits of retail level de-regulation always seemed a bit less well defined, but de-regulation of that level was no less staunchly pushed as a "potentially good thing" by early adherents of de-regulation. Arguments that retail de-regulation would drive down price were augmented by claims that it would also bring forth a wide range of extra services and innovative options for consumers (because competing companies would have to be creative in order to gain competition advantages). Anyway, giving consumers a choice about which company sold them power could not be a bad thing, de-regulation proponents argued.

De-Regulation: What Emerged

A good deal of this book discusses "what went wrong" with de-regulation, or more properly, "what went different" because de-regulation did not fail as much as evolve into something slightly different than originally envisioned. The discussion here will not dwell on many of those details or the numerous variations and subtle differences in goals and structures tried in various countries. In a few places around the world, something *very close* to the structure shown in Figure 2.2 was implemented and has been made to work. However, Figure 2.3 shows what is perhaps the mainstream approach, that toward which most of the US is evolving. Again, it is important to realize that goals, policy, and structures vary greatly from state to state and even more widely outside the US. That said, Figure 2.3 depicts, as much as any one drawing can, the current structure of the electric industry operation throughout most of United States, and in many other countries, as it would affect Big State Power and Light.

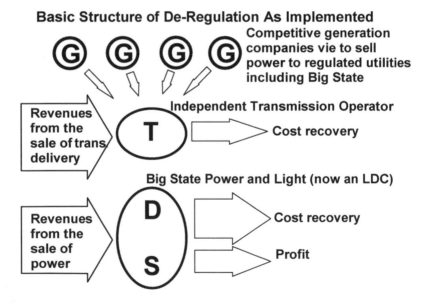

Figure 2.3 As implemented, regulated utilities still do business as local delivery companies, in charge of distributing and selling power. The generation level is unregulated companies doing business in a very competitive marketplace for power. Transmission operation has been taken away from utilities and is operated regionally by an independent grid operator to assure the power grid is used "fairly" by all concerned.

Here, the generation level is de-regulated, very much as originally conceived. The manufacture of electric power for public consumption is no longer the exclusive domain of electric utilities. Anyone who has the money and will to own and operate a generating plant can buy or build one, run it, and attempt to compete, on the basis of price, availability, flexibility of business terms, or whatever other advantage they can tout, against everyone else who owns and operates generators and who is also selling power.

The biggest difference between the original plan for de-regulation (Figure 2.2) and the scheme shown in Figure 2.3 is that retail competition has not developed. As a result, Big State is still a regulated utility that is doing business with the end consumers of electricity in a way broadly similar to how it has always done business with them, rather than as a "wires company" (T&D company). Under the original scheme (Figure 2.2) Big State would have no longer done business with those end consumers. Consumers would have bought power from the ESCos, and Big State would have been only a T&D company, its customers being the ESCos and generation companies, its revenues earned from transporting power from their plants to their customers. While this de-regulated picture represents a major change for Big State, it is considerably less radical than Figure 2.2 would have been: Big State still deals with the end users (as opposed to competitive generators and ESCos as it would have in Figure 2.2); it still sells power as its product (rather than transmission and distribution services); and it still collects all its revenue from the sales at the bottom of its system (customer level).

Big State no longer has generation or transmission (more on that later), but it is still the sole seller of electricity in the region, and it still earns all its revenues from the sale of power to end users of electricity: that hasn't changed. Perhaps most important, from the standpoint of energy consumers: They still have one and only one "local electric company" in their area. Basically, Big State is a very big version of the many small municipal and co-operative local distribution companies (LDCs) discussed in Section 1.2 (Figure 1.4). And for those many small LDCs that existed prior to de-regulation, not much in the industry has changed except a few rules about how they have to buy power now.

Figure 2.4 gives a wider view of the industry structure, a view that will be used to explore how the industry works through the rest of this chapter. Three former large vertical utilities are shown: Western, Big State, and Eastern. These three neighboring utilities are now very large LDCs. Each has disaggregated its generation – split its generation division off as a separate *unregulated* company. How and why disaggregation is done will be discussed later in this chapter.

The important point is that the new generation companies, and the LDCs, are completely independent of one another. Big State LDC is allowed no more ties to Big State generation than it has to any other unregulated company. Here, to make illustration of several points about how de-regulation works, each of the

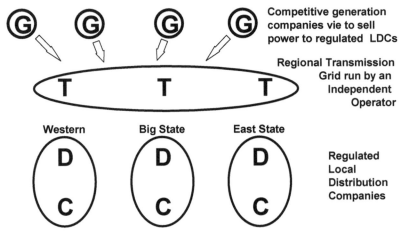

Figure 2.4 Bigger picture of de-regulated electric industry structure shows three former vertical utilities, now all large LDCs. Their former generation divisions, now competitive generation companies, and a fourth new generation company vie for business selling power to the LDCs. The transmission grids of the three former utilities have been tied together in a regional grid, run by a "neutral third party" – an independent system operator.

three generation companies spun off from the utilities have the same name as its former utility – Big State's former generation division is called Big State generation, etc. In the real world, many of these companies rename themselves to put a bit of distance between them and their former regulated owners.

A final and very important element of de-regulation shown in Figure 2.4: the transmission systems of the three utilities have been combined into a regional grid, and that is not operated by them, but by an independent organization – an Independent System Operator. This is the critical aspect of de-regulatory structure and will be the subject of much of the rest of this chapter's discussion.

How Things Work: An Overview

In this industry structure, Big State, now an LDC, buys power from a series of generating companies competing to sell it power. One of those shown is composed of some of the generating units Big State used to own and is run by the people it used to employ in its generation division. Others are former generation divisions of neighboring utility companies. Still others might be independent, never-were-part-of-a-utility companies. More generally, Big State or any former vertical utility might have broken up its generation assets when it broke them off, i.e., there might be more than one "Big State Generation." It could have (and some utilities in some countries almost did) set up nearly every separate generation plant as a separate company. Regardless, in Figure 2.4 the LDCs have four generation companies from which they can buy power.

Big State no longer *operates or controls* its former transmission system. In some states or some countries, it might still own and maintain those transmission lines, being paid somehow (methods differ and are not critical to understanding the major issues here) to maintain those lines and keep them in good working order. In other states, Big State was either forced to, or decided to, sell them or break them off into a separate company, so that it no longer has anything to do with them. For example, in Michigan, Consumers Power, a former vertical utility serving much of that state, sold its transmission to a newly formed company: Michigan Electric Transmission Company (METC).

Regardless, concerning the *operation* of these transmission lines – control of how and where power flows over them – Big State LDC has no authority. A new entity, an Independent System Operator (ISO), has authority to decide how that transmission grid is used. This development is a bit of a surprise given the original picture in Figure 2.2, because that vision had T and D coupled together. This chapter will discuss why this had to be done, and how it was done, and will show why an independent transmission grid is the heart of de-regulation.

The next three sections discuss the new de-regulated structure using versions of Figure 2.4. In order to best explain the hows and whys, the authors will tackle the various elements of this structure in the following order: the retail and distribution levels, the generation level, and then the transmission level.

2.3 NO COMPETITION AT THE RETAIL LEVEL

The original concept of de-regulation (Figure 2.2) envisioned by many governments around the world included competition at both the generation and the retail levels. But in the US, and in many other countries, while competition at the generation level is nearly universal, retail competition is rather rare. Certainly there are places where the retail level has been de-regulated, and the system made to work adequately well (the UK). But retail competition has not developed in most of the US, and in many other countries, and the authors believe it is unlikely to anytime soon.

One very important reason is purely due to political jurisdictions. In the US, the federal government, through its FERC (Federal Energy Regulatory Commission), had the authority to order de-regulation at the transmission level (and therefore at the generation level), because so much of the transmission falls into the category of "interstate trade" over which the federal government has jurisdiction and regulatory authority. The most important aspect here was probably not as much the "federal clout" that FERC had, as much as the fact that a federal initiative drove a "national uniformity" in de-regulation approach and initiative at the transmission level.

By contrast, de-regulation efforts aimed at the retail sales levels were always driven at a state level – because distribution and retail sales are always local and therefore under state, not federal regulation.[1] Any push for de-regulation at the state level consisted of fragmented initiatives. In the US, different states varied too much in their regulatory perspectives and policies to have any commonality on retail electric de-regulation. Some states saw merit in retail competition (California, Maine) while others did not (North Carolina, Nebraska). That lack of commonality in direction slowed even those that wanted to move ahead. And those states that went slowly, as well as those that were undecided, were taken aback by the energy supply and rolling-blackout problems seen in California in 2000. Whether due to de-regulation or not (the larger policy picture on energy and de-regulation there certainly contributed to them), the apparent failure of energy policy in California led many states that were planning to move toward de-regulation at the retail level to suspend those plans indefinitely. In North Carolina, where the authors live, the reconsideration because of California's problems killed any will to move forward at the state level.

Another subtle distinction between transmission and retail level is that states do not have complete regulatory authority over all the distribution and retail level electric sales, as FERC has authority over all transmission regulation nationwide.[2] State public utility commissions and laws did not apply to some municipal and REA utilities, or did not apply as completely or clearly. Thus, states could order retail competition over only parts of the retail electric industry in their states. This could create messy situations and confusion on the part of consumers (i.e., voters). Many politicians did not want to do that.

Beyond those regulatory issues, problems in creating the infrastructure needed for retail competition brought to the attention of regulators by utilities, caused many states to postpone ordering competition at the retail level. As stated earlier, retail competition can be made to work. But the systems, processes, new training, and equipment needed for operating tracking, pricing, metering, and billing of multi-retailer electric service delivered through one

---

[1] In fact, FERC defined "distribution" as the portions of electric utility systems over which it did not have jurisdiction. Chapter 13, Table 13.1, lists 7 differences between transmission and distribution that were given in FERC's "Mega-NOPR" 888. These rules define areas and portions of a system that FERC considered were covered by its authority. Generally transmission lines and facilities meet these criteria; distribution facilities do not.

[2] In truth, FERC's authority over some federal G&Ts, like TVA and BPA, is limited at best, but all of these are just various departments of the same level of government.

local, regulated, T&D system proved to have a long lead time. Worse, there were some "messy" elements: complicated, potentially costly, and in need of technological developments for which none of the apparent options looked particularly less risky or more workable than the others. And while de-regulation at the transmission level affects only the largest 300 or so electric utilities in the US (all multi-hundred-million dollar companies, most billion dollar companies), retail competition would impact the thousands of small LDCs nationwide, many of which could not afford to make rapid changes. Thus, retail-level de-regulation was a concern for politicians from both the infrastructure and "messiness" standpoints. And there was, and remains, doubt as to whether the added cost of these systems will ever be offset by the advantages retail de-regulation would create (this concern is based in part on the benefits and costs that have been seen in areas where it has been implemented).

But in the authors' opinion, the major reason that retail competition did not evolve is that the vast majority of consumers really didn't care one way or another. The majority of electric energy consumers care about little more than the reliability and the price of their electric service. Retail competition could really do little to improve either. Reliability of electric service is entirely a function of the quality of distribution system design and operation, which would remain a regulated monopoly function under any and all de-regulation schemes.

As to price, competition at the generation level was expected to, did, and will continue to drive down the price that consumers pay for power, in places and at times by significant amounts. But competition at the retail level could not show it would provide such improvement. The various competitive retail companies would all be buying power wholesale from the same set of generator companies, through a publicly-transparent and "fair" pricing system. It would be nearly impossible for any one to gain a significant advantage over the others in terms of the price it must pay and therefore what it could pass on to its customers. Further, the cost of retail sales is such a small portion of the total cost of power that even major differences in efficiency and pricing among these competing retail companies would make only insignificant differences in the prices they could offer.

There are exceptions. A lot of the push for de-regulation came from very large industrial and commercial users of electricity (refineries, air liquefaction companies, etc.) that wanted to be able to shop around for bargains in power. In many states, under regulation as shown in Figure 2.4, they are allowed to – although they must do so at the wholesale level, basically installing and operating the same systems that LDCs use to access energy trading marketplaces (see below) and operating their power purchasing as if they are a small, one-customer utility themselves. In a few states (e.g., Michigan) limited retail competition is permitted under a scheme something like a melding of Figures 2.3 and 2.4, which permits businesses to buy competitively: LDCs

(Figure 2.3) still serve the majority of consumers, but larger commercial and industrial consumers are free to shop around for power at the wholesale level as the LDCs can, "buying directly from the wholesaler." In a few instances where this has been permitted, significant numbers of businesses have shifted to suppliers other than the local distribution company. But in the authors' opinion, this is usually due to a situation where the LDC cannot compete "fairly" because it has grandfathered costs or regulatory limitations on the pricing it is permitted to offer.

Finally, although there was initially a lot of "buzz" about how retail de-regulation would compel competing service companies to offer "advanced services" such as monitoring of household appliances, energy conservation, packaged multi-utility (phone, electric, cable) offerings, and other advantages, the fact is that where this was tried there was little market response: most people saw little benefit in having to deal with deciding about their electric service provider.

But regardless of the reasons, in most of the US, something like Figure 2.4 rather than something like Figure 2.2 is at present and for the foreseeable future the norm within the United States, and an overall guiding structure of the industry in many other parts of the world: retail sales are done by monopoly franchise local distribution companies who work in an industry whose wholesale level has been de-regulated.

2.4 COMPETITION AT THE WHOLESALE GENERATION LEVEL

Intense competition at the generation level, run under rather tight rules (laws) to assure it is "fair," and with a level of government oversight comparable to what is applied in industries such as banking, is a nearly universal aspect of electric industry de-regulation, wherever it has been applied. The market structure that the government sets up at the wholesale level (how power is bought and sold), and the rules how buyers and sellers do business, vary greatly throughout the world, and even considerably within the US. Chapter 11 explores different possible wholesale level structures and their variations and rationale in greater detail than this summary. In some places there is one central authority that decides what it will buy for everyone concerned. It is the central buying authority for all the LDCs – each tell it what they will need and it orders electricity for all. But in the US and the trend in most other places is toward a system that requires LDCs and may permit large industrial users of power to buy power as they want through one or more "market mechanisms."

Under de-regulation, electricity is a commodity, bought, sold, and traded in much the way that commodities like wheat, coffee, and soybeans are bought and sold. Of course, there are some differences with respect to electricity as compared to those other commodities that greatly shape how it can be traded. Electricity cannot be stored, and thus not manufactured before the time of its use. Buyer and seller can agree and execute a contract ahead of time, but they must actually transact the business via an instant transfer of ownership when the power is simultaneously manufactured and consumed. Further, electrical properties of power, including both quantity and quality, are reduced or changed slightly by its transfer from one location to another: those changes have to be accounted for and the buyer and seller must agree on who will pay for them.

An important point about a commodity like power is that it is *fungible*: bought and sold on the basis of quantity, not quality. A kilowatt hour produced by Big State generation is no different or no more valuable than one produced by East State generation. As stated earlier the quality of power is almost completely determined by the way the distribution line over which it is delivered is operated. At the wholesale level, power is indeed fungible and is trading as such: a kilowatt hour is a kilowatt hour.

De-Regulated Market Mechanisms at the Wholesale Generation Level

Under a scheme like that shown in Figure 2.4, how do LDCs or large industrial plants who might be permitted to buy like those utilities contact the generation companies who want to sell them power? There are two mechanisms, along with a third option in some situations.

The Power Exchange (PX)

Very often, the de-regulated industry structure set up by the government in a region includes a Power Exchange, which works very much like a "a stock exchange for power." Most people who use it contact it remotely via telephone or computer. Buyers post offers to sell ("I have 100 MW at $28/MWhr" or "I need 85 MW and will pay $26 MW"). As demand and supply vary, the PX maintains a posted trading price, just as a stock exchange maintains a current stock price for any issue.

Buyers and sellers come to the exchange and post their offers. They "make" a deal as persons buying stock do, by accepting the current posted price, and selling and buying through the exchange. They do not know the identity of the person who they bought the power from or sold it to, just as someone buying stock in a company does not know who held it prior to their purchase through the exchange. A seller does not even know if what he sold went to a single buyer: the 100 MW he sold might have been split by three buyers. Similarly, a buyer has no idea if he bought all of one seller's lot, or only part, or from several. The fungible commodity of all sellers is mixed together without identities in the PX.

There are some important differences in the details between a power exchange and a stock exchange. A stock or commodities exchange has to list and track many stocks; a power exchange tracks only one "stock" – power. On the other hand, the PX has the very thorny element of time involved. A seller cannot offer "100 MW at $28/MWhr" – he has to specify a specific time he can produce that as "100 MW at $28/MWh for the hours from 12:00:01 AM tomorrow to 12:00:00 PM" or "I want to buy 250 MW from 4 PM to 6 PM tomorrow." Buyers and sellers each post to a certain amount of power at a certain time period.

Some buyers will want power to be available, but may not know in advance just how much they will need. In most PXs, buyers and sellers can therefore do deals that distinguish between capacity and energy. A seller might offer capacity (the *potential* to produce power – the seller will have his generator up and ready), along with a price for power if called upon to produce it ("I will stand by with 100 MW available at a moment's notice, for $15/MW, from 8:00:01 AM to 9:00:00 AM tomorrow. If you actually use any of that power, there will be an additional charge of $15/MWh for all that you use.")

Finally, the PX or an allied organization handles the transactions, taking the money from the buyers and transferring it to the sellers (in some areas, keeping a very small fee, in some cases, to cover its costs). It monitors performance (did the buyer actually produce the power? Did the seller use that power?).

Power exchanges differ from area to area within the US and around the world in many of the details of how they work and how one does business through them. First, although all operate with a posted price exchange mechanism, they differ considerably in how far ahead they let buyers and sellers do business – some list power and execute sell and buy orders for only a few days or perhaps just a week ahead. Others may permit people to contract to sell or buy power weeks or months or even a year ahead.

Second, in some regions, de-regulation laws do not permit any competition for the PX: *all* sales of power in the region go through the PX, period. But in other areas, there might be other, competing PXs. Some areas permit buyers and sellers to meet and separately execute agreements, what are called bi-lateral trades (see below). Others do not permit any contracts directly between a buyer and a seller.

Finally, all the tiny structural details of how the PX does business differ from one to another: one PX might list and sell power on an hour by hour basis, another in 15-minute increments. One PX might limit power sales to only 10 MW increments, another 5 MW. One PX might permit a buyer to post the minimum price he will take and wait; another might not offer any option except to "look and take it, or leave it."

Bi-Lateral Trades

A bi-lateral trade means doing business directly, rather than through the PX. The buyer and seller meet (it could be by phone or over the internet, or in person) to negotiate and execute a one-on-one contract for power. Unlike business done through the PX, the buyer knows the identity of seller and vice versa. Usually, both parties still consider the power fungible: having no distinct quality aspect regarding to other options for the same quantity. But a bi-lateral trade permits them to vary terms, conditions, or businesses arrangements from those they get through the PX, and they may see this as advantageous. Bi-lateral trades, where permitted, tend to have a different content than transactions done through the Power Exchange. Generally bi-lateral contracts cover much longer periods than sales through a PX will permit, and they may include far more power than a typical PX transaction. Further they may have both more commitments and lower prices than the parties believe they would see through the PX.

For example, in Figure 2.4, Western LDC and Big State generation might make a bi-lateral agreement covering all 8,760 hours in the next year, a longer time than arrangement through a PX would normally cover (some Power Exchanges have only permitted week-ahead orders). In addition, their bi-lateral agreement might call for a complicated schedule of demand – variations from hour to hour in the contracted amount of power, to follow the hourly fluctuations in demand that Western expects to see in consumer demand on its system. The contract would allow for some flexibility (Western does not know precisely what its demand will be) and would be quite detailed in all specifics on how that would be handled, perhaps having an estimated minimum and maximum amount of power to be purchased for each hour, special payment schedules or options, and other intricacies agreed to by the two parties.

Yet despite all this, the price might be lower than by either party expects if going through the PX, yet that might be preferable to Big State generation, as well as the buyer. With this contract, Big State generation has near certainty about its operational schedule and expected sales throughout the next year. It can schedule maintenance, purchases of fuel, work schedules, and other resource needs to optimize its costs against that schedule, perhaps lowering its costs more than the expected loss of revenue from the lower prices it agreed to take for this power compared to what it might be able to get through the PX.

In some regulatory structures, bi-lateral trades are not permitted. In others, they are, but must be fully disclosed (amount, price, and all conditions made public); in others the PX must be informed of all details, and the trade must be announced publicly but no details given (including price). Usually, though, the PX has to be informed of all power agreements and it or the ISO (Independent System Operator) has some ability to veto ones that are unacceptable for operating reasons that will be discussed later.

Middlemen

For centuries commodities like wheat and coffee have been traded by middlemen, who buy and re-sell, and who speculate on, commodity trades. In a de-regulated industry, electricity is no different: it is traded by middlemen, "power brokers" considerably different than the political hacks of that same name, who buy power and re-sell at the wholesale level, in much the same way that other traders and brokers buy wheat, pork bellies, or coffee. Brokers (middlemen) exist because they offer services, terms, and business arrangements that some sellers and buyers see as a more convenient way to sell or buy power.

A middleman might make a bi-lateral contract with Big State generation for, say, 300 MW for all 8,760 hours next year, then sell that in smaller amounts and for shorter-periods, but at higher price trades with LDCs and industrial consumers over the next year. The odd amounts the broker cannot sell in any hour are sold through the PX (he might have sold only 295 MW in one hour, having 5 left to sell). Similarly, the broker can make up anything needed to complete a big sale by buying through it. For example, he might need another 20 MW to make a big 320 MW sale, and buy that power on the "spot market" through the PX at a higher price than he is receiving for the power, but justifying that on the basis that he is making a good profit on the remaining 300 MW.

Commodity trading and broker/middlemen are not necessarily a part of de-regulation, but both have developed as part of the de-regulated industry in many jurisdictions. Generally, if bilateral trading is permitted, then commodity trading and brokers develop, simply because they really cannot be prevented without denying reasonable rights of ownership and business flexibility to buyers and sellers. As an example, an LDC like Big State could conceivably find that it ordered more power in a year-long bilateral trade with Western Generation company than it really needed. This could happen if Big State had contracted for the power it *might* need to make it through a hot summer (which would create very high air conditioning demands from the retail consumers) only to find that the weather was extremely mild and the actual demand 10% less than expected.

Big State would want to re-sell this unneeded power. It is only reasonable to allow it to do so because to deny it that right would create serious risks for utilities that would make them reluctant to lock down low-cost deals when they could instead buy short-term through the PX. That would raise their costs, and thus the prices they have to pass on to their customers. If it once made money on such a deal, it or another company who saw it make money might decide to routinely do that type of buy ahead/sell later tactic. Similarly, a generation company that already is set up to sell power might start buying power from others because it has the corporate infrastructure to do so, the additional effort required is miniscule, and it believes it can make money re-selling that power. Although there are ways to prevent or limit such trading, they are cumbersome and many regulatory structures just decide to set up reasonable rules and permit all such trades.

2.5 INDEPENDENTLY OPERATED REGIONAL TRANSMISSION GRIDS

Figure 2.4 showed the three transmission grids of Western, Big State, and East State as linked into a regional grid. This same regional grid and the relevant players and relationships are shown in Figure 2.5, which is a bit simplified from that earlier figure and will serve as a platform for examples throughout the rest of this chapter. A separate company (not shown), called an Independent System Operator (ISO), operates this regional transmission grid. The ISO is composed of expert engineers, technicians, and managers working at an operations center fitted with computer and control equipment. They make certain the electric grid stays up and running and that it is able to transport power from the generating plants selling it to the people buying it. Many of the people now working at the ISO might be the same people who used to work at Big State or its neighbors, operating their transmission systems.

Frankly, Big State and its neighbors performed transmission operation functions well in the past, but the ISO performs one additional function needed under de-regulation, one that no utility can do: it makes certain that the transmission grid is operated "fairly" – without favoritism toward any company.

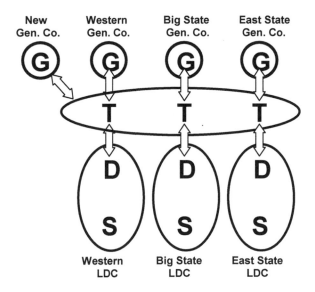

Figure 2.5 A bigger picture of de-regulation, showing Big State and two of its neighboring utilities in the region, along with the new competitive start-up generation company. Their transmission systems are interconnected into one regional grid.

Why Transmission Had to Be Integrated into a Regional Grid

Suppose that the three transmission grids were not integrated, and the situation was like that shown in Figure 2.6 rather than as in Figure 2.5. The fact that "competition" has been declared at the generation level is a moot point to all concerned. Without transmission interconnection into a regional grid, buyers and sellers cannot do business together. Power cannot cross the dotted lines in this diagram because the individual transmission systems are not connected. Both Big State and East State LDCs see only one generation company they can do business with, even though there are four that want to do business with them. Big State may wish to buy power from the New Generator Company because it might be offering a lower price than the generation it can access from Big State generation. But although New Generation Company might also want to make that deal, it is impossible. That *transaction* (buyer-seller agreement) can't be executed because the transmission system in Big State's area (its former transmission system) cannot *access* that New Generation Company's generators. The fact that competition has been declared is a moot point to all: it really doesn't exist.

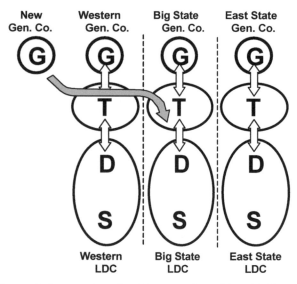

Figure 2.6 De-regulation won't work without a region-wide grid. Here, the utilities are "de-regulated" but their transmission grids are not interconnected (the dotted lines indicate borders across which power cannot flow). See text for details.

Very likely, among all the players in Figure 2.6, only Western LDC is content with the arrangement shown, for as long as the dotted lines limit access, it can probably drive down its generation costs. The New Generation Company and Big State Generation can only sell power to it. As the sole buyer Western LDC can aggressively drive down price by bidding one against the other. Obviously, both New Generation and Western Generation are going to believe they have been "screwed" in this arrangement. And *all* other players in this diagram (including the two other LDCs, who see Western getting low prices for power and want a "piece" of that low pricing) are going to be somewhat unhappy with the arrangement.

This illustrates a key point about electricity de-regulation:

> *The level and nature of competition at the generation level is dependent upon, in fact almost defined by, the transmission system's capability to link potential buyers and potential sellers.*

The word "potential" is important in the above statement. If one links all the players in the figure together with a regional grid so that any one of them *could* do business with any other, a large market has been created – one where multiple buyers and multiple sellers can potentially do business with each other. Even if, ultimately, Eastern LDC decides to buy from the Eastern generation, and Big State LDC from Big State Generation, and Western to split its buying between the two closest generators, *the availability of other buyers and sellers creates the market and makes its competition more efficient and viable.*

A regional grid, as depicted in Figure 2.5, creates that *potential* competition. To do that in this example, the transmission systems formerly owned by the three utilities are interconnected into one large regional transmission grid, so that all players have "access" to the others. More generally, the transmission in a region of many utilities and generators needs to be interconnected so that all parties have a potential access (can do business with) quite a few others. It is worth noting that while one often hears the term "transmission access" used in conjunction with this capability, that term does not mean access to transmission capability, but access, *via transmission, to people with whom one can do business.*

Why the Regional Grid Has to Have an Independent Operator

Suppose the three individual transmission systems had been connected together, and that Big State LDC has been put in charge of operating this regional transmission system (perhaps since it is the utility in the center of the system, it makes a compelling case that it is in the best position to run the grid). Big State LDC could control the transmission lines in this system so that it had an unfair advantage in buying power from the generators, and over the other LDCs, with whom it is competing to buy power at a low price. For example, its transmission operators could adjust power flow and system characteristics so that power just would not flow along the lines from, for example, Big State Generation company to Eastern LDC, or to anywhere else other than Big State LDC. That would give Big State an advantage in negotiating price with that generation company. Adding to the temptation: in some circumstances some types of favorable adjustments would go unseen by everyone except Big State.

Likewise, if any of the other LDCs or generation companies were allowed to operate the regional grid, they could give themselves the same type of advantage in dealing with the parties across the negotiating table, be they buyers or sellers. Perhaps Big State, or Mid State, or New Generation Company could be trusted to act ethically if given responsibility for operating the grid. But the temptation would be strong to make slight adjustments in one's favor – millions of dollars would often be at stake. And of course, even if there were no "cheating," inevitably some other player(s) would feel slighted and complain that they had been.

Further, and of more concern to many in the industry, operating limitations – legitimate operating decisions to back down or stop flow in certain parts of a system – are sure to happen on any grid due to emergencies or unusual circumstances. These are most likely to occur near the times of peak demand when prices are highest and the most money is riding on operation of the grid. Anyone who has a financial stake in the performance of the grid might decide to "push" too hard in such a contingency situation, rather than take prudent measures to avoid a possible blackout, if that means operating at reduced profitability for a short period of time.

Universal Transmission Access Makes the Wholesale Generation Market Competitive

The large regional grid created as part of de-regulation is a feature in nearly every de-regulation scheme, no matter how it is otherwise implemented. A wide grid area creates a big wholesale "market" by making sure large groups of buyers and sellers can do business among one another. Independent operation assures that this platform upon which they do business (the transmission system) is a level playing field for all.

With the independently-controlled, regional transmission grid in place, Big State LDC now has a way to move power from the New Generation Company into its system, or from any other generator, as illustrated in Figure 2.7. The situation for it is far better – it has four times as many generation purchase options as it had when transmission access was restricted (Figure 2.6). The independent generator also sees this as quite an improvement: the market for its products and services had tripled, from one to three LDCs.

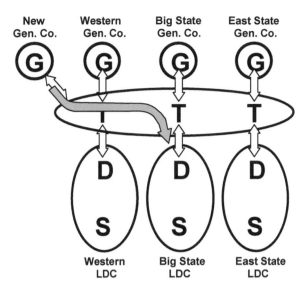

Figure 2.7 By consolidating the three transmission grids into one region-wide grid, competition at the generation level is enabled, at least in theory. Here, the regional grid permits Big State LDC to move power from the New Generation company to its system, meaning it can now buy power from the independent generator company and it has more options in buying it. Each generator company and each other LDC also sees more options.

In fact, Big State LDC is probably the only player who is not happy with the improvement that the regional grid brings. It will likely see the prices it must pay for power go up because local generators now have more options and more bidders for their business. But the situation is "fairer" as far as all other players are concerned. This brings up a second point about transmission access, limitations, and the controversies that often erupt over policy or changes in operation:

Transmission access limitations usually favor somebody at the advantage of someone else. This is another reason why independence of policy and approach is essential, and why opinions and lobbying for and against certain de-regulation issues might be hotly contested.

Market Power and Competition

A wide regional transmission grid creates a big marketplace, and that is simply better, at least according to economists who evaluate "market power" – the ability of any one player to dictate prices and control them. When buyers and sellers are mixed into a larger pot and free to do business, the market is more efficient, and everyone benefits. There are ways to evaluate if a regional grid is "big enough," and/or the various players numerous and small enough, for competition to work well. Competition is working well when no one player can dictate, limit, or drive price in opposition to the others.

Economists use an index called market power to measure whether any one player (buyer, seller) in a particular market has control over so much of the supply, or demand, that he can basically dictate price. Government policy makers in charge of setting up de-regulation frameworks use this concept to assure that: (a) the regional grid is sufficiently large that all players interact in a way where no one has too much market power; or (b) the generation companies (sellers) and LDCs (buyers) are broken up into smaller groups so that they have to compete against parts of themselves, so that again no one has too much market power.

In this regard, the situation at the generation level and the situation at the transmission level interact. A big regional grid is necessary to assure good competition, but not sufficient to guarantee fair competition. Policy makers also must make certain no generator or no LDC in the buyers-sellers mix that that grid enables (i.e., that market) has so much size or other unique characteristics compared to the others that it can dictate price all the time or in some circumstances. For example, suppose in Figure 2.7, that all three LDCs are "evenly matched" (have about the same demand) but that Big State generation has somehow acquired such a large part of all the generation assets in the region that one entire LDC will go without power unless someone buys from them. Then Big State generation can at least partly dictate the prices it gets and might totally dictate the price that one of the LDCs will pay: and by varying who and how it does business, determine which LDC it will "screw" by holding out for a high price. Many of the factors here are identical to "restraint of trade" issues considered by government regulators when they decide for similar reasons if they will approve mergers or acquisitions in any industry.

Why Regional Grids Were Easy to Implement – At Least in Theory

Fortunately for de-regulation, the transmission systems of large utilities were already linked together before de-regulation was implemented. In fact it is doubtful if de-regulation would have ever occurred if that had not already been the case, for the cost of building and connecting separate grids would have been formidable. But most of the regional grids currently running in the US were already in place.

Beginning in the mid 20th century, the larger electric utilities voluntarily linked their transmission systems together, so that, in the event of an emergency, one utility could borrow power from another, to keep the lights on. Such *power pools* were voluntary (see Chapters 12 and 16 for more history and details on power pools) and operated rather informally: A utility would simply telephone its neighbor when it faced an emergency and ask if it could help by sending power; it would "pay back" that loan with a like number of kilowatt hours at some later time, when it was convenient for the other utility to accept it. Thus, as de-regulation proceeded in the US, policy makers (mainly FERC) found a set of regional grids already in existence: individual utility transmission systems were inter-tied, and there were organizations in place (power pool committees) whereby these pools were managed and operated. Sometimes these regional grids were state-wide: utilities in both Texas and California were pretty much linked together in pools which covered each state but did not stray too far over state boundaries. Sometimes the pools covered multiple states, as for example the power pool that interconnected utilities in New York and New England, or the Southwest Power Pool, which was formed by utilities in Oklahoma, Arkansas, Louisiana, and parts of surrounding states.

Regardless, the existence of power pools prior to de-regulation did two things. First, it proved beyond a doubt that region-wide transmission grids were possible. Second, the existing power pools formed the basis for the regional grids that would operate under de-regulation.

Transmission Congestion and Other Transmission Limitations

Figure 2.8 shows a second region-wide buyer-seller agreement added to the one shown earlier in Figure 2.7. In a perfect transmission system, both of these transactions and numerous others could be done at the same time: Perhaps in addition to the two transactions (arrows) shown, Eastern LDC is buying from Big State Generation, etc. If the regional grid formed by the union of the three "T" systems has enough capacity to move all this power in the pattern that results from these contracts (i.e., simultaneously), then buyers and sellers can make their agreements and the power can flow over the network as desired. Everyone will be satisfied. Competition is not only theoretical, but real: buyers and sellers can do business as they want.

But in some cases, this is not always what happens, particularly at times of peak demand. The regional power pools that utilities set up around the US (and elsewhere in the world) were never designed with this level of complete interchange capability in mind. Power pools had limited capability to exchange power across utility boundaries, enough that regulated vertical utilities could provide mutual support during emergencies (as when a major power plant failed) and so exchanges could promote certain types of regulated efficiencies. A basic problem that stands in the way of *effective* de-regulation is that the transmission grids do not have complete capability to handle all possible transactions at the same time. This means that transmission capability limits how much competition actually exists at the generation level.

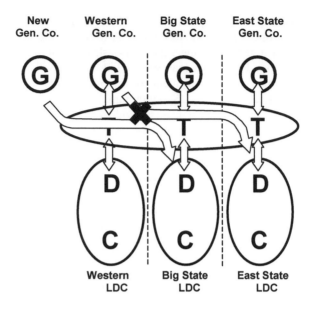

Figure 2.8 Two "transactions" (buyer-seller agreements) conflict in this picture. East State LDC wants to buy from Western generation. But Big State is already buying from the New Generation Company, as was shown in Figure 2.7. The grid cannot support the required power flowing from left to right in this diagram for both transactions. Therefore, in spite of the supposed "competition" that exists, buyers and sellers in this picture are not completely free to do business as they want, because of limitations caused by the transmission system.

For the sake of this example, assume that the only conflict in this region, at this moment, is between the two transaction arrows shown: whatever the other players are doing does not enter into consideration. The New Generation Company and Big State LDC want to do business. Big State Generation and Eastern State LDC also want to do business. But the grid will support only one of these two flows. The system is said to be *congested* – there is some point(s) in the system that will be overloaded or under voltage if these two requested transactions are allowed to occur at the same time. If permitted, this situation could either destroy equipment (overloaded equipment might overheat and fail) or lead to a cascading blackout, or both. That cannot be allowed to occur.

At this point, no doubt the reader sees even more reason why an "independent" system operator (ISO) is needed. *Someone* has to resolve these and other conflicts over usage in a "fair" way, according to the rules, and without favoritism for any one company or group. No one involved in de-regulation really disputes that. But much difference of opinion and some controversies revolve around just what the rules will be. These discussions are exacerbated by the fact that the way the system is operated, and the way the Power Exchange interacts with and communicates to the ISO, can affect if and how such conflicts arise, and limit how they can be resolved. All of this is further complicated by the facts that there are usually far more than just one conflict occurring at one time, more than just one limitation (congestion area) in a region, and that very likely one congestion situation will affect another, and vice versa.

How are transmission limitations to be handled when telling buyers and sellers – in this case the four companies involved in the two transactions shown – "You can't do that?" Should the rule be: "First come, first served?" That would mean the transaction crossed out with an "X" in Figure 2.8 would be denied. Should the rules require a compromise between conflicting transactions? That would mean that the ISO and/or the PX would tell each LDC (Big State and East State) that it can buy only half (or whatever amount the grid can support) of the power it wants from its preferred supplier (New and Big State generation, respectively). They would be ordered to buy the rest of what they need from generation companies in a pattern the grid can handle. Either of these two solutions severely limits the "market" from both buyers' and sellers' standpoints: there isn't a completely open market and competition doesn't really exist to the maximum extent it could among the four generation companies.

Then how are transmission limitations to be handled? Specifically, how is available transmission access and capability to be allocated among a grid's users when there isn't quite enough. The system needs to be: a) workable, b) fair to all concerned, c) compatible with system security needs.

The last requirement, system security, requires some elaboration for those unfamiliar with power system engineering and operation. Here, "security" has nothing to do with terrorism or concerns about external causes of damage or danger or loss in value. It refers to interconnected security, which the reader can think of as the dependability of the regional grid operating mode.[3] The ISO's number one goal is to keep the grid up and running: there will be cases where two competing transactions, such as those shown in Figure 2.8, could be simultaneously accommodated, only if the grid is run "so close to the edge" that if any major unit (generator, transformer, transmission line) failed, a blackout would occur. The ISO needs, at all times, to be allowed to intervene, to back off from what is possible to what is prudent, leaving a margin for system security.

---

[3] System security is slightly different, to utilities and power engineers, than reliability, but from the standpoint of the energy consumer, the effect is the same: a failure in either security or reliability leads to the lights going out. Chapter 16 discusses these concepts in more detail.

Congestion pricing

Limitations of a grid like those discussed above and illustrated in Figure 2.8 – transmission congestion – are a sort of electrical traffic jam: Too many people want to use the electrical interstate system at the same time. But unlike the case with traffic congestion, one cannot let electrical congestion occur and just wait while things slow down and take care of themselves over time. If a portion of a transmission grid becomes too congested, a cascading blackout begins and the operator might not be able to stop its extending over the entire region. The Northeast blackout in August 2003 occurred because a particular transmission grid area became overloaded.[4]

Opinions have and will no doubt continue to vary on how congestion is best handled. But the guiding principle that has evolved in most first-world economies is to solve this problem in a way that is compatible with the "open competition" at the generation level, by addressing congestion through pricing. The basic concept is as follows. Someone, the PX or the ISO, will charge a *congestion fee* to anyone whose transaction(s) creates or contributes to a conflict. That fee accomplishes two things. First, it discourages anyone from creating a conflict. Knowing that they would have to not only buy the power, but also pay this congestion fee, the LDCs involved in both transactions shown in Figure 2.8 might instead seek to buy from other generation companies offering to sell them power in a way that did not cause congestion. Second, if someone does go forward with such a transaction, and pays the fee, the money collected can be paid to those who "were screwed" because they were denied simultaneous use of the system by the ISO, and they will feel they were dealt with more fairly.

That is the basic concept behind congestion pricing. There are many details that are difficult to get right, including:

Congestion fees. One must be able to determine what the right congestion fee should be. Too high and it over-limits competition. Too low, and it has insufficient effect. Just right, and every one is happy: those deciding to pay the fee, those who don't get access, and the system operator, who is caught in the middle if buyers and sellers don't "cooperate." The authors will not discuss how congestion prices are set in this chapter (see Chapter 12's section on marginal locational pricing), but it can be done close enough to the "sweet spot" that it will work well. Some structures of grid operation and ISO or PX pricing do not even use "congestion pricing" although they include it implicitly. The approach called locational pricing

---

[4] Not, it turns out, strictly because of congestion, but the point is that an overload caused by congestion would have caused exactly the same result. Congestion causes overloads.

(or marginal locational pricing) sets a transmission cost on every location and path in the grid: areas near congestion (heavily used) have a high cost; little used pathways have a very low cost. Costs change as demand changes relative to capability. This sends a "price signal" that spreads out usage proportional to capability and prevents most congestion from occurring.

Anticipation and Planning. The parties considering a transaction must know in advance that they are likely to encounter congestion if they go through with their deal. This approach won't work unless the LDCs and the generators can look at a potential transaction and say "That's likely to cause congestion and to create a congestion fee of about $X." It will also not work unless they can change their minds and the PX and the ISO and all concerned can work out a change "on the fly." Why? For a number of reasons, including accommodation of emergencies. When a transmission line or major transformer unexpectedly fails, that event almost always reduces system capability and very likely creates a congestion point in what was a perfectly functioning part of the regional grid. The ISO must quickly resolve it by changing the pattern of power flow through the grid. And while electric equipment is very reliable, and the failure of any one line or transformer is rare, and unexpected, this is a fact of life: in a large power grid, composed of thousands of elements, something fails every day: "emergencies" happen frequently. The mechanism to handle congestion caused by failures or expected problems must be routine, because the problem will be routinely met.

Integration of Congestion Fees with Pricing. Very clearly, some integration of the ISO function, and the PX function, is required to make congestion pricing (whether explicit or through locational pricing) work well for both users of the grid and the grid operator. The PX is the place LDCs and generators go to do business together. At the very least, then, the PX must be made aware of possible congestion problems and the price that would be charged by the ISO, and be able to pass that on to potential buyers. Some early efforts at setting up workable de-regulated systems in the US had weaknesses here. The ISO is required to operate the grid, and to keep it free of any congestion, yet it could not implement or communicate as satisfactorily as all wanted the constraints it saw in pricing communications to the marketplace.

Complexity in the Details. There are a host of intricacies and details that will not be covered in detail here, relating to how the market works. As just a few examples of the simpler ones: How far ahead can buyers and sellers make a deal to buy/sell power? An hour? A day? A month? Is the PX exclusively the only way power can be bought and sold, or are LDCs and generators free to meet by other means and agree to long-term contracts?   How do these transaction/business deals interact with congestion pricing?   Should the ISO try to anticipate operating configurations days or weeks or even months ahead and post "expected congestion" problems and prices, so that buyers and sellers can anticipate and accommodate them in advance? Can that be done? If so is it fair to all concerned? These are only some of the messy details.

Transmission expansion

There is another way to "fix" congestion in the regional grid – by really fixing it. Someone could expand the system capability, adding new line(s) or upgrading facilities so that the grid can support both transactions shown in Figure 2.8. Sensible as this is, and necessary as it will ultimately be (because growth of electric demand continues slowly just about everywhere and even un-congested grids will eventually become congested as a result), the problem facing the power industry in the US is that *currently there is no regulatory or business mechanism to enable this in most states.*

Who should build the upgrade and own the facilities and recover (through some mechanism) the money they spent to solve the congestion problem? Since the limitation seems to be in the part of the grid inside the former regulated Western system (it will be assumed that is the case for the sake of this discussion), should it be Western LDC? Should it be expected to build this line, perhaps after authorization by the ISO or agreement by all parties involved in the regional market? Several barriers and one sound objection stand in the way of doing that, at least at present.

First, state regulators (the PUC) can only authorize something be put into a regulated LDC's rate base, and paid for by its customers, if it benefits or is needed by the customers the LDC's monopoly franchise covers. The grid upgrade probably won't benefit Western or its customers. In fact, one could argue that solving the congestion problem shown in Figure 2.8 will make things worse for Western LDC and the electric customers it serves. As long as that congestion exists, there will be situations where Western generation and/or the independent generator cannot sell to anyone except Western LDC, which means it might get lower prices it can pass on to its customers.[5] The upgrade would

---

[5] Some utilities in the south, and some PUCs in southern states, have concerns that mirror this example. Generally, southern utilities see prices that are much lower than those in parts of the northern US, something that they, their PUCs, and voters in these states feel is due to superior management of the power policy during the decades prior to de-regulation. Limited region-to-region transmission capability means that at the moment, people looking into the south from outside see good prices for power but cannot do anything to access that market because of transmission limitations. Southern utilities

mostly benefit everyone *except* Western LDC and its customers, so asking Western's customers to pay for it (the net result if the cost of the upgrade is put in Western LDC's rate base) is a bit unfair.

The four parties that will benefit most from any "solution" to the congestion problem in Figure 2.8 are the two remaining LDCs (Big State and Eastern) and the two of the generator companies (New Generation and Western Generation). A state PUC is not likely to authorize either LDC to build the grid upgrade for several reasons. First, in a regional grid, often those "other LDCs" will be located to be in another state(s): the PUC in the geographic area where the upgrade must be built may not have jurisdiction over their rates. Second, and more fundamental, letting either LDC put that cost into its rate base (meaning its customers pay for it) is somewhat unfair to those customers, because the line benefits other people beyond *just* those customers.

What about letting the generation companies build the line? This gets a bit more complicated, but is worth exploring. First, any one of the independent generators would really not want to build the line and "own" it for its own use. Look at the situation from the standpoint of the New Generation Company, who is constrained by the congestion. Suppose it went ahead and paid for the grid upgrade and then owned the upgrade "for its own use." It now faces no congestion limitation and can do business as it wants. But, the upgrade it paid for freed up the existing system, and since there is no longer any congestion, so the parties to the other transaction (Western generation and East State LDC) "win big," they can now proceed as they want and yet they have paid nothing to solve this problem.

And, even if the New Generation Company decided to go ahead, letting an independent generator build the line and use it as it sees fit is not an acceptable solution. The ISO cannot really permit an independent generator or any other user to operate a transmission line for its own use. To work in the grand scheme of things, and to achieve secure regional transmission, every line must be included in the system operation by the ISO, which can still decide to limit its usage and distribute its costs as needed for the common good.

Building the line to fix the problem will really only work if some mechanism is found that will permit:

a) The ISO to operate the line as it sees fit, not any differently than any others,

b) Recovery of the cost for the owner through some mechanism that spreads the cost "fairly."

---

see no reason why they should be asked to build lines to change this situation since in effect, they see no benefit from the investment, and worse, in effect, it will raise the prices they have to pass on to their customers. Their PUCs and state legislatures agree.

The second almost certainly means some sort of *regulated* or imposed charge or fee distributed through the independent ISO based on usage of the grid. The PUC has no jurisdiction over generation companies and the prices they charge or what they must pay (they are, after all, *unregulated* companies). Further, as unregulated companies, they probably do not want to, and cannot really be allowed to, get into the "regulated side of business." Therefore, this rather involved set of logic leads to the fact that: a generation company would not want to pay for the upgrade and would not be an appropriate party for regulators to order to pay for it.

What about the ISO? Could it be asked to finance, build, own, and recover the cost of the new facilities required to upgrade the grid? The ISO can't own any lines (it would not be "independent" if it had a financial stake in the network it was operating), so it can't be responsible for owning/operating new lines and facilities, (On the other hand, it could, perhaps, be party to the planning, assignment of priorities, and authorization for any new lines.)

All these barriers and problems could be solved by some regulatory agency or authority (probably one with an authority that covered the whole region that the grid serves), setting up a mechanism that does two things:

(1) Permits someone to authorize Western LDC or another party (Big State LDC or even new party) to build the line when proper approval by the ISO (who will be saddled with operating it) has been given; and,

(2) Creates a mechanism that permits whoever pays to build the line to recover their costs in doing so over time in a way that provides then with an acceptable business case, from some sort of revenue stream based on its usage or the benefit it provides. This last can be done by charging some cost for the use of the new line (operated as the ISO sees fit) to compensate the owner.

The savvy reader will have picked up on a detail here that seems inconsistent: some mechanism already exists to compensate transmission owners (someone owns the three "Ts" in Figure 2.8, although who is not really important to this discussion). Regardless, the de-regulated system clearly has some mechanism built into it to pay the transmission owner to keep the lines maintained and to recover whatever investment they have made in them).

To an extent, it does, but that mechanism does not cover new investment of the type discussed above, for some of the reasons discussed above, and the main economic reason goes to what might be regarded as the heart of the entire matter. Every line in the grid, all built prior to de-regulation, was justified and built and put into the (now) LDCs' rate bases because it *did* benefit the local utility customers in its system. Recall the commentary earlier about the power pools that existed prior to de-regulation, and the fact that their interconnections were rather limited.

> *Adding lines or facilities to solve regional transmission congestion*
> *problems means spending money for what is a fundamentally new*
> *reason in the electric utility industry, something not covered by*
> *existing rules, regulations, or concepts.*

This is really the crux of the matter. Facilities needed to overcome congestion and improve the "market capability" of the grid lie outside the traditional paradigm of the industry and its regulatory concepts and structure. Cost recovery and regulation frameworks and jurisdictions don't really fit what is needed. What must be created is a mechanism that permits someone to recover their costs in solving a congestion problem, and make a reasonable profit. Regional grids need a mechanism for building and operating that new line; that permits it to be operated just like any other part of the system, or in a way compatible with that operation, and that "bills" everyone who benefits in order to pay the person who built and operates the lines. At the time of this writing, this issue is not resolved, nationwide.

Federal authorization

One proposal to solve this is part of the proposed 2003 Energy Bill (not passed by Congress). It would provide that someone who built a line would own the congestion rights to the problem it solved. In such a case, the independent generator might decide to build a line to solve the congestion it sees. It would not precisely own the right to demand that it have use of the line: it would own the right to be paid for any future problems it solved, from something like "congestion fees" charged to users of the line, that would go to pay for its cost.

Merchant transmission

Another approach that has some support is to let unregulated companies invest in building new transmission lines "on speculation." These merchant transmission companies would own the rights to use or re-sell the lines they use. The ISO would have yes/no authority over their operation, but the merchant transmission company would get to charge "whatever the traffic will bear" from users. In this example, perhaps a new company, Electric Transmission Speculative Holdings Inc., would pay for construction of a new line. It would then sell the right to use the line to move power to the highest bidder – maybe one or more of the parties in the two transactions shown in Figure 2.8 would pay for its use so they could complete an otherwise advantageous deal.

2.6   THE ELECTRICITY DOESN'T CARE
ALTHOUGH PEOPLE AND MONEY DO

The ways that deregulated regulatory, business, and generation-transmission systems all function under de-regulation is complicated, as the foregoing discussion, which covers only the basics, makes clear. There is an additional complication that often confuses newcomers to the power industry as well as long-term utility professionals who are new to de-regulation:

> *There is often a substantial difference between the contracted business arrangements and the actual electrical flow of power in the de-regulated grid.*

Actually this difference, which can be considerable at times, is not that important, beyond the fact that a person must understand which picture – business or electrical – is appropriate when addressing particular aspects of regional grid performance and regulatory policy.

The diagram on the left side of Figure 2.9 shows a hypothetical pattern of transactions that could occur in the regional grid used in previous examples. After shopping around and negotiating with various companies offering power, each of three LDCs has bought power from a generation company that is "relatively far away from it," as shown by the arrows leading to each. For the sake of this illustration, the amounts of each contract will be identical: the arrows indicating Western's, Big State's, and East State's power purchases flowing across the grid from their respective generation suppliers are all for identical amounts.

The actual power flow in the grid under this set of transactions would be more like that shown on the right. Each of the three generation companies producing power is pumping out the very same amount. (One has shut down and is producing nothing.) Each LDC is pulling that same amount of power into its distribution system from the grid, and because of who is near who, and the fact that electricity tends to flow in minimum-resistance patterns, the power coming into each LDC would actually be from the source nearest to it, rather than the one from which it made a business contract.

No one takes any notice of that in any regard to the business contracts and payment. That would all be done as if the contracted (long arrows) were the "correct" flow pattern, because from a business standpoint, they are. As stated earlier, electricity is fungible – a kW generated by Big State generation substitutes just fine for one contracted for and expected from East State generation, etc. All the people doing business understand this difference (or more properly, probably don't care that much about it). To them, the "real" power flow is the way the money flows, as shown on the left.

The entity that pays the most attention to the actual electric flow (that shown

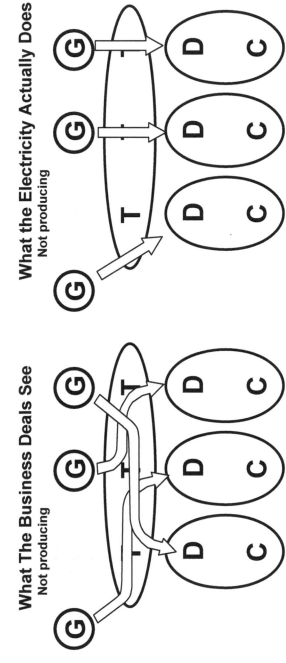

Figure 2.9 A distinction that matters somewhat less than one might expect. On the left, one possible pattern of business deals that could be executed under competition in the de-regulated grid. Arrows indicate the three transactions – purchased power agreements – between generation companies and the LDCs. In this example all three are for the same amount and one generator company is off line. As a result of this business agreement, all three generation companies would produce power and inject it into the grid and all three LDCs would pull power from the grid. From a business standpoint, everyone is satisfied: they did/got what they contracted to produce/receive. On the right, what actually happen as far as electric flow in the grid. Electricity tends to flow in a minimum effort pattern, other things being equal, which is assumed here. The point is that this difference doesn't matter to the business side: electricity is fungible, so contracts are agreed to and sellers and buyers make payments based on their contracts. Only the ISO cares about the electric flow, and assuming the actual pattern is acceptable to it (as is nearly certain in this case), it approves the transactions, the flow actually follows that shown on the right, but all parties except the ISO do business as if it occurred as on the left.

on the right) is the ISO. As explained earlier engineers and operators will receive information about the contracted power transactions in advance of their actual execution, through some means (that data will come to it from the parties involved and/or the PX through which those parties did business – how and when varies from one state or country or de-regulation system to another). The ISO's engineers and operators would analyze the set of proposed contracts *as a whole,* not individually. They would quickly determine that despite the contract pattern on the left in Figure 2.9, the electrical flow will be as shown on the right.

As mentioned earlier, if the ISO sees an operating problem(s) in the flow pattern of a proposed set of transactions it can intervene to prevent congestion or system security issues or whatever the problem might be. It would step in to prevent or limit the set of transactions, using whatever mechanism it has to do so (this, too, varies from one de-regulated grid to another). But that is very unlikely to happen in the example shown in Figure 2.9, since the electrical pattern of flow is very similar to that which would have occurred prior to de-regulation (when each LDC just used power from its own generator, that directly above it in the diagram). The grid is almost certainly built to handle this actual electrical pattern of flow.

This "fiction" might strike the reader as bizarre, and the whole system built around this difference in business and electrical flow, along with the competitive bidding, the entire de-regulation structure of regional grid, ISOs, and PXs as a waste of time and effort. Further, one can observe that, electrically, the pattern on the right in Figure 2.9 is much more efficient.

But that viewpoint misses the bigger picture entirely. The pattern on the left, the result of competitive bidding under de-regulation, is more efficient in a *business* sense. While the flow pattern on the left makes no sense from an electrical standpoint, it makes sense to the parties involved from a business standpoint, because all six parties involved were free to choose who and how they would do business and they chose to have the money "flow" in that way.

Big State LDC knew that East State generation (the supplier shown for it) was farther away than other sources. Presumably it selected East State anyway because of more advantageous business arrangements. Perhaps price was lower, but perhaps other things that were crucial to its business case also fell into line better if it did business in this way, things like contract terms, payment schedule, duration of the contracted period of power supply, and flexibility of recourse if its load changes. This deregulated system is better in a *business* sense, and policy makers (and many business leaders in the power industry) firmly believe that the "fiction" about how the power flows is more than justified. That, and the business and operating frameworks needed to permit the two patterns in Figure 2.9 to co-exist, is worth all the effort and cost. Considerable evidence says they are right, if policy-makers can get the details right.

2.7 THE ELECTRIC UTILITY INDUSTRY UNDER DE-REGULATION

Dis-Aggregation of No-Longer Regulated Business Functions

When faced with de-regulation, a traditional vertical-integrated utility like Big State Power and Light (Figure 1.2) would have to *dis-aggregate* some of its functions – splitting itself into separate unregulated (competitive) and still-regulated companies in order to fit the new structure or either Figure 1.4 or 2.2. One company cannot do business as both a regulated and unregulated business: it does not work and regulators usually do not permit it.

For example, faced with the fact that de-regulation has been implemented in the structure shown in Figure 2.3, Big State would need to "dis-aggregate" its generation holdings. An option it might consider would be to simply sell its asset (generators, facilities, etc.) to other companies (competitive generation companies). However, most utilities, and in this example, Big State Power and Light, decide to dis-aggregate: Split themselves into two or more businesses rather than just sell off assets. In general dis-aggregation is deemed to ultimately be more valuable – the disaggregated company would be worth more than the money Big State could get from the sale of its generation assets.

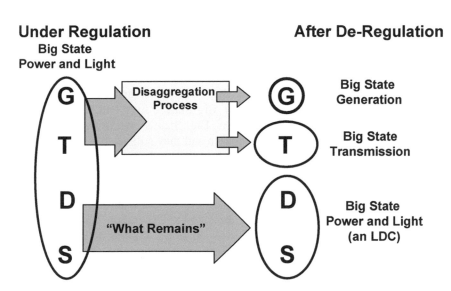

Figure 2.10   Under de-regulation, the former vertically integrated utility (left) has to dis-aggregate – break itself into independent businesses – because wholesale generation and transmission functions must be separate from its regulated retail delivery and sales operations, which will continue as a local distribution company (LDC).

Thus, as shown in figure 2.10, Big State Power and Light creates Big State Generation Company, which will competitively manufacture and sell electric power on the wholesale market. Big State's generators, generating plant sites, equipment and employees were formerly the "generation division" of Big State Power and Light. Now they are "disaggregated. Big State "cut them loose" as a separate company, one that is legally and functionally independent of Big State Power and Light.

This competitive generating company (Genco) must survive on the revenues it can obtain from the sale of electric power. However, it is no longer restricted to selling power only in its former monopoly franchise area – i.e., to Big State Power and Light. Now, it can try to sell power anywhere it would like. On the other hand, its competitors can sell power in what was formerly its exclusive territory, vying for Big State Power and Light's business by offering a lower price than Big State Generating can.

The second company in Figure 2.10, Big State Transmission Company, is formed to own and operate the Transmission assets. This company maintains a monopoly franchise on the power delivery system in Big State's service territory. It has an obligation to provide bulk power delivery and is subject to regulatory oversight of its pricing and business practices. It operates only within its franchise service territory, within which it is the sole power deliverer at the high-voltage level. In some regulatory structures, Big State does not have to dis-aggregate its transmission assets. It can hold on to them if it wants to, but as stated earlier, and regardless of whether it holds or dis-aggregates them, it may own but not be allowed to operate any transmission lines it owns – the regional ISO will do that.

The third company shown there is Big State Power and Light, the former distribution and retail sales divisions of Big State Power and Light, now an LDC, which retains the original name. Big State's executives want their customers to see as little change as possible and they still see "Big State Power and Light" on the bills they receive each month and the line trucks they see making repairs after storms.

Under de-regulation, Big State Power and Light buys power on the wholesale market, perhaps from Big State Power Generation, but also from anyone else who competes against Big State Power Generation. It resells that, at the retail level and subject to regulatory oversight and control, to its customers. To move the bulk power it buys to its distribution system, it contracts with the regional grid operator for transmission services, over the very lines it used to own, now owned and maintained by Big State Transmission. Big State Transmission does not control the lines it owns – the regional grid ISO does. Big State earns money by making them available for use. (Just who pays it and how transmission service is priced depends on the particular rules and regulatory structure in a region.)

Thus, the former vertically integrated Big State Power and Light emerges under de-regulation as three companies in this example. In addition, utilities often dis-aggregate subsets of their former integrated operations for business reasons. Many formed a *service company*. To do so, Big State Power and Light could set up Big State Power Engineering and Services Company, a separate, independent company, into which it would move all or some of its engineering, construction, meter-reading, and maintenance resources. This de-regulated company would then operate competitively, bidding on jobs that it formerly had the exclusive right and obligation to perform internally at Big State Power and Light. It could also bid competitively on similar work in other utility service areas and for other utilities and non-utilities. Big State Power and Light could also seek bids for work it needed from this company, and from its competitors.

Dis-aggregation: The Mechanism

A utility dis-aggregates its function into a separate company through a series of steps. Usually, it first moves all of the activities, resources, and people to be dis-aggregated into an organizational structure with distinctly separate human resources, payroll, accounting, and purchasing resources all within that organization and separate from its own. This is often the first sign disaggregation is planned: a division that formerly shared services with others finds it now has distinctly separate, "stand alone" services as part of its organization. Usually, this organization is moved to another physical location, to provide further business and identify separation. The owner then "dis-aggregates" (breaks off) that whole organization and incorporates it as a separate company, splitting its stock and ownership. To dis-aggregate its generation division as Big State Generation Company, Big State would have to establish:

> *Separate organization.* Big State Power and Light would take all of the people, equipment, tools, service trucks, and other assets and resources of its generation plants and generation operations and organize this, along with a portion of support services like HR and purchasing and accounting, under one executive (Fred, the Senior VP of Generation).

> *Separate location(s).* It would move all of these people, equipment, and their work activities to a separate location(s). The generation plants are all at their own separate locations already (as opposed to big State's central office building). But people working for Fred in the central office would be moved to a separate location. Anyone reporting to him in other Big State Power and Light buildings would be moved out to other, non-Big State Power and Light locations, too.

> Legally, a separation at the headquarters building would be enough if done by organizing physical location inside the building as: "generation"

takes floors 2-6 of our downtown building, and the rest of the company takes floors, 7-25. Certainly, there are many office towers that have more than one company in them. However, the utility wants to make a very visible and complete move. Generation is physically re-located into another building, perhaps in another part of town or even another county: One utility in Florida made a point of making certain the new location for a dis-aggregated unit had a different telephone area code.

Separate identity. In the examples diagrammed and discussed earlier, to make antecedents/descendants clear, the new dis-aggregated companies were all called "Big State [something], but the new company will likely have a quite different name (i.e., Inter-State Generation Holdings, Inc.) in order to further distance itself from Big State.

Separate Charter. Big State Power and Light charters the entire new organization under Bob as a separate company (Inter-State Generation Holdings, Inc). Financial instruments are enabled and a Board of Directors assigned and they and Bob given fiduciary power. Papers are signed by Big State Execs and Bob, creating the new company and transferring all assets and liabilities of the generation group from Big State to it.

Separate ownership. Stock in the new company is distributed to all present stockholders, in measures proportionate to what they owned before the split. A stockholder who previously held 1,000 shares of Big State Power and Light still has those 1,000 shares (which are each worth less because the company shed some assets and resources) and receives some shares (maybe 200) in Inter-State Generation Holdings, Inc. Theoretically, and as far as book value goes, these have the same value that the 1,000 Big State Power and Light shares did prior to the split.

At the moment of creation, the ownership of Big State Power and Light and Interstate Generation Holdings is identical. However, owners can sell or buy each, and thus these two companies are independent, and, over time, have a distinctly separate ownership.

Separate Leadership: Bob is now President of Inter-State Generation Holdings. He no longer reports to or answers to Big State's President or its board. He does not even let them be privy to business secrets within his company, and he is no longer privy to "insider" information about Big State's plans and operations.

Separate business dealings. Big State can now deal with anyone it wants to in buying generation, not just Inter-State. Interstate can offer its product (power) to anyone it wants. These two companies might do almost exclusive business together, but they are not obligated to operate like that.

Table 2.1 Types of Companies Involved in the De-regulated Power Industry

Generation companies – *Gencos,* as they will be called in the de-regulated power industry, are electric power manufacturers. They own generation units and produce electric power, which they sell "at their site" in the same manner that a coal mine might sell coal in bulk at its railhead. In each case, some transportation mechanism (transmission line, railroad) exists to move the commodity to the point of consumption, but the business concept is production and sale *at the site.* Electric generating companies produce power that is fed into an electric power system owned by someone else. This is then moved to the point of consumption over the electric lines.

Transmission utilities -- *Transcos,* as they are often called under de-regulation, are "electric railroad companies" that move power in bulk quantities from where it is produced to where it is wanted. They are regulated, and earn money by charging for transmission services. They often do not operate the transmission lines they own, letting a regional grid operator do so. *Merchant transmission companies* are similar companies that are sometimes less-regulated, with more pricing freedom (only in some areas).

Distribution company (Disco) is a regulated company that owns, operates and delivers electricity locally in an area where de-regulation as depicted in Figure 2.10 is done.

Electric Service Companies ESCos, are competitive retailers of electric power in schemes like that shown in Figure 2.10, and perhaps sell other forms of energy like gas or propane, and various services such as efficiency, backup power supplies, etc. They are unregulated. They buy power at the wholesale level and re-sell it at the retail levels.

Local distribution companies (LDCs). Under a de-regulation scheme like that shown in Figure 2.4, these are all that remain of the traditional utilities. They are, essentially, like the traditional LDCs. They own and operate only a distribution system, which they use to move power produced elsewhere to the local consumers. They provide all retail sales and services. They are regulated, have a monopoly franchise, and an obligation to serve. To their customers, they are "my local electric company."

Independent System Operator (ISO) or Regional Transmission Operator (RTO) is a non-partisan organization that actually operates the power system in a region. Its duties are to operate the system in a reliable and economical manner, and to assure all who need to use the transmission system of equitable treatment.

Power Exchange (PX) is an organization that sets up and runs the wholesale power marketplace, somewhat like a stock exchange, that permits buyers and sellers of wholesale electricity to buy and sell electric power as a commodity.

Service Companies are independent, competitive companies that include engineering, design, construction, or maintenance forces that were previously part of the utility.

Standard business practices. Regardless of whether they do business or not, all the business that Interstate and Big State do with each other will be done according to accepted practices for separate companies, auditable by various government agencies, and subject on Big State's side of the ledger to regulatory rules and oversight, etc. In particular, regulators will not allow Big State (a still-regulated company) to favor doing business with Inter-State over its competitors, just because Big State employees know and favor their former co-workers.

As a result of de-regulation, the major types of players in the new electric utility industry are as shown in Table 2.1. Chapters 11 through 14 discuss these entities, along with their duties, functions, and interactions in much greater detail. Chapter 10 gives examples of how these organizations have actually been implemented in several nations or states around the world. Table 2.2 lists a number of typical regulated and unregulated companies operated in the de-regulated power industry at the time this was written.

2.8 AN INDUSTRY IN NEED OF FINE TUNING

De-regulation does work. Not perfectly and not as well as nearly all involved would like. But for the most part, the lights have stayed on: service quality and reliability are no worse in the United States than they would have been had there been no de-regulation.[6] But as the foregoing discussion highlighted, there are a number of areas and issues where industry structure and regulatory policy in the

---

[6] Some people have pointed to the problems that California, among the first states in the US to de-regulate, had as proof that "de-regulation doesn't work." This is at best a gross simplification and in part just completely wrong. The blackouts that rolled throughout California in 2001 were caused by deficiency of generating capability, which was not due to de-regulation, but rather years of state policies, prior to de-regulation, that effectively blocked utilities from adding new generation plants, even though electric demand continued to grow.

Some of the same politicians responsible for that implicit "no new generation" policy blamed the rolling blackouts on de-regulation, because at the time it did not enable construction of transmission lines to generation sources outside the state (for some of the reasons outlined in Section 2.5). While expanding transmission capability could be viewed as solving California's problem if the state were studied in isolation, there was not a surplus of generation in the Pacific west, so that "solution" to California's woes would have simply moved the deficiency to some other state(s) nearby.

On the other hand, the bankruptcy problems of utilities like Pacific Gas and Electric were a clear example of how a regulatory/policy system that is not well designed can wreck an otherwise sound de-regulation concept system. Fortunately, over time, the system recognized most of its shortcomings and at the time of this writing seems well on its way to being fixed.

Table 2.2 Examples of Various Types of Electric Utility Companies - 2004

Type	Name	Functions*	Customers	Peak Demand	Employees
Very Large IOU	Exelon (Chicago, Phila)	T, D, S	5,000,000	24 GW	7,000
Very Large Genco	Exelon Energy	G	wholesale	32 GW	6,000
Large IOU	Centerpoint (Houston)	T, D, S	1,800,000	13 GW	4,000
Large Genco	Reliant Energy	G	wholesale	25 GW	6,000
Large IOU	Consumers Power	D, S	1,100,000	8 GW	1000
Trans. Co.	Michigan Elec. Trans. Co	T	wholesale	trans.	240
Genco	Consumers Energy	G	wholesale	6 GW	2,000
Small Muni	Hobgood (North Carolina)	D, S	365	700 kW	12
Large Genco	Calpine	G	wholesale	12 GW	3,000
Merchant Trans.	TransElec	T	wholesale	trans.	230
Services	InfraSource	services	B2B	services	7,000
Power Agency	Florida Muni. Power Ag.	G	19 wholesale	450 MW	100

* G=generation, T-transmission, D=distribution, and S = retail sales

United States in particular needs to be adjusted and fine-tuned. By far the most serious is that of recovering the costs for transmission expansion. Electric demand in the United States continues to grow, slowly but inexorably, and the regional grids are "living on borrowed time" as far as making do with the intertie capabilities that existed prior to de-regulation.

Each of the difficulties or problems discussed above has a number of suggested fixes that have their own sets of detailed complexities and interactions and over which disagreement exists as to how to best address issues. All issues interact with one another, and with the numerous differences, large and small, in local power systems design and characteristics, demand demographics, political divisions and regulatory policy, utility operating practices, energy availability and pricing, that differentiate the various regions of the US, so that sensitivities to and preferences for one solution over another vary from state to state and region to region.

In July 2003, the US FERC proposed a solution in its proposed Standard Market Design (SMD), a system built upon a common transmission pricing system which would be flexible enough, in FERC's view, to fit differing regional needs. It included ownership of congestion rights, which means people who solve or contribute to solving congestion problems "own" some of the rights to the increased system capability their "solution" provided. FERC also seems to prefer a transmission pricing system called locational pricing (which has been proven to work in many transmission regions like PJM and NY) which varies cost at points within the grid according to the cost created by demand for usage at that location. It also essentially requires all utilities to join a regional transmission grid that would not only combine operations as outlined earlier but essentially merge planning and many ownership issues, too, even if utilities retained legal "ownership rights" to their lines.

Many utilities and state PUCs immediately supported the SMD. Some others opposed it. So far, nothing has been decided. Very likely, in a modified form, something like this SMD will come to pass, and a set of regional transmission organizations will come about to plan investment in and operation of regional grids, for the reason that nothing else seems to be substantially better or even as good overall. But the details will take a while to work out.

3

A History of the Electric Power Industry

3.1 THREE INTERTWINED ASPECTS GREW SIMULTANEOUSLY

In the century from 1885 to 1985, the electric power industry grew from a series of fledgling "technology opportunists" – small companies that would be called "start-ups" today – into a cornerstone of our civilization. In its first three decades, the electric industry had tremendous hurdles to overcome. To begin with, homeowners and businessmen in the late 19th century had no way to *use* electricity. Houses and commercial buildings of the time lacked the internal wiring that is ubiquitous today. But beyond that, *there were no electric appliances available* in stores, so that even if homeowners, businessmen, and factory owners had access to electricity, they could do nothing with it. People did many things *manually* that are done today with electricity. They ironed clothes with heated irons, made coffee with percolators heated on stoves, and ran sewing machines and washers by hand. For light, they used the sun, kerosene lamps, or firelight. Finally, few people understood electricity, and even fewer trusted it.

But against these barriers to its use, electricity had several advantages. It was clean, odorless, flexible, and easily controlled. As a result, a market demand developed quickly. The electric utility industry began a process of intertwined growth in three mutually dependent aspects: electric power *usage,* electric power *technology,* and the electric power *business,* itself. Each was necessary for continued growth of the other two. For example, there would be no need for technology, and no business, if there were no demand for electricity. Similarly, despite growing demand and improving technology, without a sound business structure, the industry would not have flourished. This chapter will trace each of these three strands of electric utility history: usage, technology, and business.

3.2 GROWTH OF ELECTRICAL USAGE

The original and initially the *sole* use for electricity was lighting. Thomas Edison patented the incandescent light bulb on January 27, 1880. Electric lighting represented a staggeringly important breakthrough in the late 19th century. It provided bright and unwavering illumination, but required no fuel, had no flame, produced no odor or fumes, and presented no fire hazard. It could be controlled with the mere turn of a switch.[1]

There had been earlier forms of electric lighting. An inventor named Charles Brush, who was to compete against Edison briefly and unsuccessfully, had used arc-lights, and other types of "illuminating devices" several years earlier. Many other inventors had similarly tried to tackle the problem of practical electric lighting. But none mastered the requirements: making it small enough for room application (arc-lights were bigger than a team of horses), and making it reliable, easy to run, and inexpensive.

Edison's incandescent lighting was so utterly superior to other forms of exterior and interior lighting that *lighting companies*, whose sole business was the sale of electric illumination, had no trouble establishing themselves as viable and growing businesses. These early companies sold lighting, not electricity. *They billed their customers according to the number of light fixtures, not on the basis of actual usage.*[2] The major barrier they faced was daunting. Homes and businesses had no internal wiring! Therefore, lighting companies not only had to build electric distribution systems, but install conduit, light sockets, fuses, and switches inside their customers' homes and businesses. Still, they did a brisk business, because there was a great demand for their services. The unwavering glow of electric light through the window shades of one's home at night was a sign of personal prosperity, and the telltale conduit for wires running along the baseboards and across the ceiling to a retrofitted light fixture inside one's home were sought after signs of prestige by Victorian-era yuppies, much like a three-pointed star hood ornament would be a century later.

---

[1] Early wall switches rotated on and off, much like the switches on table lamps, rather than flipping up or down as today's wall switches do. Hence the phrase "*turn* on the lights."

[2] How and why they billed this way will be discussed in a later section on the history of the business side of electricity.

Moving Usage Beyond Lighting

For the first decade of the electric power industry, lighting represented virtually all electric usage. It still represents over 30%. Early electric companies realized that a big potential source of revenue was the sale of power to run large electric motors, which could be used for numerous industrial applications. Many of these first light companies quickly became more interested in large industrial sales than in retail sales of lighting to homeowners because of the higher revenues they could realize. They focused on selling both motors and electricity to factories, grain processing centers, water pumping plants, and the like. Electric motors were reliable, relatively quiet, needed no fuel, and did not require the plant to be located on a stream or river, as waterwheels did. More important, they were easy to operate and required relatively less maintenance than other power sources.

Another industrial application that fascinated Edison and George Westinghouse, as well as other early inventors and advocates of electric power, was electric railways. Electric-powered locomotives were particularly suited to urban use, for what is called "people movers" or "mass transit systems" today. Electric trains produced no clouds of dense smoke, soot and cinders, as did the coal-fired locomotives of the 1880s. They were quiet, controllable so they could start and stop quickly, and they accelerated much faster than coal-fired locomotives. *Electric railways and trolleys were an important element of early electrification.* Their widespread use continues to this day throughout Europe. Widely employed in the United States until eclipsed by the popularity of automobiles, they all but died out by the 1960s, but are enjoying a strong comeback in the 21st century, a modern, "low-impact" solution to urban commuting.

Although industrial usage soared to the point that many early electric utilities concentrated on it in the late 1880s and early 1890s, the residential and small business market was too big to ignore. Even before light companies began to set up shop, they recognized that electricity could perform other types of useful work besides providing light. Small electric motors could turn fans or replace the manual treadle of sewing machines, could power refrigerators, obviating the need for weekly ice deliveries, and they could also run washing machines, reducing the physical labor involved in all of these activities. Electricity could also produce heat, for toasters, clothes irons, and coffee percolators. In all such applications, it offered greater convenience and smaller appliance size than traditional methods, e.g., the alternative to a conventional oven being the far smaller electric toaster.

Not much imagination or engineering skill was needed to invent a plethora of electric appliances, like toasters or sewing machines, which basically just applied electricity to traditional needs. But it took time for manufacturers to set up shop to produce such items, and for retailers to begin marketing them. It took still more time for a few adventurous souls to buy them and discover that they worked well, and for word to spread that electricity was a superior way to get the job done. Thus, for the first quarter of the 20th century, electricity only spread slowly to other applications throughout households and businesses, where it was regarded as a *premium* means of doing work – more expensive but worth it for those who could afford it. For example, even as late as 1947, many families did not own a toaster, let alone a washing machine, the average cost of which was $240, or about $1,770 in today's dollars. Once a great luxury of the post-war era, these appliances became items that cost less today than a generation ago.

The Radio and a Change in Attitude about Electricity

Of all electric devices ever invented, the radio had the greatest "psychological" impact on society, for it permanently altered the way people regarded electricity. Before the invention of radio, electricity was just a better way of doing things, a more convenient replacement for whale oil and kerosene (lighting), coal and wood (cooking, ironing, making coffee, making toast), waterwheels (industrial motors), and other power and energy needs. Everything electricity did could be done – had been done – by other means. Even a phonograph could be powered by a manual crank. But radio was different. It worked only with electricity, and it did something never done before. *Electricity could do things nothing else could.*

Although radio communication had been discovered in the 1890s, it took until the 1920s for reliable receivers to be available at prices affordable to large portions of the public. Radios exploded into the marketplace. Americans spent an impressive $60 million on radios in 1922 – representing roughly a million home units at an average price of about $60 each. Four years later, an incredible *half billion dollars* was spent on radios. Many Americans were buying their second or even third – keeping up with technological progress. Radios evolved as rapidly as cell phones and personal assistant and entertainment systems ("Blackberries," iPods) do today, with "new" models becoming obsolete within three years, during the 1920s.

During the 1920s and early 1930s, the public attitude about electricity gradually shifted, due mostly to widespread use of the radio. Now, electricity was not merely a more convenient means to accomplish useful things that had been done by traditional methods. It was apparently essential to "modern" technology and progress.

As a result, by the late 1930s, electricity was regarded by many people as a basic utility, like water and sewers. This was an opinion that grew and hardened until it was not disputed by any credible politician, as will be discussed under the history of electricity's business side, later in this chapter. By the 1940s, prevailing opinion held that everyone should have access to electric power, and that facilities for its distribution should be built into every home and business. Thus, in roughly a half-century, from 1890 to 1940, electricity grew from a useful, prestigious service into a basic utility that was considered a necessity by both the public and the major political sectors throughout the United States.

Electricity Application Continues to Grow in Diversity

The growth of electric demand in the remainder of the 20th century was due to the broadening application of electricity, as shown in Figure 3.1. A diverse range of applications for home, office, and industry grew steadily. Although initial electricity usage was where electricity replaced traditional ways of accomplishing needed functions such as lighting, water heating, etc., most of its continued growth was driven by new applications, which, like the radio, had no non-electric counterpart (shaded area in Figure 3.1). The two most recent major new applications were the microwave oven (about 1970) and the home computer (about 1985), which by the year 2000 were in nearly 90% of US homes. While the rate of invention of new electrical applications has slowed, and for the time being focused on personal, portable applications (cell phones, etc.) there is no reason to believe it will not continue indefinitely.

Electric usage grew because electricity could perform many functions better than other energy sources, or could provide functions that could not be done by other means. Either way, usage increased because electricity offered high value.

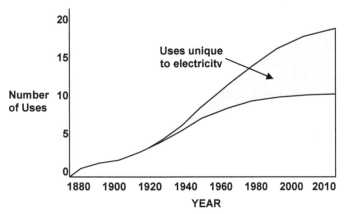

Figure 3.1 The number of different types of electric applications in a typical home (e.g., lighting, dishwashing, TV) has grown almost steadily from 1882 to 1997. Initially, electricity's sole use was lighting. Gradually, it was used for other applications, such as cooking, ironing, etc. Widespread use of the radio, beginning in 1920, initiated an era of appliances and devices that operate *only* on electricity and that have no counterpart in non-electric applications (gray area). *Source: Energy Education Specialists.*

Nevertheless, while many of these new uses for electricity contributed great value to home and business, few consumed a lot of power. Televisions, whether the crude vacuum-tube affairs of the early 1950s or the biggest-screen models of today, consume only about as much power as a large table lamp. Washers and dryers consume a lot of power when running, but operate for just a few hours a week. Then as now, the fans, toasters, coffee pots, radios, sewing machines, washers, dryers, etc., of earlier eras use much less electric power than lighting.

Air Conditioning Becomes a Major and Heavy Usage

Almost the sole exception to the trend of light electric usage among new electric applications was air conditioning. Cooling a home or office requires a lot of energy, even when the buildings are properly insulated and the air conditioning units are efficient. Although the basic principles were understood even before the advent of electric systems, air conditioning on an individual household basis really only became reliable, affordable, and practical in the early 1960s.

Increasing sales of electricity for air conditioning contributed greatly to the electric industry's continued growth in the 1950s through the 1970s. "Space cooling" increased the value that electricity could contribute to its users, and as a result it raised electric sales tremendously. Today, throughout many parts of the United States, air conditioning of one form or another accounts for up to *half* of residential electric usage. In most areas of the United States it is the only end-usage that exceeds lighting in overall amount of electricity consumed. It contributes a huge amount of revenues to electric utilities.

Cost of Usage Decreased over Time

An important accompanying trend to the broadening *diversity* of electric usage was that the *cost* for it has dropped steadily since 1880. This is partly because over the long term, 1880 to 2000, the cost of electric power, if adjusted for the changing value of the dollar, *declined significantly*. Average electric rates dropped from over a dollar per kilowatt hour to less than ten cents.

However, the cost of electricity is only *one* element of total cost, and usually not the most important to consumers. Appliance costs usually outweigh the price of electricity, and they have dropped steadily for the past century. For example, in 1925 the purchase price of a basic clothes iron was $8, that for a basic radio $65, equivalent to about $60 and $400 in 2005 dollars. Eighty years later, a basic iron costs about $20, a radio only $25. A basic television – the late 20th century equivalent of that 1925 radio – costs about $200.

Such decreases in cost occur among new inventions, too. The home microwave oven dropped from $600 in 1972 (in 1972 dollars) to about $60 today (in 2005 dollars) – in fact it dropped by about 20% since the first edition of this book was published. The cost of electricity for typical microwave home usage has held fairly steady at roughly $100 (in 2005 dollars) over the

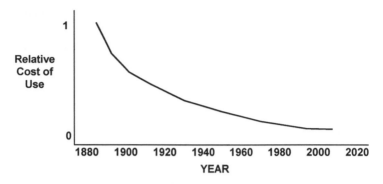

Figure 3.2 Relative cost of providing light over the 12 decades of the electric era. While part of this trend has been due to a decreasing cost of electricity, the major driving force has been increasing efficiency along with decreasing prices for lighting equipment.

fairly steady at roughly $100 (in 2005 dollars) over the average 14-year microwave lifetime. Thus, overall cost dropped from about $50 per year ($600+$100 for the electricity, divided by 14 years) to about $14 a year ($100+$100 divided by 14 years), or a factor of three and one-half to one.

But perhaps lighting best illustrates long-term trends in cost of electric usage. In 1880, Edison had calculated that in order to succeed in the marketplace, the first light bulbs had to be priced no higher than 40¢ (roughly equivalent to five dollars today). Each would use about $20.00 worth of electricity at contemporary prices during its lifetime. (This disparity between the cost of the bulb and the electricity it used was the reason that Edison focused his business interests on selling electricity, rather than light bulbs, as will be described later.)

A century later, a 75-watt light bulb costs about one dollar (2005) but still uses about $20 worth of power during its lifetime, making it now, as in 1880, one of the few applications where the electricity costs *much more* than the device. Even more-expensive compact fluorescent devices, which reduce overall energy usage for lighting by large measures, are qualitatively similar. They cost about $12 and use about $25 worth of power in their (longer) lifetime.

Figure 3.2 shows the relative cost in constant dollars to provide 2,000 lumens, the typical output of two modern 60-watt incandescent light bulbs and sufficient for a typical room, from 1880 to the year 2005. Overall, electric usage grew throughout the 20th century in both number of customers and usage. In 1900, only a small minority of homes and businesses had electric power, and the average "electric" household used less than 600 kWhr per year. Today, over 99% of households in the United States have access to electric power on a routine basis. Average usage is about 1000 kWhr per month.

3.3 THE GROWTH OF ELECTRIC SYSTEMS TECHNOLOGY

An Early Scientific Curiosity

The growth in electric usage would have been impossible were it not for the development of *machinery* that could produce electricity and distribute it to the millions who used it. Electricity was the source of experiments and curiosity among rich dilettantes as far back as the early 1600s. William Gilbert, the physician to Queen Elizabeth of England, coined the phrase *electrica* in 1600, from the Greek word *elektron* (amber, a material that could be used to produce static electricity, the only electricity known at that time). By the early 1700s, scientists throughout Europe, and a few in what was to become the United States, understood that *something* could be made to flow or jump (spark) between metal objects, and that somehow it was related to magnetism. But, the thing itself was poorly understood, and often described in different ways by different scientists. At the time, it was not believed to be capable of any practical application. Research into electricity was "basic research" – learning for learning's sake.

Many great scientists, justifiably famous today for their discoveries, helped develop mankind's understanding of electricity. Table 3.1 lists some of the most significant researchers or developers (the distinction was moot in the 18th-19th century) in electric technology. One who does not receive due credit as a scientist is Benjamin Franklin. Although often seen as a printer/politician/patriot who only dabbled in science, and whose only electric experiment involved a kite and a good deal of luck, he was, in fact, a serious experimental scientist who approached his research with intuition, careful organization, and tenacity. His contemporaries called him "the Newton of electricity," about the greatest accolade that could be bestowed at the time, and many modern scientists have labeled him "brilliantly analytical and objective."[3]

In 1747, Franklin took up the investigation of this poorly understood subject, with its competing theories and definitions. Many of the "prevailing truths" about electricity at the time were bunk, such as the widely held belief among scientists of the day that there were two types of electricity, called *resinous* and *vitreous*, which roughly corresponded to the two types of chemicals, bases and acids. Franklin knew that in order to build useful machines for electric application, mankind would have to understand electric theory correctly. However, his journals and notes indicate that he was mostly interested in understanding, rather than application. Unlike Edison and Westinghouse, and a host of engineers who came after him, and whose research was motivated by a

---

[3] Mitchell Wilson in *American Science and Invention,* Crown, New York, 1960.

Table 3.1   Persons Who Contributed Significantly to the Early Development of Electricity, its Technologies, and the Electric Utility Industry

Charles Brush (1849 – 1929) was an American inventor/businessman who developed and successfully marketed an alternating current (AC) arc-light system for outdoor (street) in the 1880s, which made him a wealthy man. His AC system, similar to but different from Westinghouse's, was the basis for the creation of several early electric companies in which he had a large financial interest, including Cleveland Electric Illuminating. The possession of several key Tesla patents, and perhaps better salesmanship and access to financing, gave Westinghouse and his equipment standards an edge which eventually overshadowed Brush's contributions.

Thomas Edison (1847 – 1931), American, was an inventor more than a scientist. Famous in his own time for a series of inventions including the phonograph, the microphone, and the light bulb, he did early research into radio (patenting a basic concept in 1891) but did not make the key breakthrough (see Marconi). Perhaps more important, Edison can be said to have invented the corporate R&D lab: institutionalized commercial, invent-for-profit effort focused on with commercial potential. Edison's careful control of image tended to focus most public attention on himself, and although he was the key researcher he employed many who contributed a great deal, including for a while, Nikoli Tesla – far smarter if possessed of far less business acumen and ego.

Benjamin Franklin (1706 – 1790) contributed so much to American independence, folklore ("A penny saved is a penny earned," etc.), and myth that it tends to overshadow his role as a serious and committed scientist whose publications contributed greatly to understanding of electrical phenomena. Franklin carried on an orderly line of research into electricity for several decades in the mid 18th century. Mostly interested in knowledge for knowledge's sake, he nonetheless was a prolific inventor, his most noteworthy electrical invention being the lightning rod and the concept of grounding facilities like buildings and signal towers.

Guiglielmo Marconi (1874 – 1937) was an Italian physicist who looked into "Hertzian waves" (radio waves) late in the 19th century and received a patent for "improvements in transmitting electrical impulses and signals" in 1897. He began manufacturing the first radios for commercial sale in 1898, and by 1901 had successfully sent signals across the Atlantic. His early radios transmitted only Morse or similar codes, not voice, and his development and business were aimed at two-way communication for ship-to-shore and similar needs. Commercial radio broadcasting, which led to mass consumer sales of radios and changed the image of electricity forever (see text), took nearly two decades, and a host of new patents by many others, to develop.

Nikoli Tesla (1856 – 1943) was a Serbian American scientist who received over 700 patents during his lifetime, including three key patents for equipment used by all electric utilities today: an improved transformer which proved practical and efficient, multi-phase AC electrical power, and AC power transmission. He invented florescent lighting, largely to prove that he could invent a lighting system that did not infringe on Edison's patents. Tesla sold his early patents to George Westinghouse, which permitted Westinghouse to build his empire, and Tesla to set up his own research lab in New York.

George Westinghouse (1846 – 1914) was an American engineer businessman who was among the first, and certainly the most successful, of those to recognize that alternating current (AC) had substantial system advantages over DC in terms of business potential. He bought rather than invented his key patents but drove considerable innovative development including hiring, for a time, Nikoli Tesla.

desire to build practical machinery, Franklin was a scientist, who focused on obtaining knowledge, regardless of its usefulness.

Franklin created electrostatic generators, Leyden jars, an early form of condenser, or capacitor, and other devices to test each theory, and gradually evolved a coherent and demonstrable explanation for electricity. He created a simple but accurate theory of electricity that has lasted to this day. Along with that he devised much of the modern terminology: *battery, conductor, charge, discharge, armature, electrician, positive* and *negative* poles, and more.

Practical Breakthroughs in the Victorian Era

Although Franklin was at heart a very practical man, ultimately he found no way to apply electricity to any useful purpose. Subsequent work by other scientists proceeded at a slow pace. Throughout the first part of the 19th century, electricity was considered an interesting curiosity, but nothing more. Three inventions changed that. The first, in 1867, was the "self-exciting" *dynamo*, or power generator, developed by Z. T. Gramme. Generators at the time were inefficient, requiring considerable mechanical power to produce just a small amount of electrical energy. Gramme's device used a portion of the electricity it produced to create and magnify the magnetic field inside itself, thereby dramatically increasing the electricity it created. *Useful amounts of electricity could be produced at affordable cost.* There was still little of practical value for the electricity to do, but there was now a reason for researchers to try.

The second development was the *light bulb*, mentioned earlier in the discussion of the history of usage. At the time of the light bulb's invention, electricity had been used to make light for several decades. But prior "illumination devices" were arc lights – huge, crackling, and smelly affairs that created an intense white light from the open discharge of an electric current. In the 1860s, several arc lights without any lens to focus their light into a narrow beam, as in lighthouses, were installed on eight-story-high lattice towers straddling major intersections in cities such as New York, Cleveland, and San Jose. At night they produced a "blinding light" that illuminated streets for several blocks around.

But such devices were unsuitable for interior use. They were as big as an average room, produced thousands of times more light than needed, and were noisy. Edison's light bulb changed that. As discussed earlier, it revolutionized interior lighting, providing something useful for electricity to do.

The third and final development that created the modern electric industry was the *transformer*, a device that can change the character of electric energy, taking power at low voltage and high current, and turning it into the same amount of power, but at high voltage with low current.[4] The transformer was

---

[4] *A transformer does not change the amount of power.* One might take ten amps at ten

based on a principle discovered in 1832 by Joseph Henry, an American who went on to head the Smithsonian Institution and the National Academy of Science. However, it was a Frenchman, Lucien Gaulard, with the financial backing of an English businessman, John Gibbs, who turned it into a useful gadget, obtaining a patent for a basic AC electric system in England in 1882.

Gaulard and Gibbs weren't entirely sure what to do with the device they had invented, which they called a *secondary generator*. It was an American engineer-businessman, George Westinghouse, who realized that it held the key to efficient distribution of electricity to thousands of potential customers. Westinghouse bought the American rights to the transformer in 1885, and built an empire on it.

Most uses of electric power (lighting, turning motors) require a lot of electric current, propelled by only a modest (low) voltage. That was in fact how Victorian-era dynamos – power generators – produced electricity. But in that form – high current and low voltage – electricity could not be moved over power lines more than two or three blocks. This limitation in "transmission distance" meant that a large, bulky, and noisy generator had to be built in every neighborhood, so that it would be close enough to the customers using the lighting it provided. Further, larger generators were efficient, but a really efficient generator, powerful enough to light ten thousand homes, was worthless if one could only move power to several dozen homes nearby.

Westinghouse understood that to move power efficiently over greater distances, one needed a low current propelled by a high voltage. A transformer could change high-current, low-voltage power at the generator site into this "easy to move" low-current, high-voltage form and move it *dozens of miles* if necessary. He had the vision to realize that this meant he could build large, cost-effective generators – big enough to power an entire city – and run the power through a large transformer at the generator site to raise voltage and reduce the current to make it readily movable. At this voltage, he could then route power on rather thin (and inexpensive) power lines throughout the city, locating small *step-down transformers* in every neighborhood to provide electricity to homes in that locale.

Thus was born the concept of the widespread *power system* – a coordinated set of generators, transformers, and high-voltage lines extending over a large region – with power produced at only a few locations, but distributed to many thousands of users. *Electricity could now be efficiently distributed to the mass market.*

---

volts (one hundred watts) and transform it into one amp at one hundred volts (also one hundred watts). That transformer can be operated in reverse, too: if one amp at one hundred volts is run through it in the other direction, it emerges as ten amps at ten volts.

Competing Electricity Formats: Direct
vs. Alternating Current Technologies

Westinghouse's transformer-based power system required alternating current (AC), a form of electricity far different from direct current (DC – see Chapter 3) which is the type of power produced by batteries and early electric dynamos (see Chapter 4). Either form of electricity, AC or DC, can produce light or heat, power motors, or run radios, etc., but the two forms are incompatible. While a light bulb will give light if provided with either AC or DC power, most appliances like sewing machines, fans, and radios have to be designed to run on one or the other form of electricity.

In the late 19th century, a bitter business and technological rivalry developed between the two types of power, with Westinghouse championing AC power systems, and Edison promoting DC power.[5] Neither man was a scientist. Both were basically technology entrepreneurs, as were their counterparts in Europe – engineer/inventors with a successful track record of invention transformed into profitable business. Unlike Franklin, Henry, and others, they were interested in electricity for the money, and both were intent on parlaying their inventions and patents into a large, profitable empire. Edison invented the light bulb as the deliberate result of what would today be labeled a business plan. It envisioned creating a need for generators, too, which he intended to, and did, produce and sell. The plan worked, and for a while he sold fantastic quantities of both.

But by the turn of the century, Westinghouse's concept of a few big generators, narrow distribution lines built with thin wire, and many small transformers had established itself as the superior way to utilize electricity. Edison's DC system had advantages in certain circumstances, but to distribute power throughout a big city, it required a forest of thick overhead wires, both ugly and expensive. Thus, AC became the preferred public power system. By the dawn of the new century, AC had become the standard everywhere, adopted even by Edison's companies. However, Edison could laugh all the way to the bank. Every Westinghouse style system still had to use the light bulb he had invented!

Technological development thereafter merely built on Westinghouse's original concept, which remains unchanged a century later, although it has been greatly refined through innovation and invention. Throughout the twentieth century, power systems technology developed along several lines of progress, as outlined below.

---

[5] There is substantial evidence that Edison knew AC power was more efficient, but he didn't own the patents on the essential element required for AC power, the transformer. Thus, he stayed with what would make him the most money, DC, as long as he could.

More Efficient Generators

The cost for an electric utility to produce electric power dropped, from nearly a dollar per kilowatt hour in 1880 to less than two cents in 1997, both figures in 1997 dollars. This improvement was accomplished through better design and materials, but mostly through economy of scale: the bigger the generator, the less expensive the power produced. As a result of pursuing efficiency, even a small power station by today's standards – 100 MW – produces more power than was consumed worldwide in the first decade of the electric industry.

More Efficient Voltage, Equipment, and Design

The use of increasingly higher voltages made power distribution ever more efficient, halving the cost of moving power in the first three decades of the 20th century. Progress continued – transformers built in the 1990s are less expensive and more efficient than those of the 1960s, which were themselves vastly superior to those of the 1930s. But beyond this, by the mid-1960s, the concept of hierarchical voltages – using a series of transformers and different voltages as needed in the power system, each optimized for its particular use had led to a further halving of cost. Cost of moving power (in real terms) dropped roughly six-fold from 1900 to 1990.

Larger Power Systems

Edison originally promoted the use of neighborhood power systems. A small generator was equipped with power lines to run electricity to homes in only a small area of a city or town, where a high density of usage would justify expenses and provide a business profit. Westinghouse's AC systems could distribute power over increasingly wider areas, which led to entire towns, and small cities, where a single power system could provide electricity throughout. Yet in the early part of this century, nearly every city and town had a separate system.

Gradually, over most of the 20th century, as steadily higher voltages permitted power to be moved on transmission lines over ever greater distances, cities and towns were interconnected into increasingly larger *power pools*, or shared power grids. This was done mainly to make electric power more reliable. If one community's generator failed, it could "borrow" power over the grid from its neighbors until repairs were made (see Power Pools in the Glossary). Interconnection also led to lower costs. For one thing, it allowed cheap hydro power produced in the Rocky Mountains and New England to be moved to cities like Los Angeles and Portland.

Today, electric utility systems in the United States are connected into only three gigantic power grids. The Eastern grid extends roughly from Kansas and Nebraska, proceeding east to the Atlantic coast. The Western grid extends from

Colorado westward toward the Washington and California coasts. Texas, which is caught in the middle, operates its own grid with only "weak" interconnections to the other two. In each-interconnected system, equipment owned and operated by the different electric utilities there works in a synchronized manner. In a very real, if technically limited, sense, it is possible to route power produced in northern Maine to western Kansas, or vice versa.

Energy Efficiency and Conservation

Viewed as a long-term trend, there has been a growing emphasis since the 1970s on energy efficiency and conservation, driven by a sense of social responsibility. Efficiency improvements throughout the early part of the century were driven by a desire to reduce cost, with a motive of actually increasing power usage – "if it costs less, people will buy more."

However, since the late 1970s, while the move toward "energy responsibility" has waxed and waned several times, the long-term trend has been toward increasing awareness that even if power is cheap, it shouldn't be wasted. By 2005, appliances such as refrigerators and air conditioners are more than twice as efficient as they were 30 years ago; doing the same job with half the power. Compact fluorescent light bulbs are increasingly used in place of incandescent bulbs, cutting power used for illumination. In a similar manner, technological progress will probably cut usage in half, for the same end result, over the period 2005 to 2055. However, if historical precedent continues, new applications for power, newly innovated, will develop to fill the gap, resulting in ever greater use of electricity overall.

Distributed Generation

Edison's original concept of "a generator for every neighborhood" had several advantages over even the most advanced evolution of Westinghouse's big-system concept. But in many ways, the concept was too advanced. It was too difficult to get all those small generators to coordinate their operation with one another. It was only at the end of the 20th century that a host of messy, but important problems associated with that idea were solved. Fuel cells, absolutely silent and almost non-polluting, and MTGs – micro-turbine generators, almost silent, inexpensive, and virtually non-polluting – can both be built as small as a step-down transformer, and can produce electric power for anywhere from five to fifty homes. They eliminate the need for further construction of transmission and distribution lines, with their often-accepted but always unwanted esthetic clutter. Both fuel cells and MTGs require a natural gas line to be run underground to them, or fuel delivery to a nearby tank, and both are more expensive than traditional ways of producing power, but not enough to preclude their being viable alternatives in many cases.

Automation and Power Electronics

A combination of modern computerization and high-power electronic devices can control electric power in ways that Edison and Westinghouse could not have foreseen. Electricity moves, literally, at lightning speed, fast enough to confound the quickest mechanical switches and control systems. The best circuit breakers and control relays and line switches that Edison, Westinghouse and three generations of engineers after them could devise were mechanical devices, which, while excellent in many ways, could not track and react to electric flows fast enough, and precisely enough, to handle many potential contingencies at full load. As a result, electric equipment throughout the first eight decades of the 20th century had to utilize conservative and often awkward designs – basically utilization levels far below actual capacity – in order to avoid these situations.

But fast as electricity is, modern computers and electronics can react fast enough to control, and even to anticipate and modify, the behavior of electricity on the power grid – a sort of "anti-lock braking system" for the electric grid, which can sense and shut down problems before power can outdistance control. This means that many power grids can have their capacities upgraded at relatively low cost, to the point where they operate safely and dependably at loading levels that would have been "at the brink of blackout" only a decade before.

3.4 THE RISE OF THE ELECTRICAL UTILITY INDUSTRY

In the United States, three technological entrepreneurs, Charles Brush, Thomas Edison, and George Westinghouse (see Table 3.1), all capitalized on their own inventions as well as patents they purchased or licensed from others (e.g., Nikoli Tesla) to largely create the electric power industry. Each created his own large commercial-industrial empire to manufacture electrical equipment and do further research. Eventually each also promoted, through funding and ownership, start- up financing, or sales of franchises, the development of electric utility companies – companies that sold electric lighting, power and appliances directly to homeowners and businesses. All three took this route largely because each saw that only a lively retail demand for electricity usage would create the large commercial opportunities for their inventions that would lead to business success.

All three were very much rivals, competing fiercely with tactics that would be considered quite "dirty" by modern standards. Initially, each of their companies sold not only electricity, but also everything required for its use, including light bulbs, light fixtures, switches, fuses, and internal wiring for homes. They really had no choice. Since there was no other supplier of these appliances and equipment, they had to provide them. But in addition, the sale of these items was profitable in itself. As a result, there were three competing "standards" or types of electric utility systems and appliances in use in the United States during the late 19th century.

Edison founded his Edison Electric Illuminating Company in 1880, along with several ancillary companies to manufacture switches and distribution equipment. Edison sold and was strongly committed to direct current (DC) technology and systems. Although he proclaimed widely a firm belief in DC and was quite vocal in pronouncing his rival's alternating current (AC) systems as dangerous, a good deal if not all of this was probably due to the fact that he had invented most of DC's technology and held few patents on AC equipment.

Charles Brush's company was founded about the same time as Edison's and initially marketed indoor and outdoor lighting. He had invented and patented a type of AC generator (dynamo) and had perfected a practical, low-maintenance arc light (a type of lighting device that uses an open electrical arc to produce light). His arc-light produced a brilliant light superb for outdoor street lighting, etc., but was not as convenient for indoor and small applications as light bulbs (arc lights make a continuous crackling and buzz). Nonetheless Brush was quite successful, branching out into other electric uses, and setting up a number of early utilities: Cleveland Electric Illuminating Company still does business under the same name as when founded by Charles Brush.

Westinghouse followed Edison and Brush by about five years in founding a large industrial electrical company (he had, however, had an earlier start, and great success, with Westinghouse Air Brake which made him a fortune in railroad locomotive brake systems). Westinghouse had a slightly different business approach than Edison or Brush, who both dominated and directly led in their company's research and development, and built businesses largely around their own inventions. Westinghouse bought patents and hired good scientists to work for him, and concentrated more on just "running the business" than his rivals. Westinghouse built his business around technologies that made the most business sense: If he didn't own the patents, he would buy them or a license for their use. Furthermore, he had what proved to be a far superior technology in alternating current (AC) compared to Edison's DC. And while Charles Brush also used AC, it was a slightly less efficient system, not as well aimed at fulfilling residential needs.

All three men sold electric distribution franchises and packaged power plants and distribution equipment along with license rights to their technology, each in intense competition with the other two. Each was interested in the sale of the electricity itself, because as mentioned earlier, it represented the major portion of the dollars involved in lighting. But all three were, by nature, mostly engineers and tinkerers who found the devices that produced and applied electricity much more fascinating than the actual business of selling power. (Westinghouse was a bit more disciplined and realistic in his approach, but even he let the "fun factor" distract him from the business at times.)

As early as 1890, the business structure of the industry began to split into "equipment manufacturers" and "electric utilities," the former building the machinery and equipment used by the latter. All three industrialists tried to stay involved with both parts, by diversifying their companies. For example, Edison founded more than a half dozen companies, including the Electric Light Company (light bulbs), Bergmann and Company (electric switchgear), Sprague Electric Railway, and the Edison Electric Illuminating Company. Westinghouse similarly diversified his activities. Both basically divided their companies into a series of regional sales (electric utility) companies, and equipment manufacturing companies.

To some extent, all three focused too much on maintaining control over the industry they had created, rather than on assuring sound finances for their various companies. All three, more engineer than businessman and more committed to invention than business management, demonstrated a lack of financial acumen which led to cash flow problems. When business difficulties forced them to sell off part of their enterprises or to seek additional investors, each chose to sell the electric business and keep the equipment manufacturing as his continued focus. That proved a poor choice, for electricity was where the low-risk profits were in the late 1890s. All three ultimately lost control of most of their companies, although none did particularly badly from the standpoint of personal wealth.

Edison's empire eventually became the General Electric Company, which survives under that name and is a major supplier of electrical equipment to this day. Westinghouse Electric was taken over by its investors and operated as a major equipment supplier into the early 1990s. It merged with a European conglomerate, ASEA Brown Boveri – itself the result of a merger of several large electric equipment firms – in the early 1990s, to form ABB Inc., now perhaps the largest global supplier of electric equipment and services.[6]

The Birth of Electric Utilities

A host of electric companies, financed by the Victorian equivalent of venture capitalists, sprang up anywhere there was a city or town large enough to provide high density market for electricity. A few of these companies were owned directly by either Edison, Brush, or Westinghouse, but most were franchises that bought franchise rights, patent rights, and licenses, as required from one or more of the three. Both Edison and Westinghouse made many sales, gradually edging

---

[6] Westinghouse, as a corporation, had by that time diversified into a variety of other businesses, such as broadcasting and defense. It sold only its electrical power businesses to ABB and continues under the name "Westinghouse" as a major player in many other markets.

out Brush and making the early three-way race into a two-man contest for dominance of the electric industry. With his AC technology, which provided lower operating costs as a utility grew, Westinghouse gradually pulled ahead. AC was so vastly superior to DC that he had nearly an unfair advantage in the long run. The two are basically equivalent if one wants to wire a single home, or a group of houses or offices that are very close together. But Westinghouse's system had an economy of scale as a utility grew: As it added more and more customers and gradually expanded its service over a wider area, cost per customer would actually decrease. By contrast, Edison's cost per customer stayed the same or even increased slightly as system size increased. Eventually, the AC approach won over DC. In 1889 Westinghouse sold 1,100 generators, only about 25% more than Edison. By 1900 he was selling more than twice as many. Except for some unusual niche markets,[7] all electric utilities had converted to a "Westinghouse" AC approach by 1920. And Edison's General Electric Company converted, too. It came to be a major competitive player in designing and building "Westinghouse" type equipment.

Many of these early electric companies – franchises purchased from Brush, Edison, or Westinghouse – were financed and run by businessmen who had no interest in the engineering and focused only on making a profit. One of the most successful was Charles Coffin, whose Troughton-Houston Company got its start with electric trolleys and railroads. He diversified where ever there was money to be made from electricity, and ultimately took control of General Electric from Edison as well as many elements of the former Brush Electric Company.

Whether they used AC or DC power, and regardless of who owned them, the earliest electric utilities thought of themselves as companies that sold lighting, not electric power. In fact, most early electric utilities billed their customers on a monthly basis according to how many light fixtures they had installed in their home, or in some cases on a "bulb evening" basis.[8] This early identification with being a *lighting* company as opposed to an electric utility persists: the term "light company" is still widely used as a synonym for "power company."

---

[7] Early elevators worked much better with DC than with AC. As a result, many utilities continued to provide DC power to buildings in the downtown core of major cities, along with AC for everything but the elevators. In the late 1960s, one of the author's (Willis) worked on what was among the very last of these systems left (downtown Houston, Texas) and subsequently designed the AC replacement circuit when it was retired in the late 1970s.

[8] Edison invented the first practical electric meter, but for a decade or more such devices were expensive, costing much more than the power being monitored. As a result, many late 19th century electric companies had "meter readers" who walked the streets every night with a clipboard, noting how many rooms had lights on in each home. There was only one light fixture per room and usually only 3-5 lights per home. Customers were billed according to this count – basically per "light-bulb evening."

Brush, Edison, Westinghouse quickly developed a host of other appliances and machines to use electricity – fans, clothing irons, electric stoves and washers, and large motors for industry. They pursued this approach both because it was what they did best – developing and commercializing new products – and because the use of such devices would expand the demand for electricity. In the very early 20th century, applications for electricity, other than lighting, began to be more important, at least in the view of some utilities.

Changes in the Focus of Utility Companies

Interestingly, the names of utility companies and the history of their names demonstrates the change in identity and focus that electric companies have undergone throughout the 20th century. Electric utilities founded in the 19th century called themselves "illuminating companies." Few of those names survived to the end of the 20th century, most being the victim of re-organizations and mergers somewhere along the line. One of the few remaining is the Cleveland Electric Illuminating Company – a part of First Energy Corporation today, but still operating under its original name.

Electric utilities founded or re-organized during the first half of the 20th century usually featured the word "power" in their names, with its prominence compared to the word "lighting" a sign of how late in the century they were organized. Early in the century they saw themselves mainly as *light* companies. Later, this view changed to *power* companies, whose main product was electric power, whether used for light or other activities.

Thus, companies established in the first few decades of the 20th century called themselves "Lighting and Power," for example, Houston Lighting and Power. Those established a bit later reversed the order of those words, as in Dallas Power and Light, and those founded later than about 1930 usually forgot light altogether, e.g., Empire District Electric, Nevada Power. After 1970, utilities merging or selecting a new name usually made no reference to power or light – Southern Company, PacifiCorp – or at best referred to "energy" instead, e.g., Entergy. These changes mirror the image companies had of their purpose. In the 21st century, electric utilities see themselves as service companies, and many have chosen names like Xcel Energy, or Progress Energy, etc., which subtly stress the quality of their service or commitment to a region.

Utility Regulation

As described earlier, by the late 19th century, the companies founded by Edison (General Electric) and Westinghouse (Westinghouse Electric) bowed out of direct retail electric sales and concentrated instead on building and selling power production and distribution equipment to retail electric companies. They manufactured and sold generators, transformers, meters and "switch-gear" (the circuit breakers and other apparatus needed both to control power flow during normal operations and to protect against short circuits should equipment fail).

Without a doubt, a big reason for the move out of retail sales was that Edison and Westinghouse along with the engineers who staffed their companies were more comfortable with the equipment business and its engineering culture than retail electric sales and the transactional culture it demanded. But an important factor was that by the early 20th century, government regulation began to reduce profit levels from electric sales to reasonable if not outstanding levels. Both Edison and Westinghouse, as well as many other early pioneers attracted to the new industry, were by nature risk-takers intent on achieving high profits. Since regulated and guaranteed profits were simply not what they wanted, they went elsewhere and left electric sales to a different type of businessman.

Electric utility regulation began slowly, driven partly by the change in attitude toward electricity as it came to be regarded as a necessity and not a prestige item, and partly by the desire of utilities to reduce risk, as will be discussed below. Many city and county governments, in areas that already had electric companies as well as in areas still without power, began exercising control over the future of the lighting industry in their jurisdiction, by passing laws that gave the right to grant franchise rights for electric distribution. A particular company was guaranteed exclusive rights to be *the* local electric company, but only if it accepted certain terms. The most important was that it had to accept an *obligation to serve* every home and business that wanted electricity, not just those it thought would be most profitable. Second, the rates it could charge were limited, so that when the cost of serving all those customers was taken into account, it would make a reasonable, but not excessive, profit.

While regulation and franchise rights limited profit levels, they took much of the risk out of the retail electric business, and most early electric utility businessmen viewed it as a good compromise. In particular, the exclusive right to be the *only* electric retail company in an area eliminated a big risk that worried early electric companies: What if a competing electric company, using a newer and superior technology, enters the local market. With its newer technology it could under-cut price and take over the market. With the guarantee of sole franchise rights, the utility could be certain that the investment it made to serve every home and business would pay off – maybe not with exorbitant profit levels, but at a profit nonetheless.

Expansion of the Urban and Suburban Utility Industry

In the first four decades of the electric era, from 1880 to 1920, over a thousand investor-owned utilities (IOUs) were formed to take advantage of the opportunities to make money through the retail sale of lighting and electric power. Most of these were "one-franchise" companies formed to serve a single city or town, often financed by local businessmen who had worked out a franchise agreement with local elected officials. In aggregate, these investor-owned utilities distribute power to over eighty percent of the population in the United States.

Mergers: Bigger is better

The vast majority of the original small, local, investor-owned power companies disappeared by 1950, having merged with their neighbors into larger utilities. Size conveys many advantages to an electric utility: the potential for slightly more efficiency, and the ability to operate at less risk from storms and unexpected emergencies.   As a result, various investor-owned utilities have been merging with one another for more than a century. Today there are far fewer, but far larger, electric utilities than there were a century ago, or merely a few decades ago, for that matter.

Many people unfamiliar with the power industry's lengthy history believe that mergers among electric utilities are a recent trend caused by de-regulation and performance-based rate pressures, but in fact *mergers and acquisitions among investor-owned utilities have been a continuous part of the electric business for more than a century.* Merger activity was actually highest during the first half of the 20th century.   There were originally over one thousand investor-owned electric companies in the United States. By 1980 there were only 238. Ten years later there were only 206, and a decade later that had dropped to less than 190.

Almost all modern IOUs are the product of many mergers over many decades. For example, the areas in New York state and New England served at present by National Grid USA were at one time broken into more than forty small IOU service territories. Modern investor-owned electric companies are quite large. Most serve well over a million customers each, and the largest at the time of this writing, Exelon, delivers power to seven million customers. The number of IOUs is also likely to continue to drop. In a manner similar to what happened in the airline industry, investor-owned electric utilities that cannot compete will cease to exist: the equivalent of those airline "routes," i.e., service franchise rights, lines, facilities, and franchise rights, will go to the successful companies who bid the most for them.

Municipally Financed Utilities

By the early 20th century, local electrification was a sign of prestige for a community, a business advantage for local industry, and a sought-after convenience for homeowners. It quickly came to be viewed as a civic necessity by many local governments – a city or town could not afford to be without "electrification" or it might be left behind as the 20th century progressed. Communities too small, or otherwise unable to make a satisfactory franchise arrangement with an electric company, often took matters into their own hands and built their own municipal electric system. They had constructed and managed their own roads, water, and sewer lines for years; they would do the same with power. A few counties did likewise, forming county utilities or Public Utility Districts (PUDs) along similar lines. Slightly less than 2,000 such municipal or public power district utilities still exist in the United States today, distributing power to about 14% of all electric consumers.

Similarly, during this period, cities and towns throughout Europe were converted to "full electric service" in much the way that North America converted. A mixture of state-owned, municipal, and investor-owned corporations focused mostly on cities and towns, where the high density of customers made electrification affordable. Slightly different standards for electric system design and operation evolved in Europe than in America, but both used a "Westinghouse" AC approach. To this day both the system frequency and nominal voltage level are different, but otherwise the electric systems are incredibly alike.

Rural Electrification

By 1930, fifty years after electric power had first been commercially viable, most cities and towns in the United States had electric power, but most rural areas still lacked it. It was a matter of business economics. In the large cities there were enough customers per mile to cover all the costs of lines and equipment at a price low enough to be affordable to nearly everyone. In small towns and the suburbs of large cities the situation was not quite as good, but retail electric sales were marginally profitable and still quite popular.

But rural electrification was a losing business proposition, even in 1930. There were too few customers in rural areas to pay for too many miles of needed line. Even if someone did provide funds for the investment in lines and equipment, recovering that investment over two or three decades would make the resulting electricity prohibitively expensive. The Europeans did not face this problem to the same degree: rural areas in Europe were packed compared to the sparse populations of rural Tennessee or western Kansas, to say nothing of North Dakota, west Texas, or New Mexico. Rural electrification in Europe proceeded slowly, but made enough business sense that government and private companies took on the task directly. In America, the countryside had remained dark at night for the first quarter of the new century.

The U.S. government realized that rural electrification would have advantages beyond those of immediate benefit to people living in remote areas. It would increase farm productivity, and it would reduce the flight of people from the countryside to the cities; while a certain amount was an acceptable sign of a growing industrial economy, too much of it would leave the "breadbasket" of the United States empty and abandoned. The nation's manifest destiny required that agrarian areas compete with the cities on equal terms.

Thus, in 1935, as part of President Roosevelt's New Deal, the United States government created the Rural Electrification Administration (REA) to bankroll rural utility companies. Boiled down to its basics, its cooperative concept for agrarian "utilities" worked much like the shared community grain elevators in many farming communities:

(1) The government would lend money to the farmers and villagers in a region to finance creation of a local electric company, and construction of a power system feeding their homes and businesses.

(2) Jointly, they would be co-owners of this local utility, called an electric membership cooperative (EMC). The cooperative was the official borrower of the money – farms and homes were not at jeopardy if the deal fell through. That was unlikely to happen, though, because the loan had a very low interest rate and a very long payoff period.

(3) Everyone would buy electricity at rates the REA set according to a formula that was aimed at paying off the loan eventually, but not quickly. These rates were perhaps twice as high as those in the cities, but still much less than one would have expected had an IOU, or even a municipal utility, been involved.

Nearly 1,000 utilities were formed in this way. Almost all of them exist to this day. (Government procedures, the financing methods used, and the EMC charters make it virtually impossible to arrange mergers or business restructuring of these companies.) Rural membership electric cooperatives distribute power to about ten percent of Americans.

While the government helped finance rural electrification, American innovation helped remove another barrier in front of it. Traditional ways of building an electric power system (by this time the industry was nearly 50 years old and indeed had "traditional" ways) would have been too costly for many rural areas, even with government assistance. A new type of power system design, called *rural single-phase*, was invented and standardized through design guidelines established by the REA. A rural single-phase power system could not distribute very much power, but then these areas did not need a lot – there were usually only four to sixteen farmhouses per square mile. Rural single-phase power systems cost *much less* than they would have if they had been built along the lines of systems in the cities. Conveniently, even though these rural systems were different in design, they used the same types of utility lines and equipment (as manufactured by General Electric or Westinghouse or the other suppliers), and they could run the same types of appliances used in the cities.

De-Regulation

As mentioned earlier, and as will be discussed more thoroughly in Chapter 10, utility regulation brought a lot of benefits to the early power industry. Viewed from the larger perspective, de-regulation, or "re-regulation" as it is more properly called, is occurring because regulation of the electric industry has served its larger purpose. In the early and mid 20th century regulation assured that adequate investment in utility infrastructure would be made. Its promise of return-on-investment led to the achievement of a universal power grid reaching throughout all cities and towns and into all rural areas. It provided a stable industry and stable prices during times when technology, systems, and prices were expanding. It led to the growth of a healthy power industry. None of that would have happened without monopoly franchises, regulatory assurances of return on investment, and rate caps.

But by the early 1990s, the technology, customer bases, and usage levels in the power industry had been fully developed and reasonably stable for decades. The original investment required to build the "universal electric system" had been recovered decades earlier.[9]   Most of the reasons that originally prompted utility regulation no longer exist to the same degree that they did at the beginning and middle of the 20th century.

Meanwhile, regulation and a complete lack of competition had led to a type of stagnation. Despite a minor facade of applied R&D and progress, regulated vertical electric utilities had little incentive to do anything other than what they have done in the past: Regulation made them afraid to take business risks, and there were no rewards for bold new ideas.

Very clearly, the electric utility industry did not keep pace with the progress of society and technology as a whole in the last half of the 20th century. Certainly, there was some progress and change, but not nearly enough. In 1995, the year de-regulation began, the electric utility industry was offering essentially the same products, services, and billing options that it had been providing to its customers in 1945. Had a similar lack of innovation occurred in the airline industry, jet planes and computer technologies would have been used by the airlines to offer passengers the type of service given in 1945: top speeds limited to 200 mph, only short hops between adjacent cities, with transcontinental travel requiring many frequent stops, low altitude operations that make flights bumpy and cause frequent delays due to storms, and prices affordable only for the wealthy or businesses in dire need of "fast" travel.

In 1997, only two years after de-regulation's start, the range of products and services, and pricing options for consumers began increasing rapidly, with no

---

[9] At least for IOUs and many municipal utilities. Many REAs are hopelessly in debt, but that is as much due to poor government policy as it is to poor management.

end in sight. Competition promotes innovation, wider choice, and lower prices. It will reward those power companies that think and do better, and remove from the marketplace those who cannot keep up. A controlled form of competition is what is being sought, from what is really "re-regulation" – a change in the rules. (See Chapters 2 and 10 for more details.)

Downsizing: One trend of the '90s

Proof that utilities can do better is evident in the downsizing trend that swept the industry in the early and mid-1990s and that continues at a slower pace today (as utilities asymptotically approach the best they can do). As de-regulation began to be a possibility, utilities prepared by cutting staff and capital spending. They did so partly to reduce cost, so that they would become more competitive. They also knew that all of them – regulated and unregulated – would face increasing investor expectations to improve their bottom line, as any and all companies in any competitive, unregulated industry do. But a more important reason was that lowering employee count lessened their fixed costs, which reduced their business risk, a prudent move in times of uncertainty.[10]

Despite what naysayers had predicted, the lights stayed on. In fact, some utilities reduced their employee count by half, trimmed capital spending by 33%, and still managed to find a way to improve customer service levels. Basically what happened is that the drastic cuts forced utilities to re-examine traditional methods, to focus more on the basics (customer value) while foregoing other issues, and, most important, to innovate.

There might very well be another round of downsizing in the early 21st century. Some industry observers point out that compared to a TV cable company, an electric utility of equal size – as measured by miles of line, dollars of revenue or in some other way – has nearly twice as many employees. The TV cable industry is the only "utility industry" born in the modern era. With no tradition or historical precedent, it organized itself to take maximum advantage of modern technology and practices, and has roughly half the employee count of equivalent electric utilities. Certainly, it is difficult to compare one type of industry to another, but most such comparisons, when they can be done, usually show that many downsized electric utilities are still not "lean" enough.

---

[10] One way to reduce business risk is to reduce the fixed portion of the business's costs. This can make sense even if the company's variable costs rise, as they would, for example, if an electric utility had to hire outside firms to do construction because it didn't keep enough full-time employees to do the work. While its overall costs might be higher, the company can react more quickly to changing conditions because it has a smaller portion of its costs that it cannot rapidly change.

De-Regulation and Debacle

As stated in Chapter 2, electric industry de-regulation can work very well when implemented with soundly designed rules and in a framework that accommodates the dual nature of utility transmission flows (see Figure 2.9 and accompanying discussion on the difference between electrical and "money" flow in a power system). However, early de-regulation efforts in many nations were less than perfect, and in some, including the US, poorly thought out and just plain clumsy in their execution. In the United States de-regulation driven by federal policy but interpreted at the state level in many different (and occasionally, nearly inept) ways created very high levels of stress on utility finances and customer service quality.

For example, in California, price volatility caused by de-regulation pushed several utilities to the brink of bankruptcy, among them one of the largest and arguably best run utilities in the industry (Pacific Gas and Electric). Utilities were caught between escalating prices for power at the wholesale level that they had to buy, and frozen, regulated rates at which they could sell that power at the retail level. A lack of perfection in de-regulated rules was partially to blame for a series of rolling blackouts that were common throughout the state in 2001.[11, 12] (The reader should see the comments and footnote at the end of Section 2.7 on the root causes of the California energy deficiency: de-regulation did not create the root cause of the problem, but did make it worse).

In August 2003, a blackout occurred over the entire northeast US and into southeastern Canada. It started with the failure of several transmission lines in northern Ohio (a situation any power system should be able to tolerate well) that cascaded in a matter of a few moments over multiple states and regions, putting more than 50 million people in the dark, some for a day or more. Whether this was caused by de-regulation is a matter of interpretation, but the very wide interconnection of the power grid in the Midwest with that in the Northeast, along with the heavy, long-distance power flows through the grid, were both

---

[11] A rolling blackout occurs when a utility or regional grid operator knows their power system will not have enough electric power to meet all demand. To avoid an uncontrolled (cascading) blackout that might take down the entire grid, it shut down portions of the grid, deliberately turning off power to perhaps a million consumers at one time. To spread the pain evenly, it "rolls" the blackouts over the region: one hour without power here, then one hour without power there, etc.

[12] It has become quite popular to blame large de-regulated energy suppliers like Duke and Enron for manipulating the market and causing these blackouts. In the authors' opinion, that is missing the point: California set up rules and incentives for its energy market that encouraged a type of look-the-other way "cheating" or even worse, that provided no financial incentive for running generators at some times. California's rules were simply not a balanced, sound system.

due to de-regulation.[13] These contributed to the problem (see Section 16.3).

Eventually, policy makers at both the federal and state levels adjusted their electric industry rules and structure, utilities adapted, and the competitive wholesale market, operating as it was expected to do under the open competition created by de-regulation, brought forth more energy supply and "solved" most of these problems. And politicians responsible for some of the most egregious errors in both pre-deregulation energy policy as well as mis-guided de-regulation policy were voted out or recalled from office. At the time of this writing, de-regulation is not entirely fixed (see last section of Chapter 2), but it is working better, and with more fine-tuning, will work quite well.

Fraud and Frustration

Chapter 2, Section 2.4 discussed the development of the trading market in wholesale commodity power that quickly grew under de-regulation. In the late 1990s several very large companies developed around buying, selling, and re-marketing electric power. Among these, the company that certainly cut the widest swath in the power industry for a short time was Enron, a Houston-based company that owned generation, wind power, gas resources, and numerous other assets, and that did an increasing volume of what appeared to be very profitable commodities trading in electric power. Ultimately, Enron collapsed due to over extension of its finances and allegedly fraudulent practices in its financial reporting and management, a spectacular scandal which brought financial ruin to thousands of its investors and disgrace to a number of executives who deserved that and a jail term.

At the time of this writing, the jury is still out (literally) on many parts of the Enron scandal, but in the authors' opinion, the "commoditization" of electricity through de-regulation created a situation where cheating was amazingly simple and quite tempting. As discussed in Chapters 2 and 10, de-regulation made

---

[13] See Chapter 2. High voltage transmission grids covering multiple states and even several regions were necessary under de-regulation in order to create efficient wholesale power markets. Heavy, long-distance power flows became common due to the competition those markets fostered. The power grids in both the Midwest and the Northeast US, which were, along with southern Canada, all tied together, were originally designed as contingency-support power pools by regulated utilities, not for such heavy flows. This combination of wide regional grid and high power flow was caused, to a great extent, by de-regulation. But the Northeast blackout was also due to the fact that utilities were operating a bigger and more complex power system than anyone appreciated prior to the blackout. Complexities and problems that could develop in such a grid were not anticipated and had not been studied in sufficient depth by anyone in the industry, nor had equipment for precise, regional-wide timing coordination and other needed control systems been put in place. It is difficult to blame these failings on de-regulation, even if it did lead to the conditions that exposed this industry weakness.

electric power into a commodity, but it has significant differences from traditional commodities. It cannot be stored as can other tradeables like wheat, coffee, and soy beans. It cannot be seen as can other commodities. It was new, and not easily understood by investors and regulators and Wall Street analysts. Therefore, it was easy to "pump up" a business's appearance through circular sales and mis-representation of trading volume: one could not count grain silos or warehouses to help check a company's math or provide a sanity check on what it was reporting in its annual report.

Many other companies including some traditional utilities jumped into the power trading "game" in the late 1990s, many in a big way. Most adhered to sound, legal, and ethical standards of corporate conduct, although a few were infected by the temptation to bend the rules. Regardless, quite a few pushed their financial leveraging past the limit of prudence, and ultimately shared the same fate as Enron (which would have very likely collapsed, scandal or no scandal).

Briefly, the claimed success of Enron and several other energy trading companies gave the appearance that they were in a vastly profitable business. In the period 1995 – 2001, the industry as a whole focused on power trading and wholesale generation with respect to investment and business expectations. But the fact is that electric power is a very competitive, low-margin, and potentially high-risk "game," in which it is very easy to over-leverage a business and quickly "cash and burn."   By 2003, many utilities and quite a few non-utilities had realized that operation of a T&D system along with retail electric sales offered more stable and perhaps ultimately a more profitable (certainly less volatile) business opportunity. One of the frustrations facing many utilities that had moved heavily into the "trading game" for a while is that while it did not destroy them, it left them with heavy debts and an inability to invest in T&D today, while they pay off those debts.

3.5 LOOKING TO THE FUTURE

Aging Infrastructures

The last sentence in the preceding section emphasized how some utilities just cannot invest in their T&D systems. But the simple fact is that most modern utilities need to invest heavily in T&D over the next two decades, because much of their present system is nearly worn out. Looking to the future, the "aging electrical infrastructure" problem is one of the greatest challenges facing the still regulated part of the industry.

All electric utilities in the Unites States, and most in Europe and many other places throughout the world, have transmission and distribution equipment in their systems that has been in service for more than fifty years. In some cases, a utility may have facilities (a substation) where a good deal of the equipment is more than 80 years old, and the authors have equipment in "front line service" that pre-dates World War I. A few utilities have equipment that *averages* between 40 and 50 years of age.

This aging infrastructure problem is most prevalent in the still-regulated, T&D portions of the power industry. There *were* many very old generation plants, but the competition created by de-regulation tended to weed them out very quickly (but not without a good deal of financial loss to some of their owners). Basic electrical T&D equipment like transformers, transmission towers and conductors, wooden poles, breakers, and control replay panels is designed to be very robust. Much of it can survive fifty or even sixty or more years in service, if well cared for. But a considerable portion of the industry's in-service stock is approaching that age.

This problem grew very gradually. For what appears sound reasons if viewed on a short-term financial basis, electric utilities almost always practice a "run to failure" policy with regard to equipment: as long as repair will keep it in service they do not replace it even if it is quite old. Proper maintenance can keep electrical equipment in good repair for remarkably long lifetimes, one reason that large portions of many utility T&D systems are composed of so much old and sometimes rather inefficient, high-maintenance designs. This older equipment breaks down frequently and can require high levels of repair effort.

As a result, the problem is getting worse every year. Most utilities do add new equipment as their systems and customer base expand each year, and they have to replace perhaps .5% to .66% percent of their equipment each year because it fails. But in spite of that "new blood" added each year, the average utility system ages about 10 – 11 months per year.[14] Failure and "must-replace" rates are gradually creeping upward: one can keep old electrical equipment in repair for a long time but eventually it just starts wearing out no matter what one does.

Although they will be challenged by these increasing failure rates, most utilities will be able to manage the increasing rate of failures well for quite some time to come, preventing undue customer service interruptions due to those higher failure rates, and finding way to absorb the higher maintenance costs those failure rates cause through improved efficiency in their management and O&M practices.

But the big problem is more fundamental: *utilities can't afford to replace this equipment.* Sooner or later, most likely in the next two decades, all of that older equipment *will* fail and they will be compelled to replace it. But they will

---

[14] A system in which no equipment was replaced and none was added would age at a rate of one year per year. One in which equipment was replaced at at a rate inversely proportional to its age (e.g., a 2% replacement rate for the oldest equipment in a set averaging 50 years old) would hardly age at all. Every utility system the authors know is much closer to the former than the latter situation.

not have the money. Utilities generally borrow money to buy electrical equipment and "finance it" for about 30 years, even if the average equipment lasts far longer than that. Today, most of the equipment that the average regulated utility has in its system was "paid for" long ago. They find themselves in a situation analogous to what many young people create during the first years of employment: They buy a new car and finance it for a period (say five years) less than its expected lifetime, making the payments without problem. When paid off, the car is still usable so they keep it and use it for several more years. During that time they gradually adjust to the lack of a monthly car payment, spending in other ways, until after several years that becomes the standard routine. Then the car fails and has to be replaced. They saved no money and thus have no cash to pay for a new car, and they can no longer afford car payments within their budget.

While many utilities face similar problems, the reason in many cases is not that they were "foolish" as much as that they are regulated. Although this is an oversimplification, regulated utilities are not really permitted to "save for the future." Since the equipment lasted longer than its depreciation period (analogous to the period of car payments) their costs dropped. Their rates are regulated based on costs. Many have gone four decades or more in situations that are not sustainable: *eventually they must spend more* than they currently take in. The amount of cost increase is significant but not staggering: on the order of 15% overall.

The problem is that the utilities cannot raise their prices to pay for needed equipment replacements unless they receive regulatory approval, and regulatory commissions, utilities, and utility customers alike have become accustomed to the present rates. Utility regulation is a political process that is reluctant to raise prices, period.

Chapter 15, Section 15.2, discusses aging infrastructures in much more detail. The authors have no doubt that this problem will gradually escalate in importance until it could dominate industry attention and utility business prospects and stock prices for a number of years. There can be no doubt, however, that the power industry will get through this problem: eventually regulatory commissions will have to recognize the need to increase replacement rates and customers will have to accept somewhat higher costs for power, but probably not before there is a considerable amount of controversy and stress for all concerned.

Technology and Automation

Despite the fact that the electric industry is well over 100 years old, there continues to be a brisk pace of development in many of the technologies that electric utilities use. As a result, gradual improvements in cost, efficiency, environmental impact, and reliability can be expected for the foreseeable future.

A good deal of this improvement will be due to progress in "ancillary" technologies such as computers, software, and data communications. Utilities and power systems, by their nature complicated entities with a widely distributed asset base, benefit particularly from a combination of computerization (which can handle the complexity) and good data communication (which can address the dispersed nature of the systems).

Although utilities have long relied on computers for analysis, billing, facilities databases and control systems, computer technology is hardly yet mature, and data communications costs continue to drop. Two areas of development will improve business and customer performance. First, integration of "enterprise-wide" systems will consolidate business processes, achieve coherency and immediacy of function and focus, and drive an overall improvement in the productivity of people. While this may sound like prime, management-consultant BS, there is every reason to count on continual gradual increases in overall productivity of people and business equipment to a cumulative 10- 25% improvement.

The second area of expected improvement is in power system automation. Automation of equipment permits a utility to work the equipment harder. It can operate the equipment in "closed loop mode" (monitoring it in real time and allowing operators to push it to its prudent limits) and to do "when needed" maintenance and service, which improves equipment condition and lifetime, driving down the utility's maintenance costs.

Use of improved data communications technologies, including the internet, makes the ability to monitor and control dispersed equipment less costly. The major impact of this gradual reduction in automation cost is not that it reduces the utility's costs for control. Rather, it improves the bang-for-the-buck of automation, which *broadens* the base of equipment that the utility can afford to automate. Whereas today only major equipment can be automated, over time it will spread to smaller distribution equipment. Power systems will become more reliable and less costly.

Fine Tuning De-Regulation's Regulations

Over a period of about 10 years beginning in the mid 1800s to the mid 20th century, electricity and the electric utility industry developed from a nascent and not-entirely understood curiosity into a trusted technology that was a cornerstone of mankind's civilizations. Regulation and monopoly franchises helped shield the industry's early development and channel its energy into productive growth, rather than competition and in-fighting. But by the late 20th century, many policy makers believed it was not just a mature, but a stagnant industry, innovation averse and unresponsive in some ways to the culture's larger needs.

While that view might be overly harsh, the fact is that the industry was de-regulated to reduce greatly the degree of monopoly involved. Many readers are no doubt struck by the complexity of the de-regulated industry structure, as compared to the traditional way electric power was produced, transmitted, distributed, and sold. But despite problems and its many complexities, de-regulation works better than most naysayers said it would, although not nearly as well, or as cleanly, as proponents hoped.

But while the industry is perhaps not in chaos at the moment, it is in a great deal of turmoil. In the US, the federal government (FERC) and the state regulatory agencies are often at odds on just who has what authority over which utilities, and there appear to be gaps, as well as overlaps, in policy and jurisdiction. What makes all of this so difficult is that there is little clear pattern to the differences and variations in policy, rules, and structure: a particular type of generation structure or policy in a region does not imply it will be accompanied by a particular type of transmission policy or marketplace policy or regulatory approach. As one looks across the industry, one can find all combinations of these different approaches, and a few not covered in this simple explanation.

There are efforts to unify national and state policies into a more coherent whole. At the federal, state, and local levels, different efforts are aimed at fixing different problems with different solutions. Most of this activity is proceeding with a speed, efficiency, and rationality no worse and no better than the political process displays any time there is money, power (literally) and blame to be apportioned. Eventually, the industry will be "fixed" and the solution will be acceptable if not elegantly simple and rational.

FOR FURTHER READING

J. A. Cazassa, *The Development of Electric Power Transmission,* IEEE Press, New York, 1994.

J. Jonnes, *Empires of Light,* Random House, New York, 2003.

M. Wilson, *American Science and Invention,* Crown, New York, 1960.

H. L. Willis, R. R. Schreiber, and G. V. Welch, *Aging Power Delivery Infrastructures,* Marcel Dekker, New York, 2001.

4

Electric Power

4.1 INTRODUCTION

Electric power is a *natural* physical phenomenon, a fundamental type of energy which mankind has learned to create and control for its benefit. Electricity is *always* energy produced by converting some *other* form of energy (heat, mechanical motion, solar light, or moving wind, etc.) into electric power.

Electricity has two advantages over other forms of energy that have led to its wide popularity. First, it is *flexible*: it can be transformed into heat, light, mechanical motion, radio signals, television images, and stereo sound. Second, it is very *controllable*: it can be turned on and off in a millionth of a second, and metered out precisely, from an amount so little that it would hardly move one grain of sand a tenth of a millimeter, to quantities that can power entire nations.

This chapter is a basic "layman's tutorial" on electricity, electric power, and the basics of electrical power engineering. It discusses electricity, electric power, and some of the fundamental concepts used in electric utility systems at an introductory level. Occasionally it glosses over messy details if they are not needed to explain the fundamental "big picture." Persons with an electric utility company who do not have an engineering background may find it useful in understanding the basics of their company's product. Engineers and others who know power may find it useful in helping explain basic power concepts to those new to the field and for preparing non-technical presentations to community groups, etc. It begins with a discussion of the basic concepts of voltage, current, power, and electric power flow in Section 4.2. Section 4.3 then discusses some of the characteristics of electric power that most shape or constrain its use.

4.2 VOLTAGE, CURRENT, AND POWER

Electricity has two fundamental components, the *current*, or amount of electrical flow, and the *voltage*, or electrical pressure pushing the electric flow. Together, voltage and current determine the amount of power – the rate at which useful work or light can be produced:

$$\text{Power} = \text{voltage} \times \text{current} \tag{4.1}$$

Units of Measurement and Typical Values

Voltage is measured in *volts*, and current in *amps*, arbitrary units of measurement for each, invented when the electric industry was in its infancy, and that are now standardized worldwide. High voltages in a power system are measured in terms of kV – thousands of volts, e.g., ten kV is 10,000 volts.

Power is measured in *watts*. One watt equals one amp times one volt. The amount of power required, in watts, can be produced by any combination of voltage and current that gives the desired product. For example, one horsepower equals 746 watts. It can be produced by one amp being forced at 746 volts; two amps at 373 volts; 7.46 amps at 100 volts; or 746 amps at one volt. Large amounts of power are measured in kilowatts (1,000 watts) abbreviated kW, or megawatts (1,000,000 watts), abbreviated MW, or gigawatts, GW (1,000,000,000 watts). For no discernible reason, the "k" used in kilowatt's abbreviation is not capitalized, whereas the M and G in megawatt and gigawatt are.

Energy used is measured in terms of *power* times the *duration* of use. One kilowatt used for one hour is a kilowatt-hour, abbreviated as kWh, or kWhr.

Alternating Current

There are two types of electric power, direct current (DC) and alternating current (AC). In direct current systems, the electricity constantly moves in one (direct) direction. It can be thought of as flowing like water through a pipe. The voltage is analogous to water pressure, always pushing in one direction, and in DC systems, current, like water flow in a pile, always moves in that one direction.

Almost every electric utility system in the world uses a different type of electricity: alternating current, abbreviated as AC. In an AC power system, both the voltage and the current in all electric equipment oscillate, as shown in Figure 4.1. Voltage and current alternate back and forth at the same rate, many times a second. Although it might seem that the resulting power would be worth little because it pushes and then pulls back so soon afterward, this is not the case. For one thing, electricity moves so fast, at nearly the speed of light, that it can travel more than 1,500 miles in 1/120th of a second (the standard duration of a single one of those oscillating pulses), a distance great enough to span the route from power generator to power consumer, anywhere on this planet. Beyond that, many electric appliances are indifferent to the direction of current flow. An electric water heater, for example, produces heat whenever current flows through its filament in either direction. It does fine with AC current. So do most other electrical devices. It doesn't matter to them that the AC power is fluctuating in direction, because power flow is a bit different than water flow: the oscillating nature of AC power actually provides power very efficiently. The physics works out and the equipment works, even if it is a bit difficult to fathom without a lot of mathematics involving complex-variables.

The rate of oscillation of the electricity in a power systems around the world is either 50 or 60 cycles per second (50 or 60 *hertz* – oscillation rate is named after an early electrical scientist, Hertz). "American" type systems oscillate at 60 hertz, "European" type systems at 50 hertz. Both work just as well, and neither frequency is noticeably better than the other, despite what one might hear from heavily opinionated "experts." Anytime one finds an engineer who insists one is substantially better than the other it will turn out that he or she "grew up" on that type of system and has a rather narrow understanding of the other type of system's capabilities.

Although both DC and AC power are electricity, they act very differently in some ways. Each has certain advantages and disadvantages over the other in different situations, but on balance AC power is considered the more useful, and in some ways a bit safer. With respect to safety, DC power is far harder to stop when something goes wrong – as in an accident or equipment failure that leads to a short circuit. Because AC power changes voltage back and forth many times a second, there is always a brief instant – all that is needed – when the power is not moving and a circuit breaker can effectively break a short circuit current. The same cannot be said for DC, which means circuit breakers for DC are, other things being equal, bigger, heavier, and, to the authors, not quite as preferable as using AC power.

More important, though, AC power allows the use of a *transformer* – a nearly foolproof device that can change voltage wherever it is needed (transformers are explained later in this chapter). Transformers enable power engineers to use high voltage when necessary – as when moving great amounts of power long distances from generator to city – and then conveniently lower voltage to a safer, more efficient, and more useful level for home and business use. By comparison, changing the voltage level of DC power is arduous, inefficient, and unreliable, meaning that a DC utility system could never function as efficiently as an AC power system (see Chapter 3's discussion of Edison and Westinghouse systems). For these reasons, all electric utilities worldwide operate at either 60 or 50 cycles per second (3,600 or 3,000 cycles per minute).

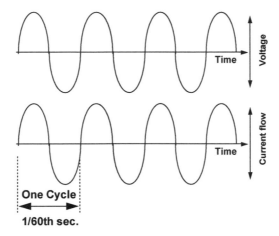

Figure 4.1. Alternating current (AC). Voltage and current in an electric system oscillate at 50 times (European) or 60 times (American) per second. AC power provides certain advantages for use in electric utility systems.

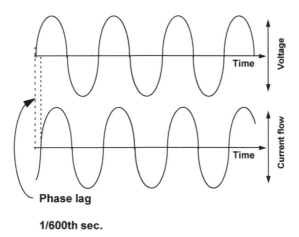

Figure 4.2 Here, current lags behind voltage by 1/600th of a second. This results in a condition called "VAR flow," electrically analogous to foam filling pipes, which diminishes the effectiveness of the power delivered by the wires.

Since voltage and current in an AC power system vary many times a second, they are usually measured in terms of the *RMS value* – root mean square – a mathematical term for computing their *average value over time* in the way that is most useful in electric measurement. RMS value is about 80% of peak value, so that when one says a typical household wiring system is 120 volts, this means it usually peaks during each cycle at more than 150 volts, while averaging about 120 volts. What is really important, though, is the RMS value, the average produced through each cycle, which is representation of the power it can provide.

Reactive Power

One of the most confusing points about alternating current is that often the *usable power* being delivered to a point in the system is much less than the product of the voltage times the current. In some cases, the current pulses fall behind the voltage pulses, as shown in Figure 4.2. This happens due to "reactive impedance," or "X" of the power lines, inside motors, and other equipment in the power system or the appliances using the power. One can think of this as delaying the current slightly, so that it lags behind the voltage.

When the current and the voltage are not in perfect phase, as was shown in Figure 4.1, but, instead, have a phase lag between them, the resulting electric power is not as useful. It will not perform as much work. Power engineers refer to the part that is not useful as *reactive power flow,* and to that portion of the flow on the line as *VARs* (Volt-Amps Reactive). VARs can be likened to "electrical foam." They take up room in the wires, but provide little substance. *Power factor* is the term given to the percent of total flow that is useful power, and not reactive power flow or VARS. A power factor of 90% is good – 9/10ths of the power being moved is useful, but one of 50% is poor. A device called a *capacitor* can be installed at strategic locations on a power system to push current and voltage back in phase, improving power factor.

Is Electricity Safe?

Any source of energy – a tank of propane stored behind a rural farmhouse, a large millwheel, a windmill, or an electric line – is a potential hazard that can injure and even kill if misused. Electricity is no different, but it is a particularly safe form of energy when handled according to standard safety precautions, such as those in the National Electric Safety Code. One reason it is safe is that it is so controllable, but a big challenge in keeping it safe is that it can act quickly, essentially at the speed of light. Automatic equipment can detect most leaks, called short circuits or faults, and "shut down" the electrical flow before substantial damage is done. Although not completely fail-safe, such *circuit breakers* and *ground fault detectors* are quite effective, and combined with sound design, good maintenance, and safe practices make electricity quite safe.

Typical Ranges of Values

A typical light bulb uses 60 watts – 1/2 amp at 120 volts. A toaster uses about 1,000 watts – a bit more than 8 amps at 120 volts; a television 240 watts – two amps at 120 volts; a large central air conditioner or heat pump 6,000 watts – 25 amps at 240 volts. As previously noted, large amounts of power are often measured in *kilowatts* – units of 1,000 watts – and larger amounts, still, in *megawatts* – a million watts. The cumulative demand of a large city or state might be measured in billions of watts – *gigawatts*. Cumulatively, metropolitan Philadelphia, Pennsylvania, uses about 8 gigawatts of power during the peak period of electric usage.

In almost all cases, whether inside a home or on the utility system, the number of volts being used in any application is greater than the number of amps. Voltage used in homes and most businesses worldwide is somewhere between 100 and 250 volts depending on local standards, but the wires for running the current through the walls to each outlet in a home will be sized to carry a maximum of about 15-25 amps.

The same is true throughout a utility power system, a large transmission line might operate at 345,000 volts while carrying 2000 amps. Again, far more volts are used than amps. The fact that more volts are used than amps is not due to any fundamental physical reason – remember volts and amps are arbitrarily chosen units of measurement. It simply turned out that the most efficient use of electricity, as engineers have learned to design it, usually calls for about one hundred times as many volts as amps. This is a useful rule of thumb.

Ranges for voltage

The power flowing through most houses and buildings in the United States is between 110 and 120 volts, in Europe about 230 to 250 volts, and in Japan 100 to 105 volts. The differences exist because these countries each established a different standard when the electric industry started there. Voltage levels in this range (100-250 volts) provide enough power for typical small appliances like TVs, microwave ovens, etc., and even large equipment like air conditioners and water heaters. What is important for good operation is that the appliances be designed to run at the voltage level standard used locally, e.g., appliances intended for 250 volt operation in France would not work well if plugged into sockets in Japan, where they would get only 40% of the voltage they need to function.

Electric utilities use a higher voltage in their systems as they move power from one location to another. *The more power that must be moved, the higher the voltage used.* Large transmission lines, built with wires hung from big, lattice-like steel towers, typically run at voltages of from 100,000 to 750,000 volts. Local distribution lines, built on poles or buried under the street, usually operate at around 10,000 volts.

As noted earlier, voltage is the electrical pressure that forces current to flow where it is needed in a power system, pushing it from generation plant to city along transmission lines, and from electric utility to a television set in a home. Voltage will also try to push power where it is *not* wanted. Thus, much of the cost and engineering effort in an electric utility system is devoted to ensuring that this does not happen. Electricity, driven at hundreds of thousands of volts, can leap several feet through the air. *Insulation* consists of materials through which electricity will not travel, and sometimes just wide spacing (distance between lines, etc.) is used so that electricity will go only where the power engineers want it to go. Much of the insulation in a power system is provided by spacing – keeping lines and equipment away from anything else, hence the use of high transmission towers.

Insulation can also be provided with *non-conducting material* to cover the electric wire. Most power cords for home appliances have a "rubber" (actually vinyl) outside jacket of insulation about 1/25th to 1/16th inch thick, more than sufficient, under normal circumstances, to contain the 120 or 250 volts used in household wiring, with a good margin for safety.

The higher the voltage, the more insulation required. Power cables for electric utility underground application at higher voltages need thicker, more effective insulation material.[1] Insulation in most utility systems is provided by air, i.e., the wires on most overhead distribution and transmission lines are bare metal, protected simply by not having anything nearby.[2] A foot of air provides dependable insulation for several thousand volts, but usually more than a foot is used. Technically, a few feet are all that is ever needed. Distribution poles and transmission towers raise power lines far off the ground to provide more than just insulation, their purpose being to get them far away from vehicles passing under them, people who forget to look up while carrying a metal ladder, etc.

Ranges for current

Most household electric appliances require from 1/2 to 10 amps. Most equipment, household or industrial, uses far fewer amps than volts: typically from 1/10 to 1/100th as many amps as volts. There are a few exceptions, e.g.,

---

[1] Some types of plastic and vinyl materials have been specifically formulated for electrical applications, and provide more effective insulation per inch of thickness than rubber or vinyl, but cost more. Such issues are important to many utilities. The space for cables under the streets in many cities is so crowded with other uses, e.g., sewer, steam and water pipes, and telephone, that the difference between a three-inch and a four-inch cable can be worth thousands of dollars a mile to the utility.

[2] The wires are held on the towers and poles with "insulators," specially designed mountings and "chains" that are made of plastic, glass, or ceramic.

unusual industrial processes like electroplating, which require large amounts of current, perhaps thousands of amps.

Usually, the current carried on electric utility lines will be in the range of 100 to 1,000 amps, but again, this is about 1/10 to 1/100 of the voltage being used.

Engineering trade-off between voltage and current

The higher the voltage, the less current is required to provide any amount of energy. Powering a toaster at 240 volts means it will need only half the amps, and the metal in the wire could be half the size. However, the higher the voltage, the more insulation (the rubber-like coating around a wire, or simply enough distance as when standing underneath a transmission line held by high towers) is required. The choice of voltage and current is determined by engineers in order to provide economy, safety, and efficiency of use. The more current that is carried, the larger the metal component in the wires must be. Wires are called *conductor* if used on overhead lines, and *cable* if used underground or inside the walls, floors, and ceilings of buildings. Only the amps being used matter in determining how much metal is needed. Most wiring is aluminum, or copper. The latter is slightly better from an engineering standpoint, but more expensive and heavier (weight matters if it has to be hung from poles). The metal in most interior house wiring is about 1/16th inch across and can carry about 15 amps. That on overhead T&D lines is usually an inch in diameter and can carry up to a thousand amps.

The amount of amps has nothing to do with how much insulation is needed: only the voltage matters. Engineering studies show that it is best to use higher voltage rather than a great deal of current. A variety of reasons leads to most applications following the rule of thumb that voltage is about ten to one hundred times amperage. This is not just for economy and efficiency, but safety as well. Most power engineers would prefer to deal with high voltages rather than high currents, large levels of which are much more difficult to switch off quickly, making safety measures harder to arrange.

Higher voltage has much more "clout" than low voltage

Voltage has a squared relation to capability: double the voltage and the power delivered increases by a factor of four, not two. As a result, slightly increased voltages have much more dramatic effects than might be expected. A light bulb designed for 120-volt operation will produce *very* intense light if connected to 240 volts, but will burn out in a matter of minutes from overheating. The 240-volt appliances used on special circuits in a home do not have access to two times the power available from normal 120-volt electrical outlets – they are being provided with the potential for *four times* the power.

To a certain extent this same squared relationship applies to the care needed when working with electricity. A macabre way of putting voltage into perspective is to look at what it does to a human being or an animal that comes into direct contact with it. Less than ten volts is not noticeable in most cases. Twenty volts can be felt to the touch. One hundred volts will sting severely, but is seldom fatal (although under certain conditions electricity in that voltage range *will* kill, making it is advisable to avoid contact at all times). One thousand volts will kill instantly under most circumstances, but leave little trace beyond a bad burn where contact was made.[3] Ten thousand volts will kill even before contact is made; current driven at that voltage will jump a short distance – an inch or less – through the air. And it will do a lot more than just burn, in some cases leaving a real mess behind. One hundred thousand volts can leap up to a foot, and vaporize many things that it contacts.

The most dangerous voltage levels from the standpoint of accidents are actually in the triple digits. One reason is that people often treat 1,000, 10,000, and higher voltages with great respect, but not so voltages in the range of 480 to 600 volts. Voltage in this range is typically used to power heavy equipment in many factories, and often it doesn't get the respect it deserves. After all, some people think, 480 volts is only four times what one has in one's home wall outlets. However, due to the squared relationship described above, it will drive *sixteen times* the energy through anybody that inadvertently grabs hold of it. Many people are hurt each year because they do not follow proper procedures with what they believe are these benign voltages. Additionally, three-digit voltages are particularly dangerous because of how they interact with the human nervous system. Voltages around 300 to 800 voltages are strong enough to hurt badly but not so strong that certain very-high-voltage effects act to "mitigate" electrocution.

Regardless, electricity is safe if the equipment is well designed, in good repair, and well operated, and if people rigidly adhere to well thought out safety procedures. But the fundamental rules in all cases, from twelve volts to twelve million volts, are: if there is no need, don't touch it, and if you must touch it, follow sound safety procedures and think before you act.[4]

---

[3] Electric fences used to contain cattle, horses, etc., often operate at voltages of one hundred or several hundred volts. However, a special circuit is used to limit to a very tiny amount the current that can flow. The high voltage assures that the electric fence will sting enough to get the attention of a big, dumb animal, but the current limiter assures that the current is curtailed so that contact will cause no permanent harm.

[4] Unwavering attention to safety whenever one is around electrical apparatus of any type should always be "Rule One." Mercifully, after 35 years in the power industry, the authors have never had an electrical accident, but we do know people who have been badly hurt. In the course of our work we have done "forensic" examinations of several deaths from electrocution. All of these accidents occurred because someone did not take his time and follow sound safety rules.

4.3 CHARACTERISTICS OF ELECTRIC POWER AND SYSTEMS

Electricity Isn't Easy to Store

One disadvantage of electricity is that it is relatively expensive to store. To be inexpensive, it has to be made and sent to the consumer at the moment of use. In fact, this is perhaps electricity's single biggest drawback compared to other power sources. Energy can be stored chemically as gasoline, propane, compressed natural gas, etc., at much less cost and in far greater density (energy per pound of storage system) than electricity.

It *is* possible to store electric power, for example, in the batteries of a flashlight. But the batteries in a large multi-cell flashlight contain only enough energy to power an average single family home for a minute or so. And the cost of that power is about $15 per kilowatt-hour, more than 100 times what utility power costs in most locations in the United States. Cost of energy storage is one reason why electric cars have not proved viable. The energy storage is so expensive, and the batteries so inefficient in terms of energy stored per pound, that inefficient storage stands as a barrier to effective design.

This concept dominates electric power system design, electric utility operations, and all thinking in the electric industry. Power systems are built so that they can sense instantly the changing demands of people and their appliances and respond literally in the blink of an eye. Equipment and systems are designed, and utilities invest, with the assumption that power must be delivered when required.

In addition, one sees that typically energy storage, whether for private or industrial use, is accomplished through some other medium. For example, most homes in the United States store the equivalent of several kilowatt hours of electrical power, in the form of the heated water inside their water heater. Many homes in Europe use "thermal storage" heaters. During the night, electricity is used to heat several suitcase-sized piles of ceramic bricks to high temperature. The next day, that energy heats the home, reducing the demand for electricity when business and industry need it.

Constant Voltage Systems

As discussed above, if a light or motor needed more power the additional power could be supplied by increasing the voltage, or the current, or both voltage and current, to the device. Although this is the case, *electric power systems throughout the world are designed to keep the voltage as nearly constant as possible, and change the power delivered by varying only the current flow.* Thus, the electric power delivered to a typical house in the United States is provided to the house at around 110 to 120 volts (with a limited amount that is at double that voltage, for large uses like electric dryers). This voltage is supplied whether the house is using no power, or a great deal of it. What varies, as power used inside the house varies, is the current drawn from the power system, zero amps if no power is used, and perhaps 90 amps if 11 kW is being used, which is about the peak demand of a typical small home with its air conditioners, water heater, refrigerator, and lights and appliances.

In fact, in modern power systems, as demand goes up, voltage actually goes down, an inevitable consequence of the rules of electrical flow. But not by much. The rules of nature make it impossible to keep voltage absolutely constant in a power system as the demand for power varies, although engineers can reduce that tendency through various means, all of which cost money. Most power systems are quite good in this regard, despite being designed to provide low cost power, but voltage will vary by perhaps as much as six percent from times of peak to times of minimum demand. Still, those extremes represent typically a six to one (600%) change in demand. A 6% change in voltage is not much in response.

The really important point is that *power engineers design the electric utility systems so that voltage at any one point stays as close to constant as possible.* Their goal is to make the power system appear electrically as a *voltage source,* i.e., an unchanging supply of constant voltage regardless of the amount of power demanded. Knowing that this will be the case, engineers of electrical equipment and appliances – motors, heaters, industrial pressure pumps, blenders, microwave ovens, garage door openers, and video games – can design them so that they will vary the current they draw as their power needs change. All of those devices, and everything else that uses electricity, would have to be a bit more complicated, and thus slightly more expensive, to make them work as evenly and safely, if both voltage and current could, or did, vary.

Synchronous AC Generation

The number of cycles of alternating power per second is called the *frequency* of the power. A cycle per second is called a *hertz.* There are two standards in use worldwide for the frequency, or rate, of cycles produced in an electric utility system. The so-called "American" system, uses power at 60 hertz (cycles per second), while the "European" system uses 50 hertz electric power. There is not really much advantage or disadvantage to either one. In fact, power systems that operated at other frequencies, such as 55, 63, or 80, could be engineered and would work just fine. Actually, there are a few 25-cycle power systems used in isolated parts of the world. And, electric power systems on military ships often operate at higher frequencies, e.g., 400 cycles, for technical reasons pertaining to their electronic systems. Furthermore, power in a few large assembly factories (e.g., automobiles) operates at 90 cycles or higher for a distressingly practical reason. Since power tools such as electric drills intended for factory use will not operate at normal power system frequencies, the employees have no reason to steal them for their own use.

The important point regarding the frequency in a utility power system is that all the equipment must be designed to work at the frequency in use. Most equipment and appliances made to work only in 50-cycle European systems will not work well in 60-cycle American systems, and vice versa.[5]

Constant Frequency Power Systems

The frequency (number of hertz, or cycles per second, usually 60 in the US) in any power system is kept constant, and to a much higher degree of accuracy than voltage. While voltage can vary several percent, depending on conditions, frequency is kept as constant as possible. The generators in an American power system rotate at whatever rate is required to produce *exactly* 60 cycles per second – not 60.1 or 59.9 – but 60.00 cycles every second. The degree of frequency regulation is truly phenomenal: 60 cycles per second is 1,892,160,000 cycles per year. Most generators will produce within five cycles of this 1,892,160,000 cycle rate each year, and a few utilities will hit this count right on the head. Many power systems run within *three ten-millionths of a percent* of perfectly uniform speed.

Frequency regulation in American power systems, and many European systems, is so constant that it is often the most exact timing signal available. Before the advent of electronics, most households used electric clocks that contained a synchronous motor, one that locks itself in step with the AC current, so that the second hand turns one revolution (60 seconds) in exactly 3,600 cycles of the power (or 3,000 cycles in Europe). Such *analog* clocks are still available, but have been displaced by digital clocks. The least expensive digital clocks have no internal circuitry to tell time. They simply have a small circuit that "counts" the power pulses of the alternating current and displays the resulting change in time, adding one minute to the display every 3,600 cycles.

Frequency is regulated this precisely in a power system for various reasons, most beyond the scope of this discussion. Precisely controlled frequency is used to regulate the power system in order to have high efficiency. Very few electric utility systems have only one generator – most have from several dozen to several hundred. Operating several dozen or more generators as a group, keeping each running at its peak performance, is a daunting challenge under the best of circumstances. This is done by *synchronizing* them – getting them to run at exactly the same rate – all of them producing exactly 60 cycles per second.

---

[5] There is also a difference in voltages – European systems use 250 volt residential voltage, American systems use 120 volts for small appliances and 240 volts for larger ones. However, a 240 volt clothes dryer made for American 60-cycle systems (designed to run at anywhere from 230 to 252 volts) would not work well in Europe, even though it would be essentially compatible with the voltage there, because the different frequency would create overheating of some parts, and generally poor performance.

Synchronization of generation and its advantages is described in more detail in Chapter 4, and its interaction with blackouts is discussed in Chapter X).

Frequency also provides advantages for power system operators prepared to measure and track it. When all generators in the power system are spinning at the same rate, the electricity from each is of *exactly* the same frequency, and power system operators can make quick analyses of problems, because they can track frequency, not power output, in order to control their system. If the frequency of power anywhere in the system starts to drop slightly – by a thousandth of a percent – it means that more power is needed there, and the system is adjusted to produce slightly more and to send it there. If frequency becomes slightly fast, the opposite is done. This turns out to be much easier than tracking actual power flow and trying to adjust on that basis.

Electrical Losses

Moving electric power through a system of wires is very much like moving water through a system of pipes. Neither electricity nor water is inclined to move unless pushed, and pushing either one of those commodities long distances can take a lot of energy. In municipal water distribution systems, large water pumps, usually powered by electricity, force water to flow through the pipe system. Without these pumps, the water would not flow. Similarly, electric power has to be "pumped" through the wires, overcoming their *resistance*, which consists of a type of electrical friction to power flow that exists inside every type of material except some very exotic "super-conductors" when they are operated at very, very low temperatures.

A water distribution system's pumps are powered by electricity, and, in a way, so are a power system's. But in the latter case, since electricity is pumping electricity, no special pumps are necessary. The electricity does the job itself, using a portion of its own energy to do so. For example, a utility might put 100 megawatts of power into one end of a 200-mile transmission line, only to have 97 MW emerge at the other end. The difference, 3 MW in this case, is called *electrical losses.*

As one would expect, the more electricity moved, the more energy it takes to move it. In fact, electric losses go up as the square of the power moved. Double the power being moved, and the electrical losses increase by a factor of four. If the utility decided to push 200 MW along that transmission line, losses would quadruple, to 12 MW, and only 188 MW would emerge at the far end.

Losses can be reduced by using a large conductor (wire). Doubling the size of the conductor in a power line roughly halves the losses. Utilities try to balance the size and cost of the conductor against losses. Install too large a conductor and losses go down, but the expense is unjustifiable. A normal conductor weighs several tons per mile, a weight that the poles and towers must hold stable in wind and rain for many years. Doubling conductor size doubles its weight. Not only does the heavier conductor cost more, but the poles, brackets, and other equipment must be designed to hold twice the weight. As a result, utilities do not try to reduce losses beyond a certain amount, called the *optimal losses level.* Typically, total electric losses on a power system will average about 6 to 10% over the course of a year, varying from a low of about 3% during times of minimum usage, to as high as 12% during periods of high usage.

Engineers categorize losses as two types. *Load-related losses* are the ones discussed above, due to the direct "pumping" of electric power from one point to another. In addition, transformers and some other electrical devices require a small amount of power simply to activate themselves. Transformers, for example, must create an internal magnetic field, which they do by using a small amount of electric power, called *no-load losses,* because it is used even if no net power is flowing to the customer.

Electric losses should be kept in perspective. They are simply the energy cost of moving power, one cost of the utility's doing business. While attention should be paid to them, some level of losses is unavoidable, and the utility's challenge is to minimize their cost, i.e., finding the best balance between the cost of using large equipment and simply living with the losses' cost.

As mentioned above, certain types of high-technology material can be made superconducting in some cases, so that all resistance to electric flow disappears and no losses occur when power moves through them. Forgetting for the moment the fact that such equipment is currently experimental and prohibitively expensive in most cases, it is worth noting that even this would not reduce losses to a minimum. Superconductivity is achieved by cooling certain types of metal or ceramics to temperatures more than 100°C below freezing. This requires cryogenic pumps that run on electricity. The energy needed to run these pumps counts as a type of "electrical transmission loss" – less than that required with normal equipment, but still a noticeable amount.

The Challenge of Moving Power Where the User Wants It

Electric power flows through a power system network in accordance with several physical laws of nature whose complexity and application are beyond the scope of this book. Their net result is that electric power will not necessarily flow from point A to point B in the electric grid just because there is a transmission line going from point A to B and the voltage is high.

Electrical conditions must be arranged to promote the flow of power – to invite it, in a manner of speaking, to flow from point A to B. Factors including voltage, phase angle of the AC pulses, and other conditions must be arranged to promote this flow of power. This is not too difficult for power system engineers and operators to arrange, except that the conditions that make power flow one place in the grid might be counter to what is best at another location.

Thus, the design and operation of a large power system is a balancing act of innumerable dimensions, as engineers arrange the system so that it will simultaneously move power throughout it, delivering the required amounts when and where needed. Some large power systems consist of over 100,000 interconnected units of equipment, all to some extent interacting with one another in a complex electrical pattern. The basic tool used to analyze electrical flow is called a *load flow*, a computer program that solves a set of equations to determine if and how the lines interact with one another and the utility's customers' demands. Engineers apply these and numerous other computerized tools to help determine how to build a system that will accommodate the demand reliably, safely, and at low cost.

Operating an Electric Utility System Economically and Securely

As might be expected, it is not only a challenge to design and build a system to meet these needs, it is equally challenging to keeping it operating smoothly and efficiently on a day-to-day basis without interruption, too. Utilities maintain large, heavily computerized control centers where they monitor the status of the system on a minute-to-minute (and for key equipment, second-to-second) basis, and also control various equipment and switches throughout their system, in order to keep power flowing and loading distributed as desired.

4.4 KEY POINTS

- The two basic qualities of electricity are current (the amount of electrical flow) and voltage (the electrical pressure pushing that flow). Voltage is measured in volts and current is measured in amps. There are two types of electric power, direct current, DC, in which electricity constantly moves in one direction, and alternating current, AC, in which both the voltage and current oscillate at the same rate, many times a second.

- Insulation, which prevents power from going where it is not wanted, is provided by spacing of equipment away from anything else, and using non-conducting materials such as plastic to cover electric wire and equipment.

- Electricity is a particularly *safe* form of energy when equipment is well designed and maintained and used, and when any activity around it is in accordance with sound safety precautions.

- All modern power systems use alternating current (AC), in which voltage and current oscillate at 50 or 60 times per second. Such AC systems provide some advantages over the use of direct current (DC), the type of electrical system used in automobiles and portable radios.

- Unlike current, which can vary widely, and voltage, which will vary by a small amount from moment to moment in a modern power system, frequency is kept as constant as possible within any one power system.

Frequency regulation in American power systems is so constant that it is often among the most exact timing signals available in a region.

- Electrical equipment must be designed to work at the voltage and frequency standard in use (120 or 240 or 250 volts, and 50 Hz or 60 Hz).

- One drawback of electrical power is that it is *expensive to store*. It costs much less if it is made and sent to the consumer at the moment of use.

- Other challenges that must be met in designing a system to deliver energy are: electrical losses, security, and economy.

FOR FURTHER READING

T. W. Berrie, *Electricity Economics and Planning,* Peter Peregrinus Ltd., London, 1992.

E. Santacana, editor, *Electric Transmission and Distribution Reference Book, fifth edition,* ABB Power T&D Company, Raleigh, 1997.

5

Using Electric Energy

5.1 INTRODUCTION

Electric utilities and their power systems exist because people want electric power. Electricity is rarely the only source of energy for a particular application, but it is the preferred single energy source worldwide. This chapter looks at the use of electricity: What it can do, as well as who uses it, how, when, and where?

5.2 A FLEXIBLE FORM OF ENERGY

Perhaps electricity's greatest advantages over other forms of energy are its flexibility, controllability, and cleanliness. Electricity can be converted easily and efficiently into an amazing range of applications, including light, heat or cold, and rotating or linear mechanical motion. It can run magnetic equipment that will grab and hold tightly a large piece of metal, or, under the right conditions, bend and shape it. It can power electronic devices that mimic human intelligence, monitor human heartbeats, track airliners through storms, or emulate the sounds of a symphony orchestra. Furthermore, the range of energy it can supply is as diverse as the types of applications it can fulfill. Electricity can efficiently power devices as small as a 2-watt bathroom night-light, or drive a ten-million watt (10,000 HP) industrial pump motor.

At its points of use, electricity is a very clean, compact, and quiet energy source. A 10 HP electric motor is no bigger than two or three briefcases, yet it runs almost silently, and produces no pollution, no leaking fluid, and very little heat or vibration. If need be, the electric motor could be located inside a home or office without causing major disturbance or disruption of quiet business routine. By contrast, a gasoline motor of similar power would be about twice as large, and even when heavily muffled would produce objectionable amounts of noise and vibration, along with considerable amounts of exhaust and heat. Unlike its many positive qualities at its points of use, at its source of production (see Chapter 6, Creating Electricity: Power Generation), electricity requires large, noisy, and in some cases mildly polluting industrial-grade equipment.

Since electricity is very controllable, it can be applied to devices such as a blinking light, or an automatic door that operates intermittently and without pre-arranged schedule, starting and stopping repeatedly, and as needed. It can be used in servomotors, which run at speeds regulated to within a thousandth of a percent, and applied in timing lasers, accurate to within one ten-millionth of a second.

Mankind has many other sources of energy at its disposal: natural gas, fuel oil, gasoline, coal, and nuclear power, to name but a few. All of these have advantages over electricity in particular applications, but disadvantages that limit their use to a rather narrow range of applications:

- *Natural gas,* where it is as convenient and inexpensive to obtain as electricity, is more efficient for water heating and for furnaces, yet it cannot power microwave ovens, stereos, or TVs.

- *Gasoline and fuel oil* are very transportable and storable, making them ideal for use in propelling cars, trucks, and ships.

- *Coal* can provide heat in great quantities, and is particularly abundant and widely available, but it is messy to use, requiring extensive pollution controls, which in many cases are affordable only in large-scale, industrial applications.

- *Nuclear power* can provide huge amounts of energy over long periods of time, in isolated locations, making it ideal as a "prime mover" for submarines and as a power source for space probes. But it requires so much care and expense that it is economical only in very large sizes.

And beyond their individual limitations, none of these energy sources can match electricity in their range of energy delivery capabilities. Nuclear energy is nearly impossible to apply unless one needs tens or hundreds of millions of watts of power. Gasoline and fuel oil are inefficient for applications of less than about 5 horsepower (roughly 4,000 watts). By contrast, most of the energy applications that households and businesses use, lamps, stereos, computers, cash registers, hair dryers, electric drills, electric garage door openers, microwave

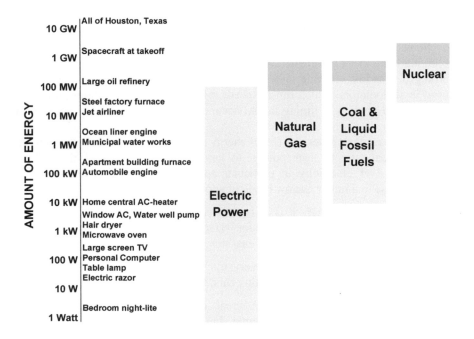

Figure 5.1 Range of mankind's energy application needs, and the capabilities of various energy sources. Scale at the left is logarithmic, with each increment representing ten times the energy of the level below, in watts (W), kilowatts (kW), megawatts (MW), and gigawatts (GW). The energy requirement of a large metropolitan area, such as Houston, Texas, would be about 12 GW, that of a single bedroom night-light would be ten billion times less, or less than two watts. The needs for various types of appliances and facilities are shown. The range of direct application of various energy types is indicated by the bars. Electricity covers needs from less than one watt to those of 100 MW oil refineries. Darkly shaded portions of the bars for natural gas, coal, and nuclear power indicate the portion of their usage that is devoted to the generation of electric power. Most electricity is created by converting these other energy sources to electricity in very large generating plants.

ovens, water well pumps, etc., require 500 watts or less. Figure 5.1 illustrates the range of efficient application of energy sources, and shows why electricity is universally popular. Electricity can *simultaneously* provide large and small amounts of power, heat or cold, and light. It can also reliably and silently furnish these on an intermittent or continuous basis. No other energy source can operate efficiently over such a wide range of applications. As a result, man's greatest use of many of the other forms of energy, such as coal and nuclear power, is to drive electric generators to convert them into electricity.

Electricity's single biggest disadvantage is that it cannot be stored efficiently. Even the most efficient batteries store less than a third of the energy per pound as coal, the least efficient fossil fuel. The cheapest battery technologies cost much more per unit of energy stored than for gasoline or natural gas. Thus, electricity is most often moved to its place of use at its time of use, and is often not competitive in applications that require storage.

5.3 FOUR BASIC APPLICATIONS

Electricity is converted into four basic essences in the course of its use. These are light, heat, motion, and electronic circuit operation. One or all four may be involved in the operation of a particular end-use device. A light bulb produces light, and heat – as a usually unwanted by-product. An electric furnace uses three: heat to warm the home, mechanical motion for the fans that blow the air throughout the house's duct system, and electronic circuit flow for the thermostat/controller, which monitors temperature and keeps it uniform.

Light

Light can be produced from electricity in at least five ways:

- Electricity can produce light directly when brought into the open, as in sparks or lightning bolts. Some types of *arc* lights use this principle to produce incredibly bright light, such as in the searchlights for spotting aircraft at night in the early part of the 20th century, but they are noisy and inefficient (See Charles Brush, Chapter 3).

- Electricity can be passed through a thin metal filament in a vacuum inside an *incandescent* light bulb. A sufficient electric flow will cause the element to heat up to the point that it will glow, creating light. Incandescent lights are inexpensive to manufacture, but convert a majority of the energy used to heat, not light.

- Fluorescent, neon, sodium vapor, and similar lighting devices produce light by running electricity through an electrically active gas, in which the flow excites atoms of the gas, which create a glow. Most *gas discharge* lights produce light of only one color (fluorescent lights are an exception), but convert much more of their energy to light, rather than to heat, than incandescent bulbs.

- Lasers use electric energy to create quantum mechanical (subatomic) disturbances in certain materials, which produces light and little waste heat. Lasers are very expensive compared to other sources of light, but the coherent light is more controllable for many industrial applications.

- Light emitting diodes are solid-state devices that give off light when a current is passed through them. They are very efficient, producing mostly light and little waste heat. Available in units that produce a number of different colors, they have a fast on-off time and are often used in computer-driven displays, etc. In recent years their cost has dropped significantly (and will continue to) and they are penetrating into former incandescent applications like automobile stop and turn signals, flashlights, and even home interior lighting.

Electricity does not have a monopoly on producing light, which can be derived from energy sources other than electricity. For example, natural gas is sometimes used as a light source, and kerosene-soaked rags are used in torches. But the vast majority of lighting, more than 99.999% of it worldwide, is provided by electric power, because it produces light reliably, cleanly, quietly, economically, and safely, i.e., without combustion.

Heat

Every electric application produces heat, even if it is unwanted. In fact, heat is the unavoidable by-product of *all* energy usage, not just electric power. An air conditioner may produce a flow of cold air inside the home, but it more than makes up for it by ejecting a stream of much hotter air outside the home. Over two-thirds of the energy used by a typical incandescent light bulb is converted into heat (only 40% in fluorescent bulbs). Stereos and TVs are warm to the touch when operating, because in the course of producing images and sound, they convert a noticeable portion of the energy they consume into heat.

When heat is the desired product of electric usage, it can be produced in one of two ways. Electric current can be forced through a *heating element,* essentially nothing more than a special wire, designed and constructed to rise to very high temperatures but not burn up or otherwise fail when great amounts of current pass through it. This wire is used inside a toaster or a hair dryer.

The second way to produce heat is with a *heat pump,* a device that merely "moves heat." Though complicated in detail, it is simple in concept. When the heat pump is operating, one end will become cool, as heat is removed from it, and the other end will become hot, as that heat is moved to it. A refrigerator applies this principle to cool the air inside it, where the "cold end" of the heat pump is in the shape of a long tube, wrapped around the inside of the freezer compartment. A heat pump used to heat and cool a home performs similarly, producing hot or cold air by selectively blowing air over one end or the other of the mechanism.[1]

Most heat pumps are quite complicated. They use motors, pipes, valves, and other mechanical parts, but are fairly quiet and reliable. However, there are solid-state heat pumps, which have no moving parts and, in fact, only one part, an electronic device, one end of which becomes hot and the other cold when electricity is passed through it. Solid-state heat pumps are quite expensive and not nearly as efficient as mechanical heat pumps, but create no vibration or noise, making them useful in some applications.

Heat can be produced by forms of energy other than electric power, and in many circumstances electric power is *not* the preferred energy source. Natural gas, coal, or fuel oil can produce heat at a much lower cost, when substantial amounts are needed. Thus, they tend to be used in large industrial applications, and even for home water and space heating, where convenient. Electricity dominates the production of heat only for small applications and where other energy sources are unavailable. For example, since most rural areas of the United States are without natural gas distribution lines, homes there often have electric water heaters and furnaces, as opposed to homes in cities, where there are extensive lines that use natural gas.

Mechanical Motion

Motion is produced by controlled magnetic fields, most commonly in an electric motor. Inside the motor, intense magnetic fields are produced by running electric current through coils of wire, and through some means, making the magnetic field rotate. This can be done with switches, or electronically, or in certain types of motors, naturally, caused by the alternating current. The rotating magnetic field turns a rotor shaft, producing rotating mechanical motion. Motors power fans, heat pumps and conveyer belts. They are available in units smaller than a beverage can, to power sewing machines and other small appliances, up to units the size of houses, which can provide thousands of horsepower.

Solenoids use magnetic force to produce mechanical motion in a straight line, usually over only a short distance, often only an inch or less. A solenoid is just a long coil of wire with a metal rod passing through its center. When electric current is passed through the coil, the resulting magnetic force through

---

[1] Cold temperatures are therefore not a direct product of electricity (or any other energy application). Cold is produced essentially by a "trick," moving the heat inside a house, or refrigerator, outside it.

the coil's center pushes the rod out one end or the other of the coil. Large solenoids can power industrial hammers so they bend inch-thick steel plates. But most applications are much more mundane. In a doorbell, when the button is pushed, a solenoid moves a metal rod about ¼ inch to strike a chime.

Linear motors have more in common with solenoids than standard electric motors. They pull continuously on a long metal bar or rail, to which they are attached. A linear motor put on a locomotive will pull a train continuously along the railroad rails. A type of linear motor called a rail gun will fire a short metal bar through a hundred-foot long magnet, up to the speed of a bullet.

But while interesting and powerful, linear motors and rail guns are not widely used, because they are not as efficient as motors and solenoids. The main reason is that motors and solenoids create their powering magnetic field *inside* their housing (windings), where it can be focused and multiplied to produce very efficient motion. By contrast, linear motors and rail guns produce their magnetic fields *outside* their windings, for example to grab a nearby metal rail, where it tends to dissipate and be partially wasted. Thus, even though electric trains could use linear motors, most use rotating motors.

Mechanical motion can be produced by energy sources other than electricity. Gasoline, diesel, and jet engines are used almost exclusively in vehicles, electric trains and buses being exceptions. Electricity is generally preferred in stationary applications, where only small amounts of power are needed, e.g., in sewing machines, or where precision or clean and quiet operation is required.

Electronic Circuits

Electrical circuits channel electric signals through them in complicated and varying ways, in order to accomplish two types of electronic performance. *Digital circuits* change voltage, or electrical charge, at points within them, called bits, between two states, high and low. Billions of such points, arranged and interconnected in particular patterns, are used in computers, robotic control systems, and similar electronic devices. *Analog circuits* use a continuous flow of electric current through various interconnected circuit elements to change radio waves into electric signals, to produce sounds, pictures, etc.

Digital circuits are usually faster, and more precise, than analog circuits, but can be applied only at very low power. Analog devices are usually more robust and capable of high power application. Many electronic devices, such as clock radios, combine both digital and analog types of circuits in what are sometimes called *hybrid circuits*.

Electric and electronic circuits have been developed for a tremendously varied range of purposes: to produce images (TV) and sound (radios, stereos, PA systems), to measure and monitor the health of human beings, e.g., heartbeats, and in machines, e.g., water pressure, and at a distance to control equipment, e.g., thermostats, robotics, etc.

The term "electronic" rather than "electric" generally means that the control or fashioning of the electric flow is the end being sought in the device. For example, a computer is an *electronic* circuit that manipulates electric pulses into specific patterns. That is its purpose. A human being interprets those patterns (dots on a screen) as written text or pictures, but the electronic circuit has no purpose other than to manipulate electricity. By contrast, electric devices have as their purpose the transformation of power into other purposes. An *electric* power system creates, moves, and controls power, but ultimately for some other end, i.e., light, heat, motion, and circuitry, to be used by mankind.

Electronic circuitry is the one domain where electric power has no competition from other energy sources. Except for very rare cases, electricity is the only format for the various functions performed by electronic circuitry.[2]

5.4 ELECTRICITY IS BOUGHT FOR END-USES

Energy consumers do not purchase electricity, or natural gas, or power from any other source for that matter, because they want energy. It is the *products* of energy use that they want: a cool home in summer, a warm one in winter, hot water on demand, cold beer in the refrigerator, and 48 inches of dazzling color image with full-bass stereo commentary during Monday night football. In every case where it is used, electricity is a means to an end, called an *end-use*. A electric utility company's customers purchase electricity in order to have these end-uses.

These end-uses span a wide range. For a few, electricity is the only energy choice. But for many, it is only one of several possible energy sources – water heating, cooking, and clothes drying being three uses that can use natural gas instead. But as discussed in Section 4.3, electricity has advantages that make it almost ubiquitous in businesses and residences, regardless of the possible application of other energy sources.

Electricity is used by many different categories of electric consumer, including residential, commercial businesses, and industry, as well as government. All of these consumers have different requirements, yet share many common needs for specific applications of light, heat, motion, or circuit

---

[2] It is possible to build a crude type of computer using flowing air instead of flowing electricity. Such "gas flow circuits" were investigated by the military in the 1970s and 1980s because they were potentially more robust on the battlefield than fragile electronic circuits. Some experimental computers circulate light instead of electricity within them, although at the nearly atomic scale being studied, there is a different distinction between the two than appears at the scale of normal life. A few devices of each type – air and light circuits, and perhaps other unusual applications – have been used in special industrial and military situations. But less than one hundredth of one percent of the "circuits" in the world use anything but electricity.

Table 5.1 Typical End-Uses in Four Major Customer Classes

Agricultural	Residential	Commercial	Industrial
Lighting	Lighting	Lighting	Lighting
Water heating	Water heating	Water heating	Water heating
Space heating	Space heating	Space heating	Space heating
Air conditioning	Air conditioning	Air conditioning	Air conditioning
Computer	Computer	Computer	Computer
Air circulation	Air circulation	Air circulation	Air circulation
Cooking	Cooking	Cooking	Filtration
Water well pump	Water well	Elevators	Fluid pumps
Grain dryers	Clothes dryers	Inventory system	Finishing dryers

performance to their particular end-uses. Electric consumers are often grouped into *customer classes*, or *rate classes,* by electric utilities, both to distinguish types of similar requirements, e.g., most home-owners' are about the same, and to distinguish how they are charged for their power. Table 5.1 shows some of the most popular uses of electric energy in the four basic consumer classes.

Appliances Convert Electricity to End-Uses

Each end-use, for example, the need for lighting, is satisfied through use of devices that convert electricity into the desired end-product. These are broadly referred to as *appliances*. In most end-use categories, there is a variety of different types of appliances, ways to convert electricity into the end-use. For instance, a wide range of illumination devices can be applied to produce light, from incandescent bulbs to fluorescent tubes, sodium vapor and high-pressure monochromatic gas-discharge tubes, and lasers. Each uses electric power to produce visible light. To heat homes, electricity can be run through heating elements in a resistive furnace, or applied to run a heat pump, or used to heat water, which can then be pumped through the house (baseboard heating). Or, other sources of energy, like natural gas furnaces, can be used.

Each type of appliance for a particular end-use will have differences from the others, that give it an appeal to some customers or an appropriateness for certain types of applications. This may have to do with the cost of the devices, i.e., electric water heaters are less expensive than gas water heaters, or their efficiency, i.e., fluorescent lighting uses less electricity than incandescent to produce the same amount of light. Sometimes the preference is cultural. For example, while in Japan fluorescent lighting is used widely in residences as well as businesses, in the United States, fluorescent lighting is restricted mainly to business applications. Incandescent lighting is chosen for home use, because many people prefer its slightly yellow glow to "sterile" fluorescent lighting.

Electric Load

The term *load* means the electrical demand of an appliance connected to and drawing power from the electric utility system to accomplish some task, e.g., opening a garage door, or converting that power to some other form of energy such as light or heat. Electrical loads are usually rated by the level of power they require, measured in units of volt-amperes, called *watts*. Large loads are measured in *kilowatts* (thousands of watts) or *megawatts* (millions of watts).

Many appliances are also rated by how much of the end-use product they produce. For example, an incandescent light bulb might be rated at 75 watts and 1,100 *lumens* of light, a lumen being a measure of light output, while a fluorescent light tube might be rated at 60 watts and 1,250 lumens output. Similarly, an air conditioner might be rated at 2,400 watts and 2,650 BTU, or British Thermal Units, an amount of heating or cooling output.

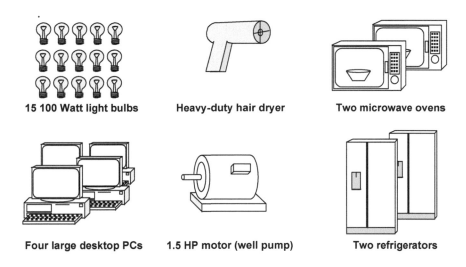

| 15 100 Watt light bulbs | Heavy-duty hair dryer | Two microwave ovens |

| Four large desktop PCs | 1.5 HP motor (well pump) | Two refrigerators |

Figure 5.2 These six groups of appliances all use roughly the same amount of energy, 1,500 watts, or one and one half kilowatts, when operating. All, *together*, use about as much energy as a two-zoned central air conditioner in a large residence (3,500 sq. ft.). Note that the production of heat is a relatively expensive way to use electricity: the hair-dryer uses as much power as all the lights that are normally on in the evening in a large home.

Efficiency is the ratio of end-use product to electric load. For example, an incandescent light bulb produces about 22 lumens per watt. A fluorescent bulb produces about 40 lumens per watt, or nearly twice as many, and is therefore more efficient. Generally, appliances of a similar type are about as efficient, but there are exceptions. For example, heat pumps vary widely, producing between 7 and 17 BTU-hours of output per kilowatt hour of energy used. Air conditioners and heat pumps are rated for efficiency according to methods set down by the U.S. federal government. The Energy Efficiency Ratio (EER) rates AC/HP units on how well they produce cool or hot air, as the case may be, from electricity. For example, a unit that supplies ten BTU of cooling from one kilowatt hour has a rating of ten. Air conditioners used in the 1960s had ratings of about 5. Those at the time of this writing vary from about 10 to nearly 18.

The Seasonal Energy Efficiency Ratio (SEER) evaluates an AC/HP unit's ability to provide the average amount of power required over an entire season of heating or cooling. This is more important to the user, as it describes how much energy the unit uses in the course of a year. Units that are very efficient, i.e., that have a high EER, may not be as efficient in day-to-day operation over a season. One reason is that some types of high-efficiency AC units are efficient only as long as they run all the time. On merely warm, as opposed to hot, days, they cycle on and off a lot and are not terribly efficient in using electricity. Almost all air conditioners cycle on and off, but some are much more efficient about how they perform in such cycling. SEER looks at how an air conditioner or heater performs over all conditions that matter.

It is often difficult to get an intuitive feel for how light, mechanical motion, and heat compare in terms of electric energy requirements. Figure 5.2 compares the electric load of several different types of common household appliances.

Market Share

The portion of all users in a particular class choosing electricity for an end-use is called the *market share* for that end-use. For example, about 34% of all home-owners in the state of Maine, in the northeastern United States, use electricity to heat their homes in winter. Electric power has a 34% market share in residential space heating.

Appliance Share

Of the several ways to heat a home, the percentage of use for any one appliance is its *appliance share*. Thus, in Maine, about 33% of all home-owners who heat their homes with electric power, use a heat pump. Therefore, heat pumps have a 33% share of the electric heating market, and an 11% share of the entire residential electric heating market (33% of 34%). Figure 5.3 shows the breakdown of market share for several appliances in the residential class of a utility in the northeastern United States.

Daily Appliance Usage

While the groups of devices shown in Figure 5.2 all have roughly the same electric load, they will not all use the same amount of power during a day, because they operate on different schedules. Of those shown, probably the pair of refrigerators would use the most energy, because they operate around the clock.

Many household appliances, such as refrigerators, water heaters, air conditioners, heat pumps, and electric furnaces, operate in an automatic "on-off" manner. For example the refrigerator has an internal thermostat that the owner sets. It then automatically starts and stops its cooling cycle to maintain its temperature at the level shown. Similarly, industrial equipment, such as air compressors, forge ovens, and water pumps, operate in an on-off manner using control equipment that maintains a pre-set level of performance.

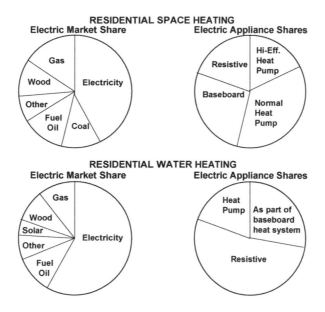

Figure 5.3   Residential electric usage for a suburban/rural power system in the northeastern United States. Top, share of home heating by energy type (left) and appliance type within electric usage (right). Bottom (share of home water heating) by energy type (left) and appliance type within the electric category (right).

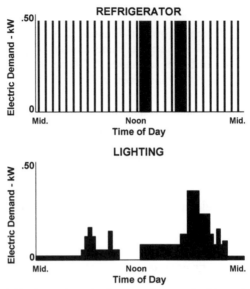

Figure 5.4 The daily demand, or load curve, for a typical household refrigerator (top) and the lights used in the house (bottom). The refrigerator activates itself whenever its thermostat decides it needs to run. During the nighttime, it has to run only a few minutes every half hour. It runs longest when it is being opened and closed during the day, or is asked to cool down warm food or beverages that have just been put inside it.

Table 5.2 Typical Duty Cycles (Percent of Time the Device Runs) of Standard Household Appliances and Equipment

Appliance	Duty Cycle - %
Refrigerator	15%
Freezer	12%
Electric oven (when on)	40%
Central space heater	35 - 90%
Central space air conditioner	90%
Electric water heater	10-25%
Water well pump	10-25%

As a result of its automatic on-off operation during a typical day, a refrigerator's demand for electricity looks like that shown in Figure 5.4 (top). It, like many appliances and equipment in a house, operates with a thermostat control based on temperature (in this case, how cold the air is inside the refrigerator). Home heating and cooling, refrigerators and freezers, and water heaters also operate in a similar manner: when temperature deviates too much from that to which the thermostat is set, the appliance switches on automatically; when temperature is at the proper level, it shuts itself off. The result is a choppy pattern of electric usage as shown in Figure 5.4. The portion of the total time in an hour or day that a particular device is on is called its duty cycle. Table 5.2 gives duty cycles for typical household electrical equipment. Most home space heaters and air conditioners are designed and sized to operate at about 90% duty cycle under extreme temperature conditions.

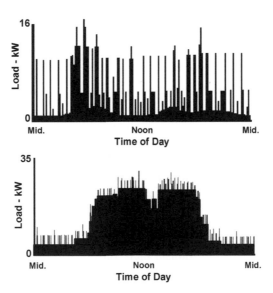

Figure 5.5 Daily load curves for a home (top) and small business (bottom) represent all the appliances and electrical demands at the home or business site. Note the choppy nature of the demand for electricity, as various appliances operate on an on-off schedule in response to thermostats, pressure sensors, etc. The residential load curve is particularly choppy, because the majority of the appliances in the house operate this way. The commercial building has a chiller-type cooling system, which runs continuously, but still exhibits many appliances that operate only for short durations.

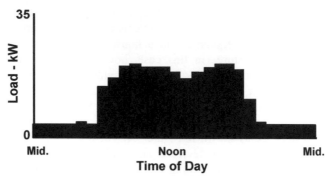

Figure 5.6 The commercial building load, shown in Figure 5.5, measured for demand on an hourly basis. The owner is billed a demand charge each month based on the peak hourly energy usage, about 20 kW at 10 AM.

Other types of appliances and devices, such as light bulbs and televisions, are turned on and operate continuously until switched off (Figure 5.4), but the majority of electric appliances, and load, operate in an on-off manner, shown in Figure 5.4. Figure 5.5 shows the daily load curve for an entire home.

Peak and Energy

A particular consumer's demand for energy is measured in terms of *peak demand* and *energy usage*. Peak demand is not the peak electrical load, but, instead, is defined as the maximum energy usage during any hour, or in some cases, any fifteen-minute period, or during a billing period, which is usually monthly. Most utilities bill commercial and industrial customers for both peak demand and energy used, but bill residential customers only on an energy basis, according to how many kWh they use each month. This is done for two reasons. *First*, meters that measure and record both peak demand and kWh usage cost a good deal more than those that measure just kWh. The average utility would have to spend millions of dollars more on residential meters in order to bill residential customers on a demand basis. *Second*, commercial and industrial customers tend to concentrate their use of electric power during mid-day, when demand is highest. Billing them for peak demand gives them an incentive to reduce their bills by spreading out their usage into evening and early morning. They can use the same total energy, but have a significantly lower bill if they reduce their peak demand by rescheduling usage.

When measured on an hourly basis, the sharp peaks and valleys of a customer's load curve tend to disappear, because these needle peaks take less than one hour and are averaged over the hour. Thus, what a commercial customer is billed for is the peak energy, and the energy usage over a month, for a daily load curve that looks something like that shown in Figure 5.6.

Power Quality

Consumers of electricity must have certain *quantities* of power, in order to meet their energy needs, as well as certain *qualities* in the power they are provided. As with power quantity, exact requirements for power quality vary from consumer to consumer. These "Two Qs," quality and quantity, are both important aspects of satisfactory electrical service.

Availability and service interruptions

The most dramatic power quality problem for most electric consumers is an *interruption* of service. For some reason, be it failure of equipment, damage to lines by a wind storm, a tree falling on a line, or a car hitting a utility pole, flow of electric power is interrupted. Lights go out, motors cease to operate, heaters stop heating, and electric circuits cease to function, causing everything from minor nuisances to major inconveniences. Service interruptions in most power systems in developed countries are rare. Throughout the United States, the average electric consumer suffers about two interruptions per year, with "lights out" lasting about one and one-half hours, total. This means average *availability* is 99.983%, equal to 8,758.5 hours over 8,760 hours per year.

Table 5.3 Classification for Interruptions Used by Utilities

Type	Definition
Instantaneous	An interruption restored immediately by completely automatic equipment, or a transient fault that causes no reaction by protective equipment. Typically less than 15 seconds.
Momentary	An interruption restored by automatic, supervisory, or manual switching at a site where an operator is immediately available. Typically less than five minutes.
Temporary	An interruption restored by manual switching by an operator who is not immediately available. Typically, about thirty minutes to one hour.
Sustained	Any interruption that is not instantaneous, momentary, or temporary. Most typically more than an hour.

Table 5.4 Annual SAIFI Events and SAIDI Hours for Six Utilities 1988-1992

Utility	Service Area	Climate	SAIFI	SAIDI
1	Dense urban area	Hot summers, bitter winters	0.13	16
2	Urban/suburban	Hot nearly year round	2.70	34
3	Suburban & rural	Lots of lightning	2.02	97
4	Urban & rural	Mild year round	2.16	122
5	Rural, mountainous	Mild seasons	1.13	168
6	Rural, mountainous	Bitter winters	2.37	220

Utilities track statistics on customer availability of power. They classify service interruptions by their duration, as shown in Table 5.4. Although there are recommended guidelines, many utilities apply slightly different definitions. The table shows guidelines recommended by the Institute of Electrical and Electronics Engineers. The most popular are SAIDI (System Average Interruption Duration Index) and SAIFI (System Average Interruption Frequency Index), which give, respectively, the average customer's time without power annually and the average number of times power was interrupted. Performance varies among utilities, partly because of climatic and geographic conditions. Lines are more prone to be damaged by ice storms in areas farther north. In rural areas repair crews must travel farther to find line damage when it occurs. Or, some utilities just do a better job of maintaining their equipment and managing service restoration or repair. See Chapter 15, Section 15.3, for more details.

Transient voltage interruptions

Power flow does not have to be completely interrupted in order for equipment operation to be stopped. An important quality that power must have is a reasonable level of *voltage regulation*. As mentioned in Chapter 4, all power systems worldwide are designed to provide a stable, constant source of voltage, from which appliances can draw power as needed. Power is of little use to customers unless it is supplied at the proper voltage, and unless that voltage is reasonably constant. If voltage strays too high, equipment is damaged. If it drops too low, equipment will stop working, and motors like those in refrigerators and air conditioners may fail. (The motors try to compensate for the low voltage by working harder, overheat as a result, and burn out.)

Voltage problems occur on a power system if the local distribution equipment does not have sufficient capacity to serve the load (low voltage), or if that equipment is not operating properly (high voltage). These types of problems are generally very rare, but when they occur, they last for minutes, hours, or even days.

Somewhat more common, and nearly impossible to eradicate completely, are *transient voltage problems.* If electric lines are struck by lightning, and various control equipment, i.e., lightning arresters, fails to stop it, lightning spikes of very short duration may flow through the power system and damage nearby equipment. If there is a sudden failure of one of the system's components, there can be a momentary *voltage sag*, lasting a second or less, before it is taken off the system and voltage is restored. Sometimes, when a large electric motor is starting, it will cause momentary *voltage flicker* on nearby circuits and houses. This often produces a flickering or brief dimming of lights.

Transient voltage deviations may not cause a problem if they are of sufficiently short duration or if the voltage does not drop too low or rise too high. However, electrical equipment, like computers, robotic control systems, and digital clocks, is very sensitive to even brief voltage sags or spikes. For example, even a 50% drop in voltage for a half second will cause many computers to lose their memory and digital clocks to forget the time.

The Computer and Business Equipment Manufacturers Association (CBEMA) has established a recommended guideline for the types of voltage deviations equipment should be able to withstand while continuing to function. As shown in Figure 5.7, the guidelines tolerate less excursion from nominal if the event lasts a greater time. A good portion of, but not all, computer and business equipment is designed to function as long as voltage stays within this envelope. A particular device may differ in its tolerance for voltage transients, as shown.

Figure 5.7 demonstrates the voltage vs. duration of 67 events at an automated hosiery factory in the southern United States. One event above the curve, attributed to a lightning strike on a nearby distribution line, led to some equipment failure. Eighteen events below the curve caused partial or complete shutdown of the robotic looms. The control computers forgot what they were doing, with loss of several hours production each time, as the equipment was re-booted, and the looms reset. Only one of these was an actual service interruption. One event, far lower right, signifies zero voltage for more than 600,000 cycles, i.e., 10,000 seconds, or more than two and one-half hours. The other seven events lasted less than six seconds, and represented voltage drops to no worse than 50% of normal, yet they caused a shut-down. Less than half of these seventeen events (seven) lay below the CBEMA curve. Most fell inside the knee of the curve, in the area where the factory's equipment did not meet the CBEMA recommendation.

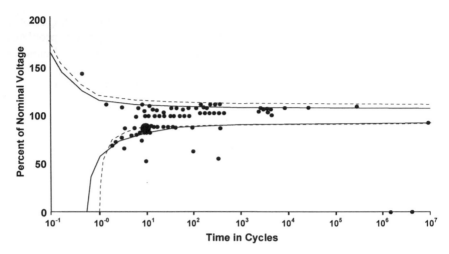

Figure 5.7 A CBEMA curve (solid lines) recommends the voltage deviation tolerance for computer and business equipment. For example, a drop to about 80% of normal voltage should be barely tolerable, if it lasts less than 1/6 second – 10 cycles or less (large black dot). Actual equipment sensitivity may vary from this curve. The dotted lines show the measured tolerance of a particular set of robotic equipment at a small hosiery factory in the southern United States which can't tolerate the voltage dip depicted by the large dot. Small dots show the worst voltage, and the duration it lasted, for 67 "events" that occurred over a 40-month period at this site.

Value of power quality

Electric power is so reliable, and used for so many purposes, that owners of both homes and businesses expect it to be available all the time. Therefore, an unexpected curtailment of power flow, for whatever reason, can cause great inconvenience. *It is testimony to the reliability of power systems that the cost of electric service interruptions is greater than the cost of the power.*

When electrical flow to a large industrial site, such as a factory, is interrupted, production has to cease until power is restored. Workers are idled during the interruption, and after power returns, equipment has to be checked and initialized, and production restarted. If this procedure is complicated, as it is in many manufacturing and chemical-processing plants, even a brief cessation of power can cause several hours of "down time" and a significant cost. Detailed analysis of the situation at a factory or large commercial building can usually provide a very good estimate of the cost of a power outage.

Table 5.5 Typical Interruption Costs by Customer Class
(dollars per customer hour)

Class	Unexpected	Prior Warning
Agricultural	1.50	0.20
Residential	0.75	0.15
Retail Commercial	11.75	6.40
Other Commercial	8.50	1.20
Industrial	13.00	4.20
Municipal	22.00	2.00

In the residential sector, costs for power interruptions are more difficult to estimate, because the values depend on the results of customer surveys, which are subject to the opinions and perceptions of individuals. Generally, between fifty cents and five dollars per kilowatt hour interrupted is given as a typical range for outage cost in the residential sector.

Cost of interruptions is always less if prior warning is possible, as it is if essential repairs must be made and the utility knows this in advance. Given sufficient time to prepare for an interruption of service, most of the cost and a great deal of the inconvenience can be eliminated for many customers. Table 5.5 compares typical cost values for unexpected interruptions with those for which prior warning was given.

But the important point to remember is: Interruptions and transient voltage events cause cessation of the end-uses that consumers want. The cost of these end-use interruptions is often much greater than the actual market value of the electric power itself.

Harmonics

Under some circumstances, electrical equipment, such as motors, computers, and other devices, as well as overloaded transformers and other apparatus on the power system, can create harmonic power flow. *Harmonics* are the electrical equivalent of buzzing or ringing, voltage oscillations that occur within the wiring at a rate much faster than the requisite 60 hertz AC power. Almost all appliances and equipment, particularly electronic devices such as TVs, microwaves, and computers, as well as electric motors, create harmonics while using power. They feed this "noise" back onto the power system. Under some circumstances high levels of harmonics can interfere with the operation of other devices, and in severe cases, cause overheating, accelerated aging, and premature failure of that equipment. However, harmonics seldom cause problems, and their importance as an aspect of power quality is vastly over-rated.

For a four or five year period in the late 1980s, harmonics engineering was the technical *"probleme de jour"* within the power industry. It was an obsession for several electric power research institutes, because harmonics are a fascinating technical problem. A great deal of attention, as well as money and effort, was focused on an aspect of power systems operation that has been with the industry since its founding and seldom, if ever, caused problems. The amount of publicity given to harmonics was completely out of proportion to its importance, so much so that there is still a tendency among some power engineers to blame unexplained operating problems on harmonics, instead of looking elsewhere, e.g., poor grounding or poor connections, etc., for the source of problems.

5.5 THE UTILITY'S PERSPECTIVE ON CUSTOMER POWER USAGE

An electric utility's mission is to serve the demands of its customers for quantity and quality of power. These two "Qs" define the goals it must set for its power production, delivery, and operations. The electric demand of its customers will vary by time of day, day of week, and season, too. Many electric uses are weather-sensitive, particularly heating and cooling, but also cooking and lighting schedules, so that demand is heavily influenced by weather conditions. Many utility executives will express hope for "a hot summer and a cold winter," because it means more sales of electricity for cooling and heating. Mild weather tends to reduce an electric utility's sales.

Most utilities experience their peak demand in either summer (summer peaking utilities) or winter (winter peaking utilities). Lighting needs are greater at times of the year when the sun sets earlier in the day, and on weekends, when activity often lasts later into the evening. Some end-uses are seasonal: heating demand generally occurs only in winter, being greatest during particularly cold periods and when family activity is at a peak – early morning and early evening. The electric load stemming from a particular application or end-use will vary as a function of time, depending on the activity patterns and demand from customers.

Customer Class and End-Use Distinctions

As stated earlier, a utility generally recognizes that its customers fall into distinct *customer classes,* each composed of consumers with similar needs and values, who are billed in the same way. These categories, which include agricultural or rural, residential, commercial, and industrial, often have sub-categories within each, for example, residential, all-electric homes, apartments/townhouses, etc.

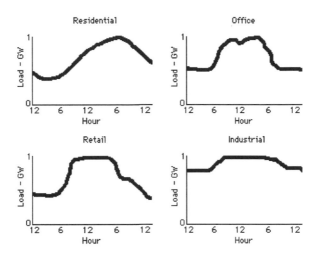

Figure 5.8 Peak day hourly load curves for four classes of load in a large utility system, showing how much the entire set of customers in each class used. Different classes have different schedules of usage, and different amounts per customer. Although industrial and residential classes use about the same total amount of power (1 GW), this constitutes the total usage of 254,000 residences (about 4 kW/home) and only 127 industrial sites (about 8,000 kW per site).

System Peak - 3,492 MW   Residential - 4.2 kW/customer

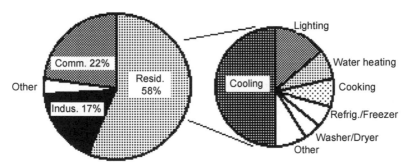

Figure 5.9 Electric peak demand of a utility in the southern United States, broken out by customer class, and within the residential class, by contribution to peak for the major uses for which electricity is purchased at time of peak.

Among other data, the electric utility analyzes the usage pattern of each class, not on a detailed second-by second load-curve basis (Figure 5.5), but through the lens of the demand period measurement, used to bill the customers (Figure 5.6). Figure 5.8 shows hourly load curves for several customer classes in a large utility system in the southern United States. The utility's peak demand, which defines the maximum amount of power production it must have, occurs when the simultaneous use of all the consumers and their appliances is at a maximum. Figure 5.9 shows the breakdown of the peak demand for a metropolitan utility in the southern United States.

Spatial Distribution of Electric Demand

The utility's customers are spread throughout its service territory, but the electric load is seldom distributed evenly over the entire region. Figure 5.10 shows a load map for a city in the eastern United States, and is typical of the electric load density and pattern in many large urban areas. In the core of the city, the downtown area has very high load densities, often exceeding 60 MW per square mile, the result of densely packed, high-rise commercial real estate development.

Outlying suburban areas have a much lower load density than areas near the center of the urban core, perhaps only 3 MW per square mile. However, the density of demand along major transportation corridors is two to five times that, due to more concentrated development of commercial and industrial buildings in those areas.

In rural regions, load density may be much lower, because homes and businesses are spread out, often being less than a thousandth of what it is in cities. However, in some agricultural areas, load density in the country actually exceeds that in the suburban areas of cities, due to the intense loads of irrigation pumps, as well as of oil pumps in petroleum fields. The spatial distribution of electric demand defines the power delivery function, i.e., the job of the utility's T&D system. Regardless of where the power is generated, it must be delivered to customers in the pattern shown in order to satisfy their needs.

Understanding Their Customers

Considering the importance of customer focus, particularly in a time of competition and de-regulation throughout the power industry, all employees in an electric utility should understand the basics of their company's power use. It is recommended that every utility employee know at least the basic characteristics of the company's sales and customer base, i.e., who buys the product, how much is bought, for what purposes, and where is it delivered, as illustrated in Figures 5.8 through 5.10.

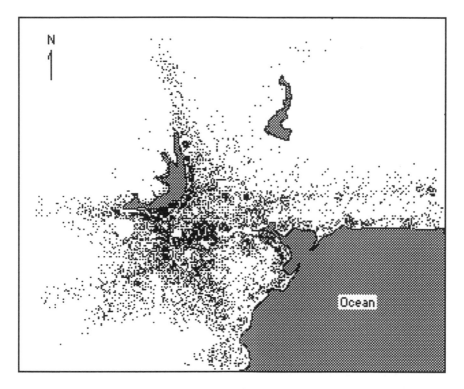

Figure 5.10 Spatial distribution of the peak electrical demand throughout a 2,000 square mile area in and around a medium-sized coastal city (population 785,000). Shading density indicates amount of electrical demand at every location. The typical characteristics of load density in a city are shown: a very high load density in the downtown core, tendrils of dense load following highways and major transportation corridors out of town, and a generally decreasing load density as one moves from downtown toward the surrounding countryside. From *Spatial Electric Load Forecasting – Second Edition,* by H. L. Willis, Marcel Dekker, 2003.

5.6 CONSERVATION, ENERGY EFFICIENCY, AND DEMAND-SIDE MANAGEMENT

It makes no sense to burn fossil fuels to produce electricity, with its attendant cost and environmental impact, transport it great distances over transmission and distribution lines at further cost, only to waste it heating a home that lets much of the heat inside it escape through poorly sealed windows, un-insulated walls, and leaky ducts. The cost to weather-seal a home or office building properly, and to install reasonable amounts of insulation so that it is energy efficient, is less than the cost of the energy those measures will save. Not only is this type of energy efficiency "the right thing to do," but it saves the building owner considerable money in electric bills.

There are dozens of other examples of energy efficiency, conservation, and effective energy control that simply make economic sense to the consumer. The tools used in achieving energy efficiency are often called customer-side, or distributed resources (Table 5.6). Each category shown includes many types of individual measures. For example, there are many ways to conserve energy simply by being aware of how to cut back waste, from installing flow limiters on showers, to reminding oneself to turn off lights in a room not being used. Under the heading of energy efficiency appliances, there are dozens of high-efficiency motors, air conditioners, refrigerators, heaters, and lighting fixtures, designed for numerous situations.

Determining how to best balance the cost of energy by using these various measures in proper combination is called *Demand-Side Management (DSM)*. The term "demand-side" refers to any energy control or management means or measures applied on the consumer (demand) side of the electric meter. DSM is hardly a simple matter.

First, the needs of homeowners, businesses, and industries vary so greatly that it is impossible to generalize about usage patterns and need even among subclasses of customer, such as high-income homeowners. DSM has to be applied in a very consumer-specific manner, with the application of the methods shown in Table 5.6 tailored to each individual consumer.

Second, determining the correct balance of the various DSM measures, and electric energy usage, is not always simple. It is possible to spend too much on insulation and weather-stripping, for example, spending more than the energy saved is worth, or worse, applying these energy efficiency methods poorly, so that they are ineffective.

Third, coordinating all the various DSM measures is difficult, for there are a lot of questions to be answered before the optimal mixture and amount of conservation, energy efficiency, appliances, control, and automation can be found. For example, which measure(s) among the many shown in Table 5.6 will

Table 5.6 Major Categories of Demand-Side Management

Method	Description
Conservation	Consumer education on how to use less energy or use it in a less-expensive manner.
Conservation voltage reduction (CVR)	The utility lowers the voltage on distribution circuits feeding homes and businesses to the lowest value that will not harm equipment or make it perform poorly. CVR tends to reduce the power demand of equipment, although it also reduces end-use performance somewhat (e.g., light bulbs are just a bit dimmer), a factor that makes heavy reliance on CVR a concern to most utilities and some customers.
Demand response	DSM under another name, but oriented toward and motivated by a desire to reduce peak demand as well as/more than energy usage as was traditional DSM. Demand response programs tend to focus on methods which the utility can control by request or operation, obtaining a "response" (reduction) when needed because it is running short of energy or equipment capacity on its system.
Efficient buildings	Weather-sealing, insulation, multi-paned windows, awnings, and other measures designed to keep a building naturally warm in winter and cool in summer.
Efficient appliances	Energy usage among common appliances such as refrigerators, air conditioners, and lighting, etc., can vary by a factor of 3:1. Efficient appliances cost more but usually pay for themselves within half their useful lifetime. Similarly, equipment used in industry can be coordinated in design to minimize its energy usage.
Load control	Control of hundreds or thousands of appliances (in many homes or businesses) such as water heaters and air conditioners can be done from a central location (the electric utility) to juggle their usage, reducing demand at peak and allowing the electric system to serve more homes with less peak energy output.
Automation	Home or business energy automation involves computerized control of all the appliances and machinery in a home or business, to schedule electrical usage among them so that they smooth out demand, resulting in lower peak usage and improved voltage regulation in the power system.
Real-time pricing	The utility installs a "smart" meter on the home or business that communicates with the utility's control center to obtain the current hour's "price" for power. The customer has installed a computer that control s appliances so that when power is expensive (as it is during peak times), more of them shut down until the price drops. The utility will then see a drop in load (from what it would otherwise be) during peak load times when energy cost rises.
Voluntary	A program in which the utility "signs up" customers who agree to reduce their demand when asked. These are usually large industrial or commercial customers who can make significant cuts for a few hours. The utility then calls (or e-mails a request) when it needs to cut back on power demand level.

make the most sense for a particular situation? How should they be balanced against one another: If a building is well insulated, it uses less heat, thereby reducing the savings from installing an efficient heater or air conditioner. How, then, are insulation and efficient heating/air conditioning balanced one to the other?

Fourth, "balancing" those resources against the cost of electricity is non-trivial, for electric utility costs and the interactions of electricity production and transmission costs with DSM methods are not always obvious. For example, during the 1980s and early 1990s, a popular method of improving residential and commercial building air conditioner and heater energy efficiency was with what was called a *variable speed heat pump*. These units included a motor that ran at "just enough" speed depending on the heating and air conditioning need at each moment, thus using only the minimum amount of power necessary.

Variable speed heat pumps and air conditioners did use less energy than the units they replaced. However, they produced copious amounts of harmonics (see Power Quality, earlier in this chapter), and they were very susceptible to power quality problems such as surges and instantaneous interruptions. As a result, performance was poor and electric utilities and building owners alike often had to install additional equipment (harmonic filters, short-term UPS backups) to achieve service equivalent to the old, less efficient heaters and air conditioners. Cost actually rose in some cases.

DSM Earned a Deservedly Poor Reputation

During the 1980s, many state public utility commissions mandated that utilities in their jurisdiction would implement demand-side management among their customers through *integrated resource planning (IRP)*. Basically, the utility's planners would determine how much of what type of the resources shown in Table 5.6 made optimal sense, and then work with its customers, through consumer education and rate incentives, to promote that type of demand-side management. Many of these programs included societal and environmental impact costs in their analysis, in order to bias the evaluation in favor of conservation and efficiency. Often, to attain the degree of consumer participation sought, the utilities were required to pay for such measures themselves or provide considerable subsidies, which required raising the rates of the rest of the customer base (those not taking advantage of the DSM programs).

While the concept was noble and the goals laudable, many DSM programs produced little or no real savings, and diluted utility focus, while providing no additional consumer value.

While there were some successes, mandated DSM programs generally failed, for four reasons. *First*, DSM was often simplified and the difficulties of proper coordination and implementation greatly underestimated.

Second, the energy savings were often grossly exaggerated, both before and after the fact. This was easy to do: Just how much does adding insulation and weather-stripping to an average home save? That is not easy to estimate under the best of circumstances, and nearly impossible to generalize over a large population. Yet many proponents of DSM, along with politicians, wanted to hear nothing about problems and were too willing to listen to simplified predictions of great savings.

Third, there was more than a bit of pure chicanery. Exaggerated promises were sometimes made by unprincipled "DSM consultants" in order to get contracts from utilities for consultation and implementation services. In some cases, later on, the amount of energy savings was overestimated by factors of more than three to one, sometimes because everyone involved had a stake in seeing that the energy savings looked good, but also because it is very difficult to determine how much energy was not used.

Fourth, consumers did not respond to mass market programs mandated for utilities. They were not that interested, despite studies and predictions that they should be. Often the savings were not that great in their eyes, or the inconveniences larger than the utility and PUC had anticipated. Perhaps the biggest flaw in regulatory-mandated DSM programs in the 1980s was that they tried to apply rather uniform types of DSM values to all of a utility's customers: DSM was applied based on assessment of its value to utility and society, not its value to consumers.

DSM: A Potential Winner in the Competitive Retail Power Market

There was and is nothing fundamentally wrong with the concepts summarized in Table 5.6, if applied well. But as mentioned earlier in this section, successful application depends on tailoring to individual customers. In a competitive marketplace, retail energy service companies (Rescos, or Escos) will not only do this tailoring in order to provide good service to their customers, but use DSM measures and the savings and conveniences they provide, to entice customers to buy power from them, rather than from competitors.

Custom Power Equipment Extends DSM
Capabilities to Improving Power Quality

An important new area of customer-side technology that has made quite an impact on demand-side management is custom power. Custom power devices include an array of different types of electronic equipment that can control power quality at a home or business site. The "surge suppressors" that many people use to protect their personal computers from lightning and power surges are simple examples of custom power devices. Much smarter and more capable devices can provide additional power quality capabilities.

For example, uninterruptible power supplies (UPS) can be installed as needed to provide power even if the "utility system" has failed, for whatever reason. A UPS that will allow a personal computer to run for about an hour or more during a power blackout is about the size and weight of a large car battery (which is, in great measure, what it contains). Larger and smarter UPS units can protect entire buildings or "power sensitive" industrial installations from the failure of the power grid, as well as from a variety of other power quality problems, including surges created by lightning, harmonics, voltage swells and surges, and other transients. The best (and most expensive) of these utilize high-speed computer control and superconducting technology to assure that power flow does not deviate from "perfection" for even a millisecond.

Custom power devices can also improve electrical performance at a particular site, by correcting the power factor to 100% on a second-to-second basis, while maintaining the voltage at precisely the optimum, with no variance over time. Integrated with plant automation and other computerized controls, they assure perfect power for ultra sensitive electrical equipment, such as the machinery used to manufacture silicon chips for computers. In particular, custom power equipment aimed at improving power quality can be coordinated in action with more traditional demand-side management equipment aimed at improving power economy, with the two working in harmony.

The Evolution of DSM Under De-regulation

Customizable power

Many custom power devices are quite costly, while others are becoming economical as volume production brings costs down. Regardless, their capabilities are usually of little value to some homeowners and business persons, but many industrial users of electricity and some specialized commercial businesses find them very valuable. And that is precisely the point of the "custom power." It provides the ability through installation of specialized equipment to provide what each customer needs in terms of power quality.

In this regard, "custom power" encompasses the full evolution of DSM. Under the traditional regulated industry structure, DSM included measures implemented en masse by the utility and regulators to improve economy as evaluated against societal as well as customer needs. Under de-regulation, and with the aid of technological advance, DSM has evolved to include measures to improve both economy of use and quality of power, customizable to the individual consumer's needs.

Demand response

Traditional DSM programs (1980s) were often aimed at and justified on the basis of *energy* reduction: they would reduce the total amount of energy a home or business used in a year. As a result, many highly rated DSM programs at the time had little impact on peak demand levels, having been designed to have an effect over the whole year, not especially during peak conditions. In fact, any impact they had on peak demand was considered of only secondary value to their energy "savings." Conservation of energy was the driver.

In today's de-regulated industry, capacity limitations are of great concern and have generated a new regard for "DSM" methods as a peak limiting resource. It is actually more appropriate to say that DSM is now used on a economic basis. Both energy and peak capability cost money, and reduction can be applied to either or both in situations where the DSM makes sense.

Focusing on peak demand makes sense in a lot of situations, and in those cases yields bigger savings than a focus on just energy. During very warm summer afternoons or cold winter mornings, an electric utility's operators might project that they will soon run out of energy (e.g., generation capacity) or that system conditions will run the load on transmission or distribution equipment past their dependable operating limits. Temporarily reducing demand at these times may cost little and create almost no inconvenience, whereas the cost for the utility to buy the equipment needed to serve "the final increment" of that peak would be expensive, particularly considering such peaks occur only a few minutes a year. *Demand response resources* include programs of any type the utility arranges so that it can reduce the level of demand during infrequent but unscheduled times where demand exceeds the resources designed to handle normal peak loads, or where there are operating emergencies on the system.

In some cases, the term "demand response" is used by utility and regulators alike to indicate only load reduction programs (voluntary or involuntary) in which customers are given a financial incentive – a price reduction – if they will agree to reduce their demand by a certain amount whenever called and asked to do so. However, in other cases the people concerned consider that it includes direct load control, automation, real time pricing, voltage reduction, or any other systems that can be controlled by the utility as needed to reduce demand level on a temporary basis.

5.7 SUMMARY

Electricity is the preferred energy source world-wide because it is flexible, controllable, quiet, and clean. It provides as diverse a range of energy as the types of applications it can fulfill.

Other energy sources: natural gas, oil, and nuclear, are suited to only a narrow range of applications, and each has significant drawbacks.

Electricity has four basic applications: light, heat, electronic motion, and electronic circuits.

Consumers are classified by utilities according to customer class, or rate class: agricultural, residential, commercial, industrial and governmental, and so forth, and within those categories by amount of electricity they use.

A variety of appliances is available to consumers of each class to convert electricity into end-uses. Fifteen 100-watt light bulbs, a heavy-duty hair dryer, two microwave ovens, four large desk-top PCs, a 1.5 HP well-pump, and two refrigerators all use about the same amount of energy.

Power interruptions, which may be caused by equipment failure, damage to transmission and distribution lines by storms, or a car hitting a utility pole, etc., are classified as :

(1) Instantaneous - less than 15 seconds

(2) Momentary - usually less than three minutes

(3) Temporary - typically 30 minutes

(4) Sustained - Normally more than an hour.

Several guidelines exist for classifying service interruptions and voltage deviations: SAIDI (System Average Interruption Duration Index), SAIFI (System Average Interruption Frequency Index), and recommendations by The Computer and Business Equipment Manufacturing Association (CBEMA) for the number of voltage events equipment should be able to withstand while continuing to function.

Demand-side management (DSM) methods balance consumer-side resources for energy economy and power quality against the cost and reliability of electricity from the electric grid and tailor the capabilities of modern power system technology to the individual consumer. Offered as services by electric service companies or local distribution companies, they will be an important part of the increased consumer value and wider customer choice available under de-regulation.

FOR FURTHER READING

A. Capasso et al., "A Bottom Up Approach to Demand Forecasting," *IEEE Transactions on Power Apparatus and Systems,* May 1994, p. 957.

H. L. Willis, *Spatial Electric Load Forecasting – Second Edition,* Marcel Dekker, New York, 2003.

H. L. Willis and G. B. Rackliffe, *Introduction to Integrated Resources T&D Planning,* ABB Power T&D Company, 1994.

6

Creating Electricity:
Power Generation

6.1 GENERATING ELECTRIC POWER

Whenever a magnet is moved while near a metal wire, its moving magnetic field induces an electric flow in that wire. The same current flow will occur if the magnet is held fixed and the wire is moved through its magnetic field. Either way, electricity flows through the wire. The amount of electrical flow will depend on:

- The strength of the magnetic field

- The speed of movement

- The direction of the wire's and the magnet's movement relative to one another.

The resulting electrical flow will be brief: As the magnet moves closer to the wire, it will produce an increasingly higher current flow, but as it passes and moves farther way, the current flow will lessen. The strength of the magnetic force acting on the wire dies off rapidly as they draw apart, creating negligible power except when the magnet and wire are very close together.

One way to get continuous electrical power is to build a rotating generator, in which a magnet rotates closely alongside a loop of wire, as shown in Figure 6.1. A continuous stream of electric pulses is produced as each end of the magnet sweeps closer to, past, and away from the wire loop.

Alternating Current Generators

All magnets have "north and south" poles, each of opposite magnetic orientation. This is true of every magnet, and is an unavoidable law of nature. As these poles alternately sweep by a wire loop, i.e., the generator, they create current flows in opposite directions: One pulls the current, the other pushes it, and the series of pulses created are a "push-pull" flow, or *alternating current* (AC). One complete set of push pull, two pulses of opposite direction, is called a *cycle*.

The diagram in Figure 6.1 shows the basic concept behind a rotating AC generator. Actual generators are much more refined versions of the one shown. The wire, *stator winding,* often has many loops, arranged in a non-circular shape engineered with extreme care, to have just the right thickness, just the right number of turns in each loop, just the right length and depth to its shape, etc., for optimum performance. Similarly, the rotor is seldom a simple bar magnet, as shown, but something closer to a round cylinder, optimized for peak performance. Sometimes the rotor is made of wire loops, and a magnet is used for the outer portion. In others, both rotor and outer portion are wire loops, since a wire loop, if fed electricity, creates a magnetic field similar to a bar magnet. Regardless, the basic concept is as illustrated.

Most generators have more than one set of magnets. Imagine a second bar magnet in Figure 6.1, perpendicular to the first, creating a rotor with an "X" cross-section. The generator would now produce twice as many pulses per revolution. Some generators have dozens of magnets. Multiple magnetic poles mean the rotor need not spin at a very high speed to produce the required number of pulses per second: A generator with four magnets in its rotor, for example, needs turn only 900 or 750 RPM to produce the 3,600 or 3,000 cycles

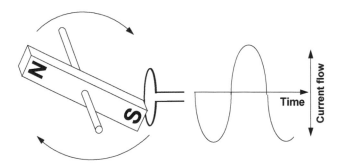

Figure 6.1 Alternating current can be produced by rotating a magnet near a loop of wire. As first one magnetic pole, then the other sweeps by the loop, sinusoidal pulses of electric flow are created, every other one of opposite polarity: alternating current.

per minute, respectively. This permits the rotor to be lighter weight and more efficient. Much the same result can be accomplished through innovative and complicated ways of arranging many wire loops in the stator. Higher speed requires more strength, which means heavier parts throughout.

A typical large "central station" generator, capable of producing 600 MW of AC electric power (enough to meet all the residential, commercial, and industrial needs of a city of 200,000 people), is actually not very large if one thinks of the amount of resulting power, as illustrated in Figure 6.2. The largest electric generators are perhaps 50 feet long and 15 feet wide, with the rotor inside being four to ten feet in diameter, spinning at 1,200 to 2,000 RPM. Generators in large hydro plants are similar in concept, but quite different in design. They have a vertical rather than horizontal shaft, are wider than they are high, with dozens of magnetic poles inside a rotor that is perhaps 40 feet across, and turn much more slowly, at only about 100 RPM.

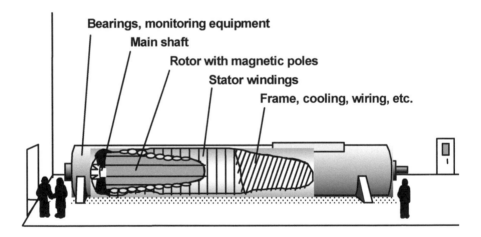

Bearings, monitoring equipment
Main shaft
Rotor with magnetic poles
Stator windings
Frame, cooling, wiring, etc.

Figure 6.2 A typical electric generator, as engineered to maximize the utilization of the principle illustrated in Figure 6.1. The portion shown is only the electric generator part of a power station, located inside a building. Not shown is the turbine required to spin the generator, the boilers required to make steam to power the turbine, the ancillary equipment required to run all that equipment, the control and operations house, or the electrical switching yard (outgoing substation), which together make the plant a very big facility.

Direct Current Generators

Sometimes, for a variety of practical engineering reasons, it is best to use direct current (DC) for a certain application or appliance. DC power moves only in one direction, instead of alternating back and forth. The generator in Figure 6.1 could be converted to DC by installing mechanical or electronic switches (not shown) that reverse the direction of current flow every time the magnetic rotor makes half a turn. Since the current flow is naturally reversing itself each half turn, due to the opposite polarity of the magnetic poles, this "reversal of the reversals" results in a series of pulses, all with the same polarity. This is a DC generator, commonly used as the small generator or alternator under the hood of an automobile: All automobiles use DC electrical systems for their lights, instruments, air conditioning, and radio, etc., which still produce pulses of power, but all with the same polarity. However as explained in Chapter 4, electric utility power systems use alternating current (AC).

In special cases a generator unit might utilize a DC generator and then transform the DC power into AC power for use on the utility power system by inputting it into a *converter*. While this might seem like a circuitous way to produce AC power (and it is) it provides a utility special advantages in some cases: converters have special "power shaping" capabilities for voltage, current and power factor that make them useful when very rapid electrical control of power quality or output is needed.

In the reverse of that process, direct current can also be created from alternating current using a DC-AC *rectifier*. Rectifiers can be either mechanical or electronic. Rectifiers are much more common than converters, but in utility systems (most of the control equipment at power stations and substations is run on DC, not AC, power converted on the spot) and particularly in homes and businesses. Computers, televisions, and microwave ovens also run on DC power. The power electronics inside them converts AC current from the wall socket to DC electronically. About the only time homeowners or small businesses use a converter is if they buy one to power, say, a small TV from their car's electrical system.

Three-Phase AC Generation

Most generators do not use one wire loop, as shown in Figure 6.1, but instead use three, equally spaced around the generator, as shown in Figure 6.3. These are called *phases*. The result is that during each revolution, the magnetic rotor produces a pulse in each of the three phases. One end of each loop is connected together, and only three wires leave the generator. Each represents one of the phases, and because they are all driven by the same rotating magnet, they carry power that is alternating at the same frequency, but the *timing* of the pulses is 1/3 revolution (120 degrees of phase) apart, as shown.

Three-Phase Generator Layout

Timing of Power Output by Phase

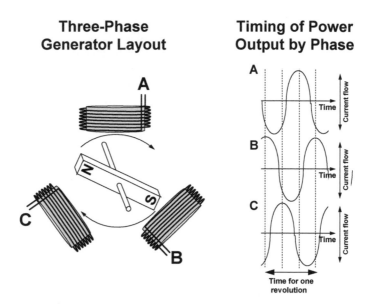

Figure 6.3 Generators actually have three loops, or *phases,* spaced evenly around the rotor, as shown at the left. One revolution of the rotor thus produces a single back and forth pulsation of current (a *cycle*) in each loop. Timing of the rise and fall of current in each phase is slightly different, separated by the time it takes the rotor to spin 1/3 revolution, as shown at the right.

Why three phases and not two or four?

There is no special reason why three phases are used in power systems instead of two, or four, or six. Sometimes, engineers or scientists will claim that the number three is the closest integer to pi (3.14159 . . .), or the base of the natural logarithm system (2.78182 . . .), as if this has some special significance, but it doesn't. The fact of the matter is that the first few electric generators, built near the end of the 19th century, had three-phase loops in them, probably because three was seen as a good compromise between two, which would leave a perceivable gap on each side of the generator, and four or more, which lead to manufacturing complexity, a real issue before the advent of standardized parts and assembly line production. Since three phases proved reasonably efficient and easy enough to work with, it became a de facto standard and then the official one, until today all power systems are standardized on three phases. In very rare cases, often for esoteric special reasons, six-phase transmission lines may be built, and special generators or equipment are used in certain industrial processes that require nine, twelve, or even fifteen-phase power. In short, though, power systems have three phases just because they do.

6.2 ELECTRIC GENERATING SYSTEMS

Power distributed by an electric utility is produced by electric generators, which convert some other form of energy into electricity. These original sources of energy can be fossil fuels, i.e., coal, oil, and natural gas, or energy released by the fissioning of radioactive materials, by falling water in hydro-electric plants, or by energy taken from sunlight, wind, or geothermal sources. The complete set of machinery for producing electricity, including an electric generator, whatever converts the original energy source, e.g., coal, into energy, and all control and supporting ancillary equipment, is called a *generating unit.* Usually, several generating units will be located together at one site, a *generating plant.*

Central Station Generation

The various generating stations in a traditional utility system are connected together by a high-voltage transmission grid (Figure 6.4). The set of transmission lines, each of which can usually move an entire generator's worth of power several hundred miles, joins all the generators electrically, so that they act as one large set. Theoretically, power from any one generator can be routed to any location in the grid. This concept is called *infinite bus generation.* Since the power from any generator can, conceivably, be routed to any customer in the system, *all the generators serve all the customers.* By contrast, *zonal generation* has generators in one part of the system reserved only for customers in that area, e.g., those in the south serve customers in the south, and those in the north serve customers in the north.

Electric utilities and private generating companies (Gencos) usually try to put two or more generating units at sites called *generating stations*, partly to reduce site acquisition and licensing costs. In the developing era of electric systems, prior to 1930, most utilities had only one such big generating facility, the *central station.* Most modern utilities have grown to where they have 12 or more such generating stations. The largest ones have almost 100, but the concept of producing power at a few large generating sites is still called the *central station generating concept.*

Modern independent generating companies usually build generators in the range of 50 to 250 MW for financial reasons. They know that bigger units have lower operating costs, but do not want to make the sizable investment for building large, single generators. Instead they tend to diversify their investment by building several medium-sized units in different parts of the country to sell into different markets, rather than one very large unit at one location.

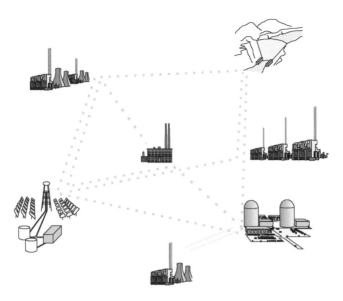

Figure 6.4 Central station generating concept involves producing electrical power at a few large generating plants, central stations, scattered throughout the service territory. Each plant has between one and perhaps six large generators, and all are connected electrically by a high voltage transmission grid (dotted lines).

Generation mix: Base, intermediate, and peaking units

Any large utility system, or power pool, will use a variety of generating unit types and sizes, for several reasons. Foremost among these is that *all the generation capacity is not needed all the time.* Demand for electric power varies as a function of time, rising and falling each day, being greater during weekdays than on weekends, and rising during summer and winter, due to air-conditioning and heating usage. As a result, a particular utility may study its annual demand curve and determine that demand will exceed 4,000 MW every hour of the year, and that peak demand will reach 10,000 MW. Analysis of usage patterns typically reveals that peak demand occurs infrequently, e.g., demand might exceed 9,500 MW (95% of peak) only 250 hours per year, less than 3/10ths of a percent.

The utility knows it will need at least 4,000 MW of *base load generation –* power production equipment that will run all 8,760 hours of the year. Given that these units will run continuously, it makes sense to spend whatever is required to buy very reliable, and highly efficient equipment. Thus, base load generators tend to be nuclear, specially designed large natural gas and coal units, and big hydro-electric plants that have a relatively high initial cost, but very good operating economy. Higher-than-average initial costs are justified because the savings in fuel and operating cost will be substantial when such units are run for many hours.

However, for the 250 hours per year when the utility needs that last 500 MW increment to meet its peak requirements, such highly efficient generators are not economically justifiable, because the units will not run enough hours to pay back their higher initial cost. A utility or power producer might buy a number of gas turbine generators as *peaking units*. These are not quite as fuel efficient as base load units, but since they have a very low initial cost, they work well for peaking applications.

Peaking and base load generators also have another difference – *load tracking*. Many of the engineering and design methods used to achieve high efficiency in a generator have a side effect: The generator must run at a very stable level of output and cannot increase or decrease its power output quickly. For example, nuclear power plants, the penultimate base load generators, cannot change their output rapidly. Many are licensed to ramp load (increase power output) only 3% per hour, meaning that it takes them nearly a day and a half to run up to full output after starting.

This inability to vary output is not too important in a base load unit that will slowly be run up to near maximum and left there for thousands of hours. However, a peaking unit will be started and stopped. In addition, demand for electricity varies from hour to hour, sometimes by as much as 10%. Therefore, peaking units are designed to vary power output rapidly – the best can go from zero to full output in less than five seconds.

Filling in between these two extremes are the *intermediate generating units*. These are intended to operate for longer periods of time than peaking units, and represent a compromise between low initial cost, good fuel economy, and operational flexibility as needed to meet the production needs of the system.

Contingency margin

Typically, a utility that expected a peak demand level of about 10,000 MW would have about 11,500 MW of power production available. Any power system needs more generating capacity than its peak demand, in order to allow for the failure of any one generating unit. The general rule followed by power systems engineers is that a utility should have enough generators to get by even if its largest unit is out of service, due to failure or maintenance, either or both of which can take up to two weeks annually.

Reserve margin is the amount of generating capacity over and above the peak load, allowed to cover maintenance and contingency (failure) needs. Generally 15% is considered a good rule of thumb. *Spinning reserve* is the actual amount kept running. *Standby reserve* is that available on short notice.

Thus, a power system that had to meet a 10,000 MW peak load might have access to 11,500 MW of generation in order to have a 15% reserve margin. Of this, 500 MW might be spinning reserve, extra generation running and connected to the system, but not producing power. This would cover power being produced by any one generator, including the largest one currently on line, should it fail. Another 500 MW might be on standby, manned and with control and other equipment checked out and even warmed up, so that it could start and be on-line within 30 minutes. This would permit the utility to bring another reserve unit on-line if the first one were needed due to a failure of one generator.

Synchronization of Generators

Almost all power systems have many generators in them, both because no one generator can be built large enough to provide all the electricity needed, and because use of many generators means the failure of any one does not cause a serious problem. Due to the magnetic fields and electric flows inside each generator, if two or more AC generators are connected by a transmission line, they will sense one another, and under the right conditions, lock into step with one another, turning at exactly the same rate. The tendency to stay in *synchronization* is only slight, but enough that if carefully used, a utility can arrange for all of its generators to turn at *exactly* the same rate.

When synchronized, the set of generators can be precisely controlled with respect to which generator produces how much power in a relatively straightforward manner: If the utility wants a generator producing 500,000 kW to increase output to 550,000 kW, the generator is instructed to rotate faster. Locked into synchronization with the other generators in the system, this particular unit will not be able to spin faster. Instead, what happens is that in trying to do so, it tries to push all the other generators to spin faster too, by pushing more power out onto the grid. The spin rate of the entire generating set does not increase, but this particular generator shoulders a bit more of the electric output demand of the entire system.

When all generators in the power system are spinning at the same rate, it makes instant analysis of problems easier for the power system operators, because they can track frequency, not power output, in order to control their system: If the frequency of power anywhere in the system starts to drop slightly, e.g., by a thousandth of a percent, it means more power is needed and the generators are set to produce slightly more. If frequency becomes slightly fast, the opposite is done. Beyond that, if the frequency in a particular location starts to deviate from that in the rest of the system, it may mean that there is a failure or some other difficulty with equipment, lines, or a customer appliance, and devices designed to sense this will operate circuit breakers, etc., carefully opening circuits to avoid any possible damage or safety problems.

6.3 TYPES OF GENERATING PLANTS

What Turns the Generator?

Something must turn the rotor in a generator so that it can produce electric power. The effort required to turn it is not trivial. *Power out equals power in.* All that the electric generator does is convert mechanical power from whatever is turning it, into electrical power. A human being can turn a very small generator with a hand crank, producing enough power for one or two light bulbs.

However, to achieve any significant amount of power, some other source of energy beyond human or animal power must be used to turn the generator. A wide variety of energy sources is utilized for the resulting generation systems. In order of their usage worldwide, these can be divided into three major categories:

(1) Fossil fuel powered generation: Coal, oil, natural gas.

(2) Nuclear fission powered generation.

(3) Renewable powered generation: Wind, solar, hydro, photovoltaic, geothermal, and bio-mass. These types are covered in Chapter 7.

Fossil-Fuel Powered Generators

The most convenient, economical, and widely used method of producing electric power is to burn a fossil fuel, be it natural gas, coal, peat, fuel oil, diesel fuel, kerosene (paraffin), or refined gasoline, in a suitable machine, using rotating power to spin the electric generator. Except in the case of fuel cells, where no rotation or combustion takes place, this is always accompanied by burning, with its attendant pollution concerns.

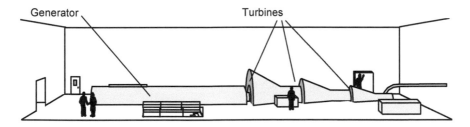

Figure 6.5 A turbine generator unit for a large power plant. This unit is inside a building to protect it from the elements and to provide a comfortable, clean environment for the people who maintain it. Only the top half of the generator and turbine is shown in this drawing, the other half being accessible from the floor below. The generator is like that shown in more detail in Figure 6.2.

Steam turbine generators

By far the most popular machine worldwide used to spin electric generators is the steam turbine. Water is heated to very high temperatures, producing steam at high pressure. The steam is then directed at a turbine, basically a series of windmill-like bladed wheels inside a metal housing, and a set of "windmills" one behind the other, designed so that the rapid flow of steam forces them to turn quickly. The turbine, including housing and all plumbing, is typically about the same size as the generator itself (Figure 6.5). Usually, turbine and generator are built as a tailored unit, the characteristics of each one optimized for the other.

The steam to drive the turbine is produced in a boiler, which can be designed to burn oil, natural gas, or coal, or in some rare cases, garbage or hay, with the heat from this combustion used to produce steam. Many boilers are engineered to operate on only one fuel type, like coal, although some are multi-fuel, able to run on natural gas or oil. Burning coal and oil can produce significant amounts of pollution. Therefore controls, similar to the catalytic converters in automobiles, but more sophisticated in operation, are a big factor in the design of any fossil-fuel generating plant. Coal plants, particularly, create challenges in this regard. However, very "high-tech" methods of burning coal, such as *fluidized bed processes*, can both increase fuel economy and reduce the pollutants, if used properly.

High efficiency and economy of operation come from using a combination of very high temperature and pressure in the boilers and turbines. Most boilers operate at temperatures of up to 1,000°F and pressures of up to 800 lb. per square inch. This means the turbine must be built to withstand both these very high pressures and temperatures.

Physically, the boiler is by far the largest part of a typical power generator, as shown in Figure 6.6. The cooling towers, if used, are the other major part, in terms of size. For maximum efficiency of operation, the steam must be cooled and converted back to water after it has passed through the turbines. This can be done with large radiators (cooling towers), or by circulating the steam through pipes laid out in a large body of water (cooling pond), or if there is lots of water nearby, like a large river, discharging hot water into it and withdrawing cold water upstream. However, for both environmental and engineering reasons, this is often unacceptable.

Very efficient turbine-generators can be designed in almost any size, from very small (50 kW) to very large (1.25 gigawatts), but boilers have a very definite economy of scale: Bigger is always better. Thus, most steam turbine generating units are built in sizes to produce 200 million watts (200 MW) or more, the most efficient size generally being 500 MW or more. Such power plants are complex and take several years to build: The turbine generator will be ordered as a custom design and the boiler will be constructed laboriously on site, taking about four to seven years overall, at a total cost of perhaps $400 a kilowatt. A 600 MW coal-fired power plant would therefore cost about 2.4 billion dollars.

Fuel efficiency

The foregoing discussion referred repeatedly to "fuel economy" and efficiency of operation. Although large generating plants are very expensive, that cost pales next to the fuel cost, which over their lifetime will be several times that of construction and maintenance. Therefore, great emphasis is put on designing the power plant to use fuel as effectively as possible. The efficiency of a generating plant is typically measured in the number of British Thermal Units (BTUs), the amount of heat needed to produce 1 kilowatt of power for one hour. This is called the *heat rate* of the unit. Like fuel economy of a car, fuel economy of a generating plant depends on good design, proper maintenance, and good "driving habits" – if run to a good schedule by expert operators it will be much more efficient. A well-designed modern steam turbine-generator-boiler system, in good condition and operated optimally, requires about 10,000 BTUs for each kilowatt hour of power produced. Older units, poorly maintained, i.e., "out of tune," and not well operated may require up to 12,500 BTU/kWhr.

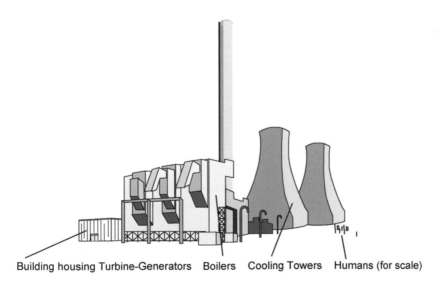

Building housing Turbine-Generators   Boilers   Cooling Towers   Humans (for scale)

Figure 6.6 A generating unit. The largest items are the boiler (foreground, about ten stories high) and the cooling towers. Turbine generator is in a building behind the boiler.

Often engineers refer to a power plant's economy in terms of percent efficiency. A rating of 30% means that the unit converts 30% of the chemical energy in the fuel into useful electricity. A heat rate of 10,000 BTU/kilowatt hour is equivalent to about 35% efficiency. Due to the way that efficiency is computed, based on a formula and conceptual model called the *Carnot cycle,* it is impossible ever to achieve anything close to 100%. Even the fusion process inside the sun falls short of that theoretical limit. An efficiency of 35% is about the best that can be expected from normal steam-turbine generators, and about 42% from very exotic types of boiler systems that burn coal at very high temperatures under computer-controlled conditions.

In general, the overall efficiency of a generator is a function of the highest temperature inside it. Certain natural physical laws decree that the hotter the temperature, the more efficient an energy conversion process can be. Thus, most fossil-fuel and nuclear engineering seeks to design equipment that works at the highest temperatures possible.

Overall efficiency is not the only criterion of operating economy, however. The type of fuel has some bearing on it, too. For example, coal is available in various grades, from low-sulfur (inherently hotter when burned and much less polluting) to lignite, often referred to as "burnable dirt," which produces much less heat per pound and much more pollution. But lignite is much less expensive, perhaps sufficiently so in some cases to outweigh the cost of more expensive pollution controls, and lower efficiency that results from its typical application. At the time of this writing (2005) natural gas is although more expensive than in the past, and still so economical it is the preferred fuel for most new generating plants. However, long-term fuel costs are uncertain. Coal may have a competitive price advantage in the future.

Gas turbines

Rather than burn natural gas to create steam to drive a turbine to spin the generator, one can use a *gas turbine. This is similar in concept to a steam turbine,* but natural gas or oil is burned inside the turbine itself, producing a hot, expanding gas that pushes against the turbine's blades, creating circular motion. Gas turbines are basically just big jet engines, like those used on large airplanes. They became increasingly popular as generating sources in the last third of the twentieth century, for several reasons:

(1) For any amount of power, they are slightly simpler, with fewer parts, etc., and noticeably smaller than steam turbine generating plants of equivalent power output.

(2) They have a different economy of scale so that they can be made into modular units as small as 25 MW, rather than only in large sizes like steam turbine units.

(3) But most important, they are mass-produced at factories and can be ordered, shipped, and set up in only a year or slightly more, reducing their lead time and their cost per unit of output compared to steam turbine plants.

(4) Finally, if run on natural gas, they produce low pollution levels.

Against these advantages, gas turbine power plants have two drawbacks. The first is that except for unusual (more expensive) designs, they run only on natural gas or gasified coal. By contrast, steam turbines can be built to run on natural gas, fuel oil, wood and refuse, or coal which is particularly plentiful. This is a not a serious drawback in an era when natural gas is inexpensive, but does limit gas turbine application to locations where natural gas is available, which makes them impracticable in many third-world rural areas where there is no gas pipeline system. The second and major disadvantage of gas turbines is that they are not as efficient as steam turbine power plants. A gas turbine is considered fairly good if it can produce power at below 11,000 BTU/kilowatt hour, a figure 10% worse than the figure attainable with good but traditional steam turbine units. Since most of the cost of producing power is fuel cost, this means that gas turbines are relatively expensive to run.

Combined-cycle generators and measures to improve efficiency

Gas turbine fuel efficiency can be improved tremendously through a complicated process of using its exhaust heat to produce more power. The exhaust from the "jet engine" part of a gas turbine generator is very hot, enough to boil water and produce steam at high pressure. A normal gas turbine simply wastes all that heat, venting it into the atmosphere. But a *combined cycle generator* uses it to boil water to produce steam, which is applied to run a steam turbine to help spin the generator in company with the gas turbine. As a result, a combined cycle generator is much more efficient than a gas turbine generator. *In fact, "world class" combined cycle units are about the most efficient type of generator possible, the very best having heat rates of about 9,100 BTU/kilowatt-hour.*

Since they use two different types of turbines – gas and steam – in one machine, combined cycle generators are quite complicated. They have many more parts than either a gas turbine or a steam generator, and they require complicated control systems, because they are harnessing two different types of mechanical power processes – steam and gas – that behave according to similar, but quantitatively different physical laws. This means they cost more initially, and are more expensive to maintain.

However, the added cost and complexity is justifiable in many cases,

Containment building (reactor inside)
Machinery building (turbine generators inside)
Electrical switching substation
Operations and Administration

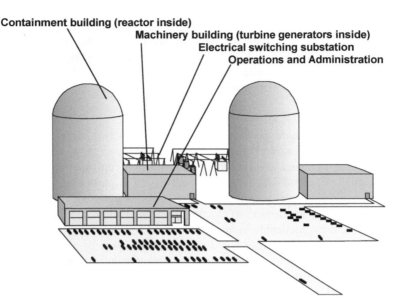

Figure 6.7 A nuclear power plant with two nuclear generators, each reactor in a separate containment building. Since this plant has no cooling towers, it has a nearby *cooling pond* or lake, from which it will draw cool water as needed, not shown.

because combined cycle generators achieve phenomenal efficiencies, up to 57% for "experimental" units utilizing every design trick known. Like gas turbines, combined cycle units can be ordered in "pre-designed" modular units, which can be manufactured and set up at the site in a relatively short period. This reduces the lead time and front end investment of the electric utility or generating company, and the high fuel efficiency reduces the cost of fuel needed by 25% for gas turbines and 15% for steam turbines. Thus, combined-cycle units are the "generators of choice" for new construction.

Nuclear Power Generation

Another way to create the heat to drive a steam turbine-generator is with the thermal energy produced when uranium is fissioned under controlled circumstances. This heat is produced inside a nuclear reactor, and used to turn water into steam. That steam is used in the same way as if produced by a fossil-fueled boiler, to drive a steam-turbine generator.

For practical nuclear power, great effort must be put into design, materials, and safety systems in the reactor/steam generator, to prevent radioactivity from the nuclear core from contaminating the water/steam, during normal operations, or during innumerable possible failure scenarios. In fact, the main cost of a nuclear power plant is devoted to these goals.

Similarly, the major consideration in the course of operation is to avoid situations or conditions during which contamination might occur were there an accident or failure of some component of the system.

Nuclear fission and the reactors that create and control it are quite simple in concept, theoretically no more difficult to control than the fires inside a fossil fuel boiler. However, nuclear power plants are among the most expensive machines built by man for two reasons:

(1) Nuclear fuel is among the most toxic substances known. Nuclear power stations are extremely safe if designed and operated well, but only because they have myriad safety features including:

> A basic design intended to be "fail-safe," so that accidents are unlikely and not catastrophic if they do occur.

> A huge containment structure built around the reactor and all ancillary equipment so that any possible radiation leakage is contained.

> A raft of monitoring equipment, along with a nearly paranoid control system to check system security constantly and *scram* the reactor, that is to shut it down instantly, should anything unexpected happen.

(2) Nuclear reactors have a tremendous economy of scale. The basic physics favors reactors well over 150 MW in output. And, since the licensing, safety, and labor costs for a nuclear plant are much the same whether it produces 150 MW or 1500 MW, the rule of thumb is that larger is better. Therefore, most nuclear power stations are in the range of 1,000-3,000 MW. Producing enough power for a major city involves a large facility (Figure 6.7) and costs dozens of billions of dollars.

Nuclear power plants built for commercial application are only in the range of 25 to 30% efficient, but that is not a critical factor. Fuel cost is relatively low, when averaged over all the power produced, and modern, "next-generation" designs would be quite cost-effective if built. Nevertheless, due to the high initial cost and the risk of accident, the lengthy and costly licensing procedures with an inherent uncertainty about both if and when, as well as questions about the future cost and liability for disposal of nuclear waste, few companies in North America – traditional utility or competitive Genco – seem willing to commit to construction of any new nuclear power plants. Despite this, the authors believe nuclear power has a good future. There aren't many other alternatives, particularly when one considers that, for all its disadvantages, nuclear power produces no air pollution at all.

Nuclear power safety and business risk

Actually, the business risk from the possibility of a nuclear accident is not primarily driven by any liability or likelihood of hurting the general population. While nuclear plants owners and operators worldwide have always been quite concerned about public safety, there is virtually no possibility of danger to the public from modern commercial nuclear power plants of the type built in the United States. These are among the safest devices ever built by man, particularly because of the containment buildings which house the nuclear machinery and keep any radiation that might be spilled inside this protective shield. The Chernobyl nuclear plant did *not* have a containment building, an unforgivable short cut on the part of the former Soviet Union. All modern reactors in the US and Europe do.

The real risk with nuclear power, and one that scares off many potential investors, is the possible catastrophic loss of their investment in a *contained* nuclear accident. As an example, had the same chain of accidents and operator inattention that occurred at the Chernobyl nuclear power plant happened at a coal-fired generating plant, it would never have received much public notice. Very likely the resulting "accident" would have blown the boiler tubes in the coal-fired steam boiler, releasing a huge cloud of steam, perhaps badly burning or killing any workers unlucky enough to have been within a few yards of the boiler at that instant. The boiler failure would have cost the utility about fifty million dollars in "minor" repairs (minor versus the near billion dollar cost of a big coal-fired power plant) and millions more in lost power production during the several months required for repairs. Several plant personnel probably would have been reprimanded or fired, and if the incident had made the news at all, it would have been quickly forgotten.

But had that same set of events happened at a US nuclear plant (one with a containment building), no one would have been harmed, but the plant would have been ruined: investors would have lost billions of dollars in just seconds. Even a less serious nuclear incident, such as what occurred at Three-Mile Island, can cause massive monetary losses for the owner. Three-Mile Island caused a nine-billion dollar asset to destroy itself in a matter of minutes. While dangerous levels of radiation were never released into the atmosphere (because the plant's containment building kept almost all radiation leakage isolated from the atmosphere), the inside of the plant was made "too hot to handle."

This particular business concern has made utilities and potential investors in power generator plants "gun-shy" about building new nuclear power plants: Can a nuclear plant be protected from its own ability to poison itself into uselessness due to a moment of operator inattention? While the answer to the question seems to be "Yes" for modern high-technology nuclear plant designs, that lingering worry, plus the gauntlet of permits and licensing procedures that an owner must endure, has made the industry reluctant to pursue nuclear power.

Hydro-Electric Power Generation

A completely different way to spin an electric generator is to use water power. In a hydro generating plant, water is held behind a dam and routed through a turbine, a water wheel inside a special housing constructed to extract the maximum power possible from the water flow, to produce the energy needed to spin the electric generator. Hydro-electric generation is the only traditional, and widely used, *renewable* energy resource. Other renewable energy resources are solar, wind, and geothermal generation. These and newer applications of hydro-electric generation are covered in Chapter 7.

The best type of hydro plant, viewed from the electric power standpoint, is one on a major river where there is a very sizable water flow throughout most of the year, with a high dam, and an extensive reservoir (big lake) behind it. A large water flow is preferable because there will be plenty of water to use. A hydro generating plant can use millions of gallons of water per hour. Consumption is so high that the water in its reservoirs is typically measured not in gallons, or even cubic meters, but in acre-feet. One *acre-foot* is the amount of water that will cover one acre, one foot deep, or about 2,000,000 gallons (Figure 6.8). The higher the water level behind the dam, the greater the *pressure head,* i.e., difference in pressure developed by the water level on one side of the dam vs. the other, and the greater the power per gallon of water used will be. The mechanical energy available at a hydro plant to spin the generator is mainly that of the water "falling" from the upstream side, through the turbine to the downstream side. Thus, a head of two hundred feet will produce twice as much power as a head of one hundred feet.

From the electric utility's operating perspective, the larger the lake behind the dam, the better. Water behind the dam is essentially stored energy. The flow down most rivers varies greatly on a seasonal basis – much more in spring than in other times of the year. An ample reservoir lets the utility store a significant portion of a year's worth of flow, and in a few cases, more than a year's flow.

Water energy "collected" in spring, during the annual snow-melt runoff, can be used later in the year, during the peak electrical periods of the summer. Having a lot of stored water behind the dam means that the generator output is *dispatchable* – power production can be scheduled on a definite, pre-planned basis, because there is absolute assurance that the energy will be available. Hydro plants without significant storage, or with environmental restrictions that do not allow use of stored water, are called *run-of-river power plants.* They can produce power only proportionally to whatever flow the river has at the moment, providing little or no output during droughts, for example. These are less dependable as power sources because they cannot be scheduled weeks and months in advance – they are *non-dispatchable,* making them less valuable.

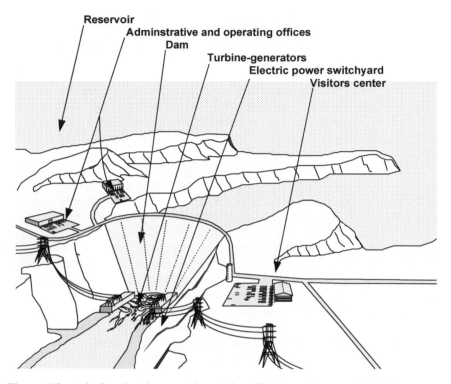

Reservoir
Adminstrative and operating offices
Dam
Turbine-generators
Electric power switchyard
Visitors center

Figure 6.8 A hydro-electric generating station. The dam gives a substantial pressure difference in water from upstream to downstream sides, providing power to water turbines that spin the generators. The lake behind the dam is stored energy, water that can be streamed through the turbines whenever needed to provide power on demand.

Hydro plants can be quite expensive to build. The water turbine and generator need a very different design than that for steam, gas, or nuclear units, but ultimately their cost is roughly the same as their equivalents in other power plants. However, the construction and materials' cost of the dam, and the land for the reservoir behind it, are much more expensive than the cost of any boiler in a typical steam turbine system. More than compensating for this additional cost, at least if viewed over the lifetime of the dam, is that the fuel is "free." In addition, hydro-electric generators are relatively simple machines in terms of the number and difficulty of manufacturing their parts, and in their operation and control. They involve no high temperatures within, but do involve very high pressures, necessitating very robust designs. Overall, they are reliable, dependable, and easy to maintain.

Against these advantages is one major disadvantage beyond their very high initial cost: One has to take hydro power where one can find it. Most of the good locations are in the foothills near mountainous areas, usually rather difficult terrain, and often quite far, 500-1,500 miles, from population centers. This means that the major construction project required for the dam and reservoir must be done at an isolated site, increasing cost, and that a lengthy, high voltage transmission line must be built to move the power from the hydro site to the population centers where it is required.

But on balance, hydro-electric generation is an attractive proposition, widely used. Hydro plants are the oldest commercial form of renewable resource electric generation, i.e., energy production that does not use up natural resources.

Since hydro-power plants involve no combustion, they produce no pollutants. Nevertheless, they do have significant environmental impacts, even if generally more benign than other types of power systems. Hydro plants create a man-made lake, often displacing hundreds of square miles of wildlife habitat, or using up land where man could grow crops. In many areas, depending on the underlying geology, this lake will leak a great deal of water due to seepage, into the rocks below. This is a considerable environmental impact: Water that once would flow to the sea now seeps into the subsurface hydrology hundreds of miles from the coast.

A large hydro dam provides some local immunity from flooding during periods of strong rains, which is a plus. But while the controlled, even flow of water that can be imposed by a large dam is more convenient for man, providing not only electricity, and water for irrigation, but also mitigation of flood effects, environmentalists are beginning to understand that the *lack* of occasional floods and rushing water creates long-term changes in the downstream ecology. These impacts are not always negative, but in any responsible application they should be anticipated and controlled. Regardless, if properly built and managed, hydro plants are among the most environmentally benign power generation sources.

Many of the best hydro-electric dam sites in the United States are already taken, and those remaining are not environmentally acceptable to all concerned, so that growth of hydro-electric generation in the United States will be slow in the 21st century. However, significant untapped potential still exists in North America, mostly at sites in northern Canada. Worldwide, the amount of untapped, available hydro potential is phenomenal. Most experts estimate that the amount of economically exploitable potential exceeds mankind's need for electric power at the present time. Other estimates put it at enough to meet all mankind's power needs through the end of the 21st century, if properly harnessed and supported by long-distance power transmission lines that can move power from those hydro sites to the cities where it is needed.

For the foreseeable future, hydro power will continue to be an important source of new generation construction throughout the world, and will contribute to a significant portion of the electric power used on this planet. Chapter 12 discusses the "globalization" of power grids and the role that inexpensive hydro power might play in the 21st century, in both providing mankind with more environmentally benign power while fueling the economic growth of third-world countries.

Co-Generation

Co-generation is the term used for power generators in which the *waste heat* is not wasted. The most efficient boiler or gas turbine leaves a large plume of very hot gases escaping out its smokestack or exhaust chimney.[1] There is a good deal of heat energy left in that exhaust. One often-used *heat recovery method* is to run this hot exhaust through a grid of water pipes, heating the water to very high, e.g., 300°F temperatures. Another common co-generation method is to apply the exhaust to heat buildings safely. To do so, the exhaust is not routed into the buildings directly, but is used to heat large volumes of water to a temperature just below boiling (about 180°F). This hot water is then pumped through baseboard heaters, etc. There are many other similarly useful benefits to which waste heat can be applied. (See, for example, Figure 8.5.)

Co-generation raises the net use of the energy released when fuel such as coal or natural gas is burned from a maximum of about 45% if only electricity is generated, to as high as 70% when considering both the electric and ancillary uses.

One problem with this type of co-generation is finding a nearby use for the "waste heat." Close to big power plants, there is seldom enough residential development to create a sufficient demand for the heat energy for home aplications: The steam or hot water cannot be piped efficiently more than about 1/8th mile. As a result, co-generation is most often used in industrial and commercial sites, such as where the generator/co-generator unit can be built into the design of a paper mill, and where the generator is small compared to large utility system generators, i.e., 10-25 MW.

A typical co-generation application at a paper mill that might involve a large boiler is designed to power a steam generator to supply the mill with electricity, while providing enough additional heat to feed the paper pulping process; converting recycled paper, or wood chips, to paper pulp requires *a lot* of steam and hot water in its initial stages. A boiler designed to produce that hot water can also be used to produce steam to run through a turbine to produce electricity to meet the plant's needs (or even more). Such dual-applications of steam have

---

[1] Power plants that meet modern air quality guidelines produce very little pollution. The "smoke" seen rising from power plant smokestacks and exhaust chimneys is simply very hot steam condensing as it hits the far cooler surrounding air.

been used in industrial plants for decades. More recently, manufacturers of modern gas turbine generators produce special designs tailored to specific industrial co-generation applications. Paper mill units are designed to produce the co-generation products (electricity, hot air, and hot water) in the appropriate ratios that a typical paper plant would use: a big plant might use three of these units, a smaller one, only one, etc. Similar designs are available to meet other industrial needs with different mixtures of hot water/electricity needs, etc.

Many industrial plants with this type of co-generation arrangement can produce more electric power than they need during most hours of the day. In the United States, throughout the 1980s and 1990s, these industrial businesses would routinely sell the excess power to their local utility company under PURPA regulations. For this reason, many long-time electric utility personnel use the term *co-generator* to indicate an industrial plant that has a co-generation unit and sells electric power to the utility.

6.4 SUMMARY

Large electric utility power generators convert rotating energy, obtained in some form from fossil fuel, nuclear power, water flow, or another manner, into electric energy. Over 95% of the electric power consumed worldwide is produced from large, central station generators that run on fossil, nuclear, or hydro power. Although newer technologies are available, tens of thousands of these big traditional power generators are already in place around the world, with their power output essential to our society. However, despite the fact that most of these units are "paid for already" and functioning productively, they have two negative aspects:

- Adverse environmental impacts

- Potential total depletion of the earth's fossil fuels.

Consequently, much of the emphasis in power generation at present is on using *less* of the finite fossil fuel and nuclear resources through more efficient power systems and electric usage, and on improving methods for harnessing "reusable" sources of power, what is known as renewable energy. They are covered in the next chapter.

FOR FURTHER READING

Institute of Electrical and Electronics Engineers, *Proceedings of the IEEE*, Special Issue on Advanced Power Generation Technologies, IEEE, New York, March, 1993.

H. L. Willis and G. B. Rackliffe, *Introduction to Integrated Resource T&D Planning*, ABB Guidebooks, ABB Electric Systems Technology Institute, Raleigh, NC, 1994.

7

Renewable Power Generation

7.1 FREE FUEL AND LOW ENVIRONMENTAL IMPACT

More than 95% of the electric power consumed on this planet is produced by traditional central station generators of the type described in Chapter 6. Two new applications in electric generating technology, far different from this traditional approach, show promise of increasing usefulness in the 21st century. Both are actually decades old concepts, but technological breakthroughs and refinements appear to have finally made both viable.

The first is renewable resource electric generation. *Renewable resources* are energy sources that do not consume irreplaceable natural resources. Sunlight, wind, and water power are renewable resources. They are constantly replaced by nature. The only renewable resource used in any significant amount for traditional power sources is hydro-electric power. Continuous research on solar and wind power has improved possibilities for those energy sources, to the point where they may well be competitive, at least in some situations. By contrast, coal, oil, natural gas, and uranium are not renewable. Once used, they are gone. At present rates of usage, these fuels will be exhausted in 50 to 400 years.

The second possible change in generating technology is *distributed generation*, which will be covered in Chapter 8. Instead of using large, central station power plants, distributed generation involves building many smaller generators which are scattered throughout the power system so that they are all close to where the power is consumed. With the generators located near the consumers, large transmission lines don't have to be built to move the bulk power from remote sites into cities and towns. This cuts cost and environmental impact.

These two concepts, renewable resource generation and distributed generation, are linked. Distributed generation calls for small, "home-sized" generators, and nearly all renewable resource generation, like wind and solar, is most conveniently built in small units appropriate for individual homes and businesses.

However, renewable energy has an additional advantage: there is no fuel cost, or, as will be discussed later in this chapter, fuel delivery cost – often that is actually the bigger factor. In all cases, "fuel-less" renewable power generators cost much more than their fossil or nuclear alternatives. But those technologies have the following unwelcome side-effects:

1) They use up natural resources that cannot be replaced. Eventually, this planet will run out of oil, natural gas, and fissionable uranium. By contrast wind and solar power tap into a virtually unending supply of energy

2) They produce pollutants, perhaps not much, but fossil and nuclear plants have an environmental impact. Renewable energy generation does not produce pollution in the strict sense of the word, although it can have undesirable esthetic or environmental impacts (wind generators are big, ugly, and noisy).

3) They cause a net increase of heat in the earth's environment. Burning fuel or fissioning uranium for power generation are processes that are not natural. By contrast, wind, solar, hydro, and geothermal are simply moving heat that is already here around on the planet's surface. This may have some small secondary or tertiary environmental impact, but it is far less.

There are numerous good arguments for and against the importance of each of these three issues. But they are arguments only of degree. No one seriously suggests that any of these side effects of fossil and nuclear power production has the least positive aspect.

Thus, renewable power generation is a sustainable alternative to using fossil fuels or nuclear power, depending on something that is restored through a natural process, like sunlight, wind, or water power. Assuming that mankind exercises good stewardship over the earth, these resources should last as long as we and our planet do. In turn, this chapter will look at small hydro, wind, solar, and other renewable technologies in Section 7.2 – 7.5. Fossil and hydrogen fueled distributed generation units are covered in Chapter 8.

Dispatchable versus Non-Dispatchable Power

Some generators are *dispatchable* – the operator of the generator can control output. When more power is needed one "pushes the throttles forward"; when less is needed, the generator is throttled back. Actually, no operator intervenes: automatic governors control the generator output to maintain output voltage at a constant level (when demand for power increases, voltage tries to drop and the generator compensates; when it drops, voltage rises and the generator will respond by backing off its power production).

Dispatchability is a key aspect of generation value, because it permits the utility to "track" customer demand: as its customers use more power, as during the afternoon on a hot July day, its generators respond automatically by producing more power. Operators at its control center have a lot to do to make this happen efficiently. While they let the generator governors control each unit, they look ahead an hour or a day, and "dispatch" generation by starting enough units to have a sufficient reserve of power as needed – enough generators spinning that there will be just a bit more than enough.

Fossil fired generators are dispatchable – they have a throttle and their output can be controlled: Put more fossil fuel in and more power comes out, etc. But most renewable generation units are not dispatchable, a severe handicap. Solar power produces power only when the sun is shining – one can "open the throttle wide" at night and it would do no good. Moreover, power output varies depending on how hazy the atmosphere is on a given day, and it can drop by 50% or more if a cloud passes overhead. Similarly wind power is also not dispatchable, and worse, subject to a very unpredictable pattern of wind speed. Practical utility systems need to be dispatchable, so that the power produced matches demand. This can be accomplished in two ways. First, a utility can mix the renewable power in with a sufficient amount of fossil power, so that it has enough controllable generators that it can match variation in demand: not all generators have to be dispatchable, only between 1/3 and 2/3, depending on the situation.

Or, the utility can install energy storage. In this case, the renewable energy is stored in some way (energy storage will be discussed in Chapter 8) in a battery of similar devices, so that it is available when needed. Energy storage is expensive, but worth the expense. For one thing, it makes the renewable generator dispatchable. And energy storage systems, and their converter and control circuitry, often bring additional benefits in power quality and system stability that are worthwhile (see Chapter 8, Section 8.3).

Non-dispatchable power generation is not worthless, but it is not nearly as valuable to a utility as dispatchable power generation, particularly in small, isolated applications: the very case for a lot of renewable applications. The ratio is roughly two-to-one.

7.2 Hydro Power

Despite being a renewable resource in the strictest sense, hydro power is not classified by the US government as a renewable power generation source. This is largely for tax incentive reasons – it is a very competitive technology without tax credits or incentives. But there is also a growing body of opinion against hydro because of environmental impact. Changes in river flow patterns can be significant. There is evidence that after many years those changes can have a marked impact on the ecology of both the river valley downstream and that upstream of the hydro site. For these reasons, and because large hydro plants are a significant part of the traditional generation base in the US and around the world, it was covered in Chapter 7. This brief section will make comments only on small hydro plant applications, generators often in the size range for distributed or private homeowner generation purposes.

Traditional hydro-electric generation technology needed a large *head*, i.e., difference between water levels on the upstream and downstream sides of the dam, in order to produce electricity. It also required great amounts of water, as in a large river, in order to produce considerable amounts of power. By contrast, low-head hydro generators produce small amounts of power from a head of only a dozen feet or so, and a flow equivalent to that of many small rivers or large streams. Low-head hydro is considered feasible and economically competitive at many locations throughout North America and, most likely, worldwide.

The majority of small, low-head hydro installations are *run-of-river* systems. This means they have no, or only a very small, reservoir behind the hydro site. When the river is flowing rapidly, as after spring floods, power output is high; during a drought it is far less. The high proportion of run-of-river as opposed to reservoir-dam units among small hydro applications is due partly to the fact that good low-head hydro locations occur where the flooding of large areas of surrounding countryside simply cannot be done. Equally important, the power production levels involved – often only a couple of megawatts, not hundreds, as from a large hydro-power plant – do not justify the purchase and conversion of land into a reservoir.

As with large hydro plants (see Chapter 6), one has to find a site that matches one's needs, even if it is somewhat inconvenient, because river flow is the key to power output. Figure 7.1 shows the relative expected annual output of hydro units in a portion of a very mountainous region. Sites farther downstream have more potential. They are also usually nearer to population centers (which are, typically, more downstream than upstream in rugged country) but tend to be farther away from mining and timber industrial sites, which are often the prime candidates for making use of small hydro power.

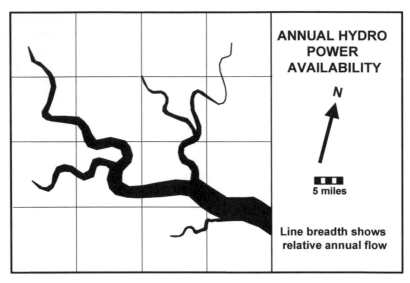

Figure 7.1 Map of average annual river flow at sites along a small river. Width of the line representing river or its tributaries shows relative potential for power. Anyone desiring a certain amount of power, therefore, has to pick a site where the river flow is strong enough.

Thus, most low-head hydro generators are *non-dispatchable*: Their output cannot be raised under operator control to meet demand. Instead, it varies seasonally and yearly, according to the amount of water upstream, meaning that one has to take what is produced, period. Low-head hydro utilization often raises environmental concerns about its impact on aquatic life, and the changes it makes on the natural "cleansing" of the waterways by spring floods. These are real issues that must be weighed carefully against the benefits provided: reliable and economical power without the production of air or heat pollution, or the use of non-renewable resources.

7.3 WIND-DRIVEN POWER GENERATION

Mankind has more experience with wind power than with any other energy source, except perhaps fire. For millennia before the invention of combustion engines or electricity, wind was the only viable form of propulsion for ocean vessels and the only practical means of driving water pumps other than beasts of burden. Despite the unpredictability of wind, its energy was adequately dependable and economical enough that empires were built upon those technologies.

Wind, which is just moving air, has a good deal of energy. Air moving at 26 mph contains 1 kW per square meter of cross-section – more than one horsepower. Wind electric generators use some form of windmill to turn a power generator. Usually these *wind turbines* are a design refined by computer to be much more efficient than the traditional "water pump" windmills seen throughout the western United States.

Figure 7.2 shows several wind energy conversion machines, illustrating the variety of designs used to convert wind energy to electric power. All are rather large. A typical wind turbine (center, top) that produces 1 MW of power would have a tower perhaps 200 feet high. The most popular design is the horizontal axis wind turbine, superficially similar to the traditional "water well windmill," but quite different in operational and design features. The most efficient designs use only two or three very narrow, long blades that rotate at high speed to capture the maximum amount of wind energy. Use of more and/or fatter blades, as in a traditional western "water pump" windmill, increases starting torque but decreases maximum power output.

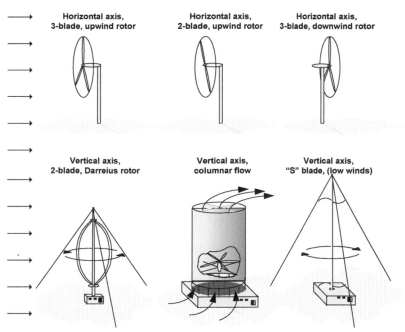

Figure 7.2 A wide variety of wind turbine designs have been tried, several of which are shown here. The three on the top row represent the most widely used types.

Wind turbines utilize only two of three high-aspect ratio blades (long, narrow) similar to the wings on high-performance aerial gliders, in order to minimize drag and turbulence. The best of these modern wind turbines can recover about two and one-half times as much energy from the wind as the traditional water well type. They spin either an AC generator, or a DC generator operating through a DC/AC converter. Horizontal-axis machines must have a *yaw control* – some mechanism to keep them pointed into the wind. This is necessary for two reasons:

1) To maintain power production – if pointed at an angle to the wind, power production rapidly declines.

2) To avoid possible operating problems. Side torque on the rotors, resonance in the blades, etc., caused by operating at an angle to the wind can quickly destroy a modern wind turbine.

If the turbine is designed with rotors *behind* the tower (upper right, Figure 7.2), yaw adjustment will be automatic. As the direction of the wind shifts, the turbine will be turned to face it. But for several reasons, most designs eschew this feature in favor of rotors *in front of* the tower, and active, i.e., computer-controlled, power-driven yaw aiming. Rotors located behind the tower have problems as the blades pass through the tower's "wind shadow":

1) Uneven rotational speed and a consequent loss of power.

2) A surprisingly loud "thumping" noise that becomes bothersome when heard 24 hours a day.

3) Blade vibration and resonance, which lead to metal fatigue and equipment failure.

Vertical axis machines have several advantages over horizontal ones. Most of their mechanical and electrical machinery is on the ground, rather than 40-90 meters in the air, making maintenance and operation simpler, and reducing structural needs. More important, vertical-axis machines need no yaw control. Since large horizontal-axis turbines have considerable inertia and require time to change yaw, they do not "track" wind shifts instantly, whereas vertical axis machines are immune to shifts in wind speed, and produce power even as the wind is changing direction.

Either way, a wind turbine has to be rather large to supply even modest amounts of power, compared to a traditional power plant. A horizontal axis wind turbine needs to have a blade diameter of nearly 180 feet, and a tower height of over one hundred feet, to generate 1 MW of power. One constraint on operation is that the energy in wind is proportional to the cube of its speed: Wind at 13 mph contains only 1/8th the energy of wind at 26 mph;

wind at 52 mph has eight times as much. To be viable, wind turbines need to be placed where they have stiff breezes all the time. Winds that are too light provide no power, and heavy winds provide too much power, with much of the energy in the wind going to waste.

Wind turbine "farms"

Substantial amounts of power are created from the wind by locating many turbines at one site, in a *wind farm*, as shown in Figure 7.3. Turbines must be separated by enough distance to avoid disturbing one another's air flow: About two to four diameters apart side-to-side, and about eight diameters apart in the direction of the wind. Extensive computer simulation and innovative thinking are needed to balance the many objectives involved in how to situate a wind farm, such as optimum wind recovery, lack of interference of one turbine with others, minimization of noise pollution, esthetic impact, operating cost and O&M (Operations and Maintenance) access.

Environmental impact of wind generation

Although wind turbines create none of the typical kinds of pollution nor contribute to global warming, they are not completely safe. Poorly maintained wind generators can throw their blades, which can be tossed over one half mile, doing significant damage. In cold weather, wind turbine blades can become coated with thick ice, in the same way airplane wings can. The ice will break off in chunks, tossed at speeds of up to 200 mph for up to a quarter mile, creating a potential hazard.

Noise and its abatement is another major consideration. The turbine blades make a slight amount of noise as the wind passes over them. Most objectionable is the audible "thump" produced as the rotor moves behind or in front of the tower. At 100 meters distance, the noise from a single large wind turbine, i.e., 500 kW, is about 50 decibels, or the same noise level inside a typical car traveling 55 mph. The turbine's noise is not distributed across the acoustic spectrum, as is the motor, wind, and tire noise in a car, but rather is concentrated in the thumping sound that some people find quite annoying, particularly if heard 24 hours a day. The noise problem with turbines is additive, in that if one makes discernible noise a quarter mile away, a dozen will jointly create a very noticeable sound, and a hundred will create quite a racket.

The other drawback of wind farms is esthetics. The best locations are on hilltops and sites with unobstructed wind, where, unfortunately, they are easily visible. While a few turbines may be tolerable, dozens or hundreds create a bizarre appearance most people don't want to see on a daily basis.

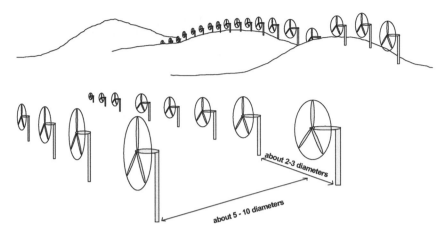

Figure 7.3 A wind farm. Placement of the turbines depends on a number of factors, including the design of the individual turbines, the terrain, and local wind conditions.

But despite the problems of icing, noise, and esthetics, wind generation is an appealing option in many cases, owing to the following advantages:

1) There is no fuel cost. The energy is "free."

2) Wind turbines are non-polluting, and available in small sizes, to fit situations where a lesser amount of power is needed.

3) They are modular: Two or more standard size units can be used where a greater amount of power is needed.

Against these advantages, wind power generation has the following disadvantages versus other types of power generation:

1) They have a high initial cost, as much as two times that of fossil fuel generation. While it is tempting to say that in the interests of the environment, everyone should just pay the additional cost, the margin could be unacceptably burdensome.

2) They are non-dispatchable, and their energy production is more difficult to predict in advance than solar or hydro units. Storage can be arranged to make them dispatchable, but this raises cost and complexity, rendering them uncompetitive in any sense.

3) Environmental impacts are often undesirable. Although wind turbines do not pollute, they use up a great deal of land, create noise pollution, and are objectionable in appearance.

7.4 SOLAR POWER

Under peak conditions at the equator, sunlight on a clear day provides the equivalent of 1 kilowatt of energy per square meter – more than 4 megawatts per acre! No known process can convert all of this solar power into electric power. In fact, it is impossible to convert more than a small portion of it into electricity. But with that much solar energy, the resulting electric power can still be quite substantial, making solar energy an appealing source of generation.

Potential for Solar Power

Under the very best of conditions – the sun directly overhead, the air as clear as likely to ever occur – the amount of solar energy reaching the surface of the earth is about 1 kilowatt (about one and a third horsepower) per square meter. That is a good deal of power, amounting to over two million horsepower per square mile. But the realizable potential for practical solar power is far less. To begin, no solar generation technology is even close to 100% efficient. In fact, none can convert even half of that energy into electric power – most convert only about $1/6^{th}$ – 16%. In fact, "solar efficiency" is usually not an important concern; cost is. In most cases it doesn't make sense to spend a lot of money to buy "efficiency" in solar energy conversion through use of advanced or exotic materials and systems: just use more sunlight, instead. Using six square meters of 16% efficient solar photovoltaic panels to produce 1 kW of energy under ideal conditions will be far less costly than using three square meters of 32% efficient panels to do the same. If one has the space to collect more sunlight, using lower performing technology is usually more economical when working with PV systems.

But even six square meters of panels each $1/6^{th}$ efficient would not normally produce 1 kW, or anything like it, because the sun is not always directly overhead, and the air is often anywhere from slightly hazy to completely overcast. And of course, sunlight is not available at night, so that basically divides generation potential in half. Figure 7.4 gives a map prepared by the authors that shows the estimated average annual gross solar power available throughout the US. Peak value (in the Arizona – New Mexico region) is 250 watts, or only $1/4^{th}$, of that 1 kW. Thus, that 16% ($1/6^{th}$) efficient panel installed in the southwest US would start out with only ¼ kW per meter of potential, as shown, meaning that on average it would produce 1/6 x 1/4 = 1/24 kilowatt, or about 42 watts.

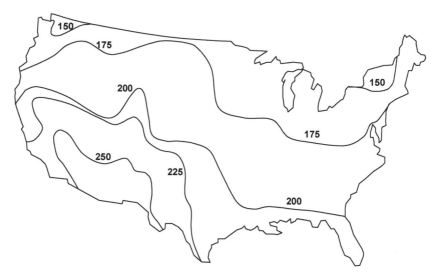

Figure 7.4 Map of the United States showing average yearly solar potential (per day) in watts per square meter, developed by the authors by averaging data and regional solar data from several sources.

Photovoltaic solar generation

When most people think of solar power, they imagine the flat photovoltaic panels or cells they see on solar-powered calculators and radios they can buy in stores and used to power the emergency telephones located on the side of many highways throughout the United States. Photovoltaic (PV) generation converts light energy directly into electric power, without any rotating or moving parts, using a semi-conductor device vaguely similar to a transistor or the integrated circuits found in computers. These "cells" are made to be as flat as possible, so that they have as much area as they can to catch light.

Photovoltaic generation is a proven and widely used technology – there are about 200,000,000 PV arrays operating world-wide – most of them in hand calculators and similar small appliances, where their output is used directly to power the devices or machinery. There are numerous types of PV cells, differing in chemical makeup and manufacturing technique and level of technology, each with advantages and disadvantages whose importance will depend on the particular situation for which solar power is needed.

Details are not given here (see Willis and Scott), but among important characteristics that one can "buy" through selection of solar PV are weight (which might be important if one were going to mount them on a rooftop, etc.), resistance to vibration and extreme temperatures, and, of course, efficiency – how much of the sunlight hitting them they convert into electricity. In particular, efficiency varies with price: A little bit more efficiency generally costs a good deal more money.

PV cells are called "cells" because their electrical output is much like a battery cell's. Each PV cell will produce a small amount of direct current (DC) power whenever exposed to sufficient light. A single PV cell may be only a square centimeter or several square centimeters in area, and will produce only a minute amount of power, less than a watt. Usually many cells are connected in series to provide higher voltage, and in parallel to produce higher current, in what is called a *photovoltaic array* or *panel*. It would take several hundred in an array to power a normal overhead light fixture.

PV power is perfect for some applications, such as in hand calculators. A calculator requires DC power at low voltage, and uses only a small amount of power, characteristics of small PV array output. Further, most calculators will never be operated in the dark, when the PV cells could not produce any power, since users would not be able to see it to use it. Millions of small, inexpensive calculators use a small PV array to augment their batteries, an array which produces at best only one watt.

PV generation is less ideally suited to run heavy machinery, such as a water well pump, which requires alternating current (AC) power, and a lot of more of it – roughly 100,000 times as the calculator, and which might need to run at night. For such applications, numerous PV cells must be grouped into a very large array in order to get the required amount of power. And, if power is needed 24 hours a day, some means of storing energy must be arranged (see Chapter 8), usually a set of batteries. The battery storage will be less than perfectly efficient (all that goes in doesn't come out) so a few extra sets of cells – perhaps 10% more – will be needed.

Finally, most water well pump motors require alternating current (AC) power, which means the DC power from the array/battery set will need to be converted into AC using a converter. Again, those are not completely efficient, so perhaps a further 10% must be added to the array size. The net result would be a fairly large (and expensive) system including several square meters of solar arrays, batteries, and a converter system. This isn't to say that such a system can't be made to work and work well. All of those factors can be addressed and in fact are in large-scale PV generation systems, but they raise the cost of the equipment and the resulting power.

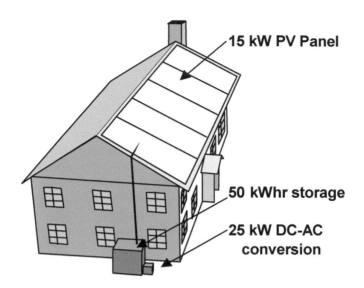

15 kW PV Panel

50 kWhr storage

25 kW DC-AC
conversion

Figure 7.5 A photovoltaic array system designed to power a typical home. It consists of 120 square meters of panels that "roof" the house and produce in peak sunlight, about 15 kW. Batteries store 50 kWh of energy for use at night, slightly more than enough to make it through one complete day, and a DC-AC converter transforms the power to alternating current for household needs, on demand. The 25 kW converter needs its considerable excess of capacity over the home's maximum (about 8 kW) in order to handle needle peaks (see Figure 5.5) of the household usage pattern.

Figure 7.5 illustrates a PV system put on a single-family home and designed to be able to provide power to this home that is equivalent in quality and availability to that from a utility system. Panels cover nearly the entire roof. A large set of batteries is required, along with a hefty converter system. But the system, which costs somewhere close to as much as the house might, if one considers maintenance costs, etc., provides plentiful, high quality, and reliable power for "free" without need of utility lines and without any substantial environmental impacts.

Advanced Solar "Cell" Systems

PV cells produce maximum output when pointed directly at the sun, and large arrays meant for generation purposes are sometimes provided with a means of aiming themselves automatically (Figure 7.6) which adds to their complexity but reduces overall cost (the cost of the aiming mechanism is less than the cost of the much larger array that would otherwise be needed).

Fixed   Single-axis   Two-axis

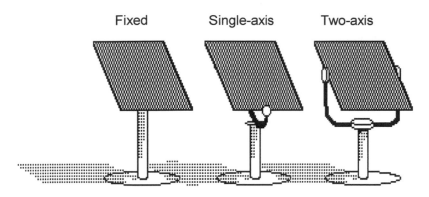

Figure 7.6 Fixed PV arrays can be put in mounts (left), or on the roof of a house, for example. In such cases, they produce power only when sunlight is falling rather directly on them. Use of motor-driven single or dual-axis tracking systems keeps them aimed more directly at the sun. Tracking provides higher output over a longer period of the day, but increases both initial and maintenance costs. Panel at the right uses a two-axis tracking system, similar to a heliostat in a solar thermal power plant.

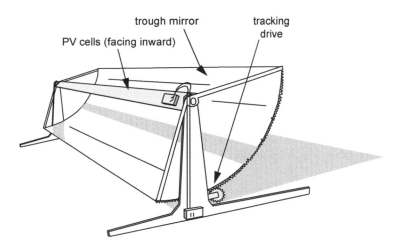

Figure 7.7 A trough type of mirror-concentrator PV generator. This uses a polished aluminum mirror to focus sunlight by a factor of twenty on a thin row of PV cells that actually face away from the sun. A tracking drive keeps the unit aimed at the sun in one axis – this is not as effective as two axis aiming, but much simpler. This unit is about the height of a typical human being, with 14 square meters of gross collector area. It produces a peak output of 2.8 kW and an average of about 1 kW.

Another "trick" to boost output and lower cost of solar generators is to use some form of concentration of sunlight. One can get more power out of a particular solar cell or panel by focusing five, ten, or even fifty times the normal level of sunlight onto it. The PV cell must be designed to handle this intensity (that much light will tend to make the PV itself quite warm), but it will produce very nearly five, ten, or fifty times as much power, as the case may be. The bottom line is that mirrors are less costly than PV cells, so a concentrator like that shown in Figure 7.7 results in an overall drop in cost for the power produced.

But the net result of using "every trick in the book" to maximize the output-versus-cost of photo-voltaic power is that it still costs about two to three times what power from fossil fuel units costs: in the range of 18-22¢ per kilowatt hour for most applications. While there is hope that continuing research may reduce this cost disadvantage, there is no reason to think that PV generation will ever be the lowest-cost form of electric generation or even the lowest-cost form of solar power generation.

But despite the higher cost, photovoltaic systems offer a number of distinct advantages over other types of power generation. They require no fuel. For applications in remote, hard-to-reach locations, the most important point here is that no fuel has to be *delivered* to the site. Often, transporting fuel to an isolated location costs far more than the fuel itself. Further, PV generators have very limited esthetic or environmental impacts. They are absolutely silent. They produce no exhaust or thermal pollution, and they do not interfere with natural water or wind flow. They are reliable, too, having no moving parts to wear out (although they do deteriorate over time). [1]

Also quite important, PV generation has no appreciable economy of scale, so that a small PV generator costs little more per kW output than a very large one, making PV useful for small-power applications. For these reasons, PV generation "owns" a small but distinct niche in the power generation market: for isolated, hard-to-reach locations that need only small amounts of power, and at which there might be extreme environmental sensitivity.

But despite these advantages, PV generation's cost disadvantage over other options, even over other solar options, makes it unlikely they will ever occupy much more than that one market niche. And although some people select them just because, "It's the right thing to do," the high cost of PV generation means it is not expected to become a mainstream generation source any time in the near future.

---

[1] Strangely, although they have no moving parts, some units have proven to have short lifetimes. One pilot test by a utility in New England found that 50% lasted less than 8 years in service.

Solar Thermal Energy Storage (STES) Power Generation

Another way of converting solar power into electric energy is solar thermal steam generation, in which heat from concentrated sunlight is used to produce steam to drive a traditional steam generator. Solar thermal power plants are robust and reliable, because they employ only proven technological components such as heat exchangers, steam turbines, etc., and they produce AC power directly using a synchronous generator.

Solar thermal power generation units use mirrors to reflect sunlight onto a gigantic boiler, to create great heat. This heat is used to raise the temperature of a fluid – usually oil or a liquid salt – to a very high level, and a lot of that hot fluid is kept in an insulated tank. The heat in that is then used to create steam to drive a traditional steam turbine generator, producing power as needed. The storage capability means that power is available at night or at other times when the sun isn't shining. As a result, solar thermal generation plants are *dispatchable*, which means they are much more useful than non-dispatchable types of renewable resource generation.

Unlike PV generators, there is a definite economy of scale with solar thermal generation plants, due to considerable fixed costs in the design of the system and its control/system interface. Units under 10 MW in net electrical output are not as economical as larger sizes, and the optimum size is generally regarded to be about 75 MVA electrical output (power for a small town of perhaps 20,000 persons), although no one has built a unit that large.[2]

STES system layout

Figure 7.8 shows the major elements of a typical solar thermal electric system (STES) generation plant. The unit depicted uses a field of *heliostats* to reflect sunlight onto a tower-mounted receiver. Heliostats are merely large mirrors, often 30 feet to a side, each mounted on its own motor-powered aiming structure, which keeps it pointed at the sun so that it reflects sunlight to a stationary target, in this case the STES's *energy collector*. If enough heliostats are used, they focus several hundred or thousand times the density of normal sunlight onto the energy collector. The resulting concentration produces intense temperatures – up to 1,200°F. This heats a primary circulating fluid that is pumped to a heat exchanger, where its heat is used to turn water into steam to power a traditional type of steam turbine generator. A good deal of the primary fluid is also heated and stored for later use in a large insulated tank.

---

[2] This figure is extrapolated from existing (1995-1997) information on solar thermal technology and power plants. No plant of this size has been built.

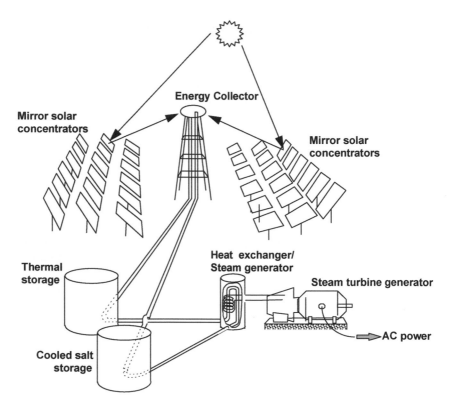

Figure 7.8 Schematic drawing showing the major components of a typical STES (solar thermal energy storage) power plant that employs both solar collection and energy storage so that it produces dispatchable, 24-hour-a-day power. The plant can use a primary cycle employing molten salt or heated oil, which withdraws fluid from a "cool" tank and returns it, at high temperature, to a "hot tank." A secondary water/steam system transfers energy from the heated salt or oil to the steam-generator.

Water is never used as the primary liquid in a good STES system. Instead an oil or a liquefied salt is used.[3] The higher the temperature of the primary fluid, the higher the net electrical efficiency of the STES plant, i.e., the more energy it will produce with the same amount of sunlight. Earlier, when discussing PV systems, it was pointed out that efficiency is not a greatly

---

[3] Certain types of salt turn into an odorless, clear, and very viscous liquid at temperatures over 600°F.

desirable trait in solar conversion. While this is generally true, the point with a STES is that high temperatures means that not only the solar conversion efficiency will be high, but also the efficiency of converting the heated liquid in the tank to electric power at night. Considerable effort has gone into heating and storing that hot liquid, making efficiency in that conversion important.

Water turns to steam at a relatively low temperature of 212°F. Steam, a gas, is not nearly as efficient for the generating cycle and storage heat exchanger process as a liquid. One can raise the boiling point of a water-based system by pressurizing the entire cycle. That drives up the boiling point which makes the system more efficient, but it requires that pipes and tanks be built with heavy and costly high-pressure materials throughout.

Instead, STES systems can use an oil or a liquefied salt instead of water, one with a boiling point far above that of water. That gives the system higher efficiency, but does not require pressurized pipes and tanks. Oil is not quite as efficient but much simpler to use: some salts solidify at room temperature, a serious design and operating constraint on the STES unit.

The foregoing discussion about the fluid selection emphasizes one element of STES design – keep it simple and basic. In this case, that means unpressurized, plain steel pipes, a very normal type of storage tank (insulated on the outside) and old-fashioned turbine generators. An advantage in the eyes of many potential users is that these equipment are reliable; the whole STES can be built using only low-cost labor and materials.

Some oils can be heated to 600°F and will remain a liquid (above that it boils or breaks apart), meaning that the tanks illustrated in Figure 7.8 do not have to be extraordinarily large to store tremendous amounts of energy. In a 10-megawatt plant the tanks are about 65 feet wide and 25 feet tall yet store enough energy to run the plant and provide electricity to a small town all night long. During the day, the heliostat-collector system both runs the generator directly and heats extra fluid for storage in the tank for nighttime use. Therefore, heliostat field and collector have to be large enough to do both at once, roughly twice that 10-MW capacity.

Features of a STES plant

Referring to Figure 7.8, the largest element of a STES plant is *the heliostat field.* Each heliostat in this type of STES layout is a large mirror with electric motor drives and a control system designed to keep the sunlight it reflects aimed at a particular point, in this case the receiver. Modern heliostats are rather large, the mirror being about ten by ten meters, or nearly 1,000 square feet. A set of heliostats is arrayed around the central receiver so that they concentrate sunlight on it during all expected operating hours.

In practice, roughly 8,000 square meters, or about 86,000 square feet, of heliostat mirror surface is required for each MW of peak electrical output. A rule of thumb: Given that peak *insolation* under perfect conditions can reach 1 kW per square meter, there is roughly an 8:1 ratio between peak solar incident energy collectable by the mirror field and realizable net electrical output of a solar plant. The net solar to electrical efficiency of a solar thermal power plant is roughly 12%, with only one eighth of the sunlight energy falling on the mirrors converted into usable electricity. While this is low, one must bear in mind that the "fuel" is free, so that if the cost of the plant itself can be kept down, the resulting electricity can still be relatively economical.

STES power plants typically involve hundreds if not thousands of heliostats covering an area of 100 acres or more. Location of the plant and heliostat field on flat terrain, with no nearby mountains or hills to produce shadows, makes sunlight available to the mirror field during the greatest possible portion of the day. Efficient arrangement of the individual heliostats and optimal usage of land will still require on the order of 7 to 12 acres of land per MW of net electrical output. This is two to three times the amount of land required for a typical coal-fired generation plant of equivalent capacity. However, this comparison does not take into account any allocation of railroad ROW land requirements for the coal transportation, inclusion of which could conceivably bring the ratio closer to equality.

Central receiver. Sunlight from the heliostats is focused on the solar receiver, located at the top of a tower. Some mirrors are partly obscured by the tower for a brief time daily, a situation impossible to avoid. The receiver is basically an array of pipes, through which a liquid is pumped to absorb heat and carry it away to the electric power conversion machinery. Typical receivers consist of a cylinder whose total area is roughly one one-thousandth of the total heliostat mirror area, covered with metal pipes, since no plastic can withstand the temperatures required, and ceramics are too expensive. The operating fluid, either oil or salt, is circulated through these pipes, where it absorbs heat from the reflected sunlight. In modern STES plants, it can reach 1,100°F.

Operating fluid. Either an oil or a liquid salt is used as the primary circulating liquid, depending on the design of the generating plant. The advantages of oil are:

Relatively high heat transfer and retention, roughly that of water.

Good chemical stability.

A liquid at room temperature, since the low-side of any heat exchange loop can be a very low temperature.

Oil is inexpensive, a significant factor considering that a typical STES plant requires about 250,000 pounds per MW net output.

A disadvantage is that oil breaks down at about 900°F, precluding highly efficient STES plants from using it.[4]

Salts, such as $NaNO_3$ or KNO_3, while solid at room temperature, liquefy above 450°F and do not break down or become gassy at temperatures up to 1,100°F, permitting higher temperatures, and greater efficiencies than with oil. Molten salts are inexpensive, last indefinitely, and are relatively non-toxic.[5] They have one disadvantage compared to oil, however. They are solid at room temperature. Therefore, some means must be included in the STES plant design to melt the salt into a liquid, before the plant can begin producing electricity. And the owners can have a real mess on their hands if the plant unexpectedly shuts down due to a failure long enough that the salt solidifies in pipes and tank.

Dual tank fluid storage system. Any viable solar thermal power generation unit must have a limited energy storage capability, so that it can maintain a regulated power output as solar flux varies slightly due to clouds, etc. Most systems have additional storage capacity, so they can run through periods of little or no sunlight. Energy is stored by using the primary circulating fluid. Most solar energy storage power plants use a "two tank" storage system, a "cool" tank for operating fluid, which might still be quite hot, and a "hot" tank that keeps only very hot fluid. During the day, "cool" liquid is taken from the cool tank and routed through the receiver, heated, and then passed into the "hot" tank. By the end of the day the "cool" tank is nearly empty and the "hot" tank is full. During the night, hot fluid is used to convert water to steam for generation, which cools it, and then it goes back into the cool tank. At sunrise the whole process starts again.

The tanks themselves are stainless steel, of a somewhat greater thickness than would be required to store an equivalent volume of water, the liquid salt being twice as dense and thus providing more weight for the tank to contain. The tanks typically have insulation on the order of 12-24 inches of thickness around them. Even a system with minimal storage, just enough to permit stable, dispatchable operation, will require a good deal of fluid. A rough rule

---

[4] The basic lesson of the Carnot cycle theory: the higher the temperature of any process, the more efficient it will be. A STES plant can be made more efficient if its mirrors are used to focus on a smaller area to raise less fluid to a higher temperature. Thus, a liquid salt plant will provide noticeably more power than an oil plant.

[5] None of the materials used in the primary loop of solar thermal power plants is as benign as water, or particularly pleasant to be around. However, to the authors' knowledge, none is considered hazardous or life-threatening if handled prudently.

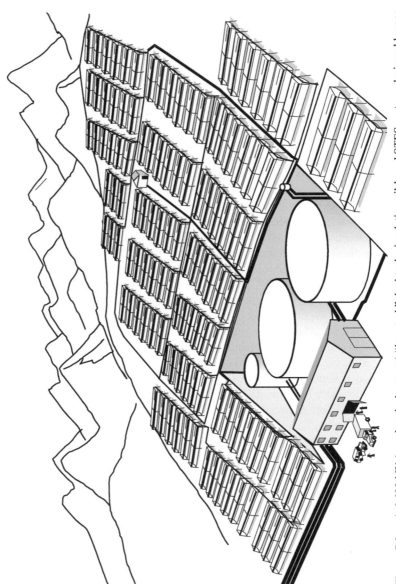

Figure 7.9 A 6,600 kW (net electrical output) "low-tech" dual tank circulating oil-based STES system, designed by one of the authors for installation in a third-world country, where simplicity was more important than high solar efficiency.

of thumb is about 2,500 to 5,000 gallons of fluid per MW net electrical output, and an ability to store several thousand gallons of fluid in each tank.

Heat Exchanger, Steam Generator, Turbine Generator, and Power System Control. Most of the remaining portions of a solar thermal power plant are identical in technology, and often equipment design, to those used in traditional steam turbine power plants. Hence, the technologies and equipment designs are well known and long-proven.[6] The steam turbine/generator and its control equipment are identical to those used in thousands of power plants worldwide. Liquid to liquid/steam heat exchangers are familiar elements of some types of nuclear power plants. Units using salt, or pure molten sodium, were perfected for use in Soviet submarines and other facilities decades ago.

STES systems: good for some applications

Figure 7.9 depicts an oil-based STES plant designed by one of the authors (Willis) for use at an isolated site in a second-world country. Robust design, low cost of materials using only minimal construction skill requirements, and simplicity of operation and maintenance were prime criteria in its design. Therefore, this system uses oil (easier to maintain and operate) which is heated by running it through pipes that zig-zag through a field of concentrator mirrors, which are individually much like the PV concentrator system shown in Figure 7.6. Each concentrator is a simple device: sheets of polished aluminum bent and bolted to mild-steel frames for mirrors, with no tracking. A pipe runs along the focal point. Run oil through enough of these mirrors on a sunny day and it is quickly heated to 900°F.

This STES is somewhat less efficient than a salt-based system, with a present worth (PW) production cost averaged over the next thirty years of about 19-21¢/kWh as compared to as low as 17¢/kWh for "high-tech" STES systems. But nonetheless it can appeal to the governments of many developing countries, particularly those with limited fossil fuel resources. Solar power avoids continuing fossil fuel costs that would drive up trade deficits. And much of the plant's cost will be money spent in the country, rather than sent overseas for high-tech equipment: a good deal of labor is required to set up the mirror field, but that is mostly un- or semi-skilled labor which can be obtained locally. Most of the materials and machinery can be manufactured locally, too.

In general, STES power plants are robust, reliable, and proven technology: only slightly more complicated than hydro power. They produce

---

[6] The only significant difference between those used in existing solar thermal plants and in traditional steam plants is size: Most steam plants are 100 MW and over, largely because boiler design is more efficient in larger sizes.

power directly in AC form, and are dispatchable. However, they have a high initial cost, and even when amortized over a 30-year lifetime, the power produced by solar thermal conversion costs about twice that produced by the most economical hydro, fossil, steam or gas-turbine, or nuclear power plants: about 16-18¢ per kilowatt hour. But this is low for dispatchable *renewable* power, beating other solar technologies, as well as wind (dispatchable wind power needs both the wind generators, and electrical energy storage).

However, for power production in isolated or environmentally sensitive areas, or where delivery of large amounts of fossil fuel is expensive, STES systems may be ideal because they are among the simplest of dispatchable solar or wind generation systems.

Solar Tower Generation Units

A solar tower generation system (STGS) uses a unique combination of solar energy extraction and wind power generation to produce dispatchable energy. No large STGS system has been built, although several projects are planned or in the early development stages. A 50 kW proof of concept plant has operated for many years

The idea is deceptively simple and quite clever, and is depicted in Figure 7.10. A "greenhouse" is built in the shape of a shallow, flat cone, using clear plastic or glass on a light framework, and with the edge of the greenhouse open. As with any greenhouse, when the sun is shining the air inside is heated quite a bit. Air temperature does not reach high temperatures, but it is warmed quite noticeably, perhaps by 10-20 degrees Fahrenheit.

Hot air rises, in this case that inside the greenhouse rises to the ceiling of the greenhouse, and because the roof is sloped, toward the center. As it does so, air moving from the outside of the greenhouse toward the center accelerates – it is being squeezed into a tighter space as it heads toward the center. Rising from the center of the greenhouse is a tall chimney. The air, reaching the center, is moving relatively rapidly. It heads up the chimney, where natural draught forces accelerate it a bit more.

Built into the chimney is a wind-powered generator. The wind created by the greenhouse and the draught force of the chimney is converted into electric power. The "wind generator" in this system produces much more power than a typical wind generator because the air speed up the chimney is higher, perhaps over 100 km per hour.

To produce any significant amount of power, a STGS must be built on a huge scale. The greenhouse of a 200 MW unit planned for the Australian desert is kilometers across. Most of it is constructed only of thin plastic film on wooden supports, but near the center, to handle the increasing wind speed and pressure, a metal frame with glass is needed. The chimney, of cast concrete, would be 1 km high.

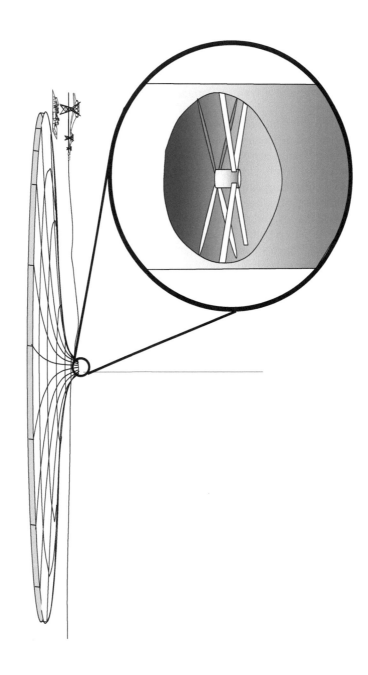

Figure 7.10 A solar tower generator uses a low-cost "greenhouse" to produce warm air from sunlight, and feed the resulting air flow into a chimney, in which a wind turbine is housed. The hypothetical unit depicted here has a greenhouse 1 kilometer in diameter and a chimney 450 meters high. It produces a peak output of about 4 MW but can average only about 2 MW of dispatchable power throughout a typical day.

Perhaps most interesting, the unit is capable of producing dispatchable power 24 hours a day, because it has a type of energy storage: under the greenhouse is the earth – sand, rocks, etc. – which is heated during the day and retains heat overnight. At night, output drops slightly as the rocks gradually cool, but not so much that useful amounts of power are not available even just before dawn. Day or night, by controlling the wind generator and a type of flapper valve on the chimney, one can control the wind speed and thus the amount of power produced.

STGS are not efficient generators – in fact they can have a total sunlight to electric efficiency conversion factor in the range of only 1%, among the worst of any energy conversion technology. But efficiency isn't an issue when the "fuel" is abundant and free. STGS make up for their inefficiency with a low marginal cost for truly large scale – the type of scale required to compensate for their inefficiency: Their energy storage is just earth under the whole facility and their energy collector just a greenhouse; one can buy a lot of both for very little money. Overall, their economics can work out well.

STGS have several disadvantages, not the least of which is that no large unit has ever been built, only a small pilot unit. Yet the technology is simple, and there is little risk that a larger unit, if well designed, would not work. An obvious drawback is that any large production unit will require *a lot* of land, but it can be real estate that is marginal, at best, for other purposes, like deserts, wasteland, etc. However, such land, while often cheap (even free), is usually far from power grids – new transmission lines might be required. And of course, the chimney is quite high and would be a navigation hazard to aircraft, and the greenhouse would be susceptible to damage from hail and violent weather. But despite that, STGS is a novel and practical design that might well lead to several large plants built around the world.

7.5 OTHER RENEWABLE GENERATION TECHNOLOGIES

Trash Burning

These approaches are closely related. The first includes burning trash to produce the heat for a boiler for what is, from that point on, just a traditional steam turbine generator. This is "renewable" in a different way than solar and wind power. Mankind renews trash, not nature. Nonetheless, almost everyone will admit mankind probably has an unending supply of trash as one looks into the future.

Many commercial trash burning plants have been built and operated in large metropolitan areas, most in the 5-25 MW net electrical output range. All are 24-hour, dispatchable units. None is particularly efficient, but the base fuel is free, although the ashes must still be processed in land fills, etc. Unlike sun and wind power, this "renewable" generation technology produces pollutants. In fact trash combustion is not particularly clean. A minor complication is possible contamination of trash with metal and other "toxic" materials burned in the combustion process, which produces some quite undesirable air pollutants (more toxic than normal pollutants). On the other hand, air pollution is controllable and the fuel is not only free, but our socieity has to get rid of it: potentially one might be paid to take it.

Biomass Plants

More intriguing, and offering great potential for third-world development, are "biomass" power plants, which burn crops specifically raised as fuel. Biomass crops are usually special breeds of very fast-growing, tropical light woods, somewhat similar to bamboo, or very fast growing, high (3 meter) grasses. Some special breeds of grass have been patented for this purpose. These crops can go from seed to harvest in less than a month and regrow after cutting, like the grass in a lawn, so that no reseeding needs to be done. In most cases, optimization of power-yield per acre calls for harvesting the crops before they mature: a plant that goes from "freshly mowed" to "full height" in 30 days will typically add only a small amount of mass during its final week of maturation. Cutting it in 23-day "harvest cycles" thus provides more fuel per acre per week.

Such power plants are proposed almost exclusively as small, "central station" facilities for remote areas in the tropical latitudes, where heat and sunlight are plentiful most of the year and the growing season very long. Perhaps a square mile or more around the power plant will be planted with the fuel crop. For example, if the crop goes through a harvest cycle in 23 days, the surrounding square mile would be divided into 23 areas of 26 acres each, leaving 42 acres of the square mile for the power plant, harvested crop storage, and access roads, etc. One crop area would be harvested every 23 days.

From an energy cycle standpoint, these biomass plants are essentially a type of "solar" power plant although one would not normally think of them that way: but sunlight is the driving force behind them. But this power generation method requires access to considerable water for irrigation – all that foliage growth doesn't occur without water. Still, abundant energy – harvested and stacked plant material – can be stored on-site at almost no additional cost, making these plants both dispatchable and able to ride through periods of several months where growth slows for seasonal reasons.

Biomass plants are less feasible in northern climes where growing seasons last only a portion of the year, although with more space made available for storage, they could be made to work. However, plants that burn crop waste or grass clippings might be more feasible, although those are very seasonal supplies of fuel, a potential disadvantage.

Geothermal

It is possible to find locations within a mile or two of ground level where substantial thermal energy exists in near-molten rocks below the earth's surface. Water can be pumped into such locations, where the heat turns it to steam, which in turn can be used to power a steam turbine generator. Numerous sites like this exist around the world, most in volcanically active or tectonic plate activity line areas.

Geothermal power uses robust and proven AC steam technology for the generation system, and is fully dispatchable. Despite concerns that such a plant might "use up" the thermal energy at a particular site (i.e., gradually draw off so much of the underground heat that it cools the rock to the point that the plant is no longer functional) some have been in operation over 30 years with no measurable degradation in output.

Tidal Power

Several attempts have been made to harness the wave and varying water-height energy action of the ocean. These have included several innovative but bizarre designs for machinery that will turn the oscillating fluctuations of wave action into power, and a type of "two-way" low-head hydro that generates power by letting water flow through a hydro unit into a large reservoir at high tide, then letting it flow back out at low tide. To the authors' knowledge, such systems produce power, but not in commercial availabilities or at competitive costs.

Ocean-Current Turbines

Underwater currents in the ocean contain truly prodigious amounts of power. Deep ocean currents flow at almost constant speeds, providing a dependable and constant base-load capability, and their far greater mass more than compensates for their slower speed as compared to wind. A generator somewhat analogous to a wind turbine, but with a blade shape optimized for capturing energy from water currents, could either be anchored to the ocean floor or suspended underwater from a tethered raft. Located offshore, in the middle of a steady mid-ocean current such as the Gulf Stream, such generators could produce a great amount of power.

While intriguing, such plants would probably have a high initial fixed cost, making utilization efficient only on a large, central station size, a scale quite beyond the range of distributed generation applications.

7.6 ARE RENEWABLE RESOURCES PRACTICAL?

While few people disagree that renewable energy is a good idea, most also recognize there are two practical barriers to implementation on a wide scale. First, all renewable energy sources, except for hydro-electric power, cost from two to five times what fossil fuel and nuclear energy costs, even when their "free" fuel costs are factored into the analysis. Many, first-world countries could afford the higher cost, thus decreasing their emissions and fossil fuel use at some slight reduction in spending on other items. But low-cost power is essential to the industrial processes that provide the basis for the world economy, not just that of rich nations. And among the first "other items" for which spending would be reduced would no doubt be charity and aid given to developing nations. In addition, those third-world countries cannot improve their standards of living unless plentiful, cheap power is available in growing amounts. As a result, while renewable energy costs will likely continue to decrease at a slow pace as technology improves, only cost competitiveness will bring about big changes and thus widespread usage.

Second, all renewable energy sources have some type of special site requirements that make them suitable for only some parts of the world and some locations. Many renewable methods, such as hydro, geothermal, and wind, can be used only at a relatively small number of locations, where the water, geological, or wind energy can be found. The power must be generated there, regardless of where it will ultimately be used. Considerable land must then be cleared and devoted to high-voltage transmission lines to get the power to where it is consumed – creating another type of significant environmental impact.

Even solar power has only regional applicability. While theoretically, solar thermal and photovoltaic power will work anywhere, both are practical only in locations where the sun shines a good portion of the day. Near the equator, solar thermal power plants receive nearly 12 useful hours of sunlight daily. But beyond 50 degrees latitude north or south, there are many days of the year when there is simply not enough daylight, leaving energy consumers in need of spending more on some other "backup" supply for those times.

Similarly, biomass plants work only in areas where there is a long growing season. Year-round growing seasons are best, of course, although biomass can also work where the added cost of considerable fuel storage for cut grass and timber is feasible.

FOR FURTHER READING

G. Cook, *Photovoltaic Fundamentals,* DOE/SERI Report No. CH10093-117, 1993.

Institute of Electrical and Electronics Engineers, *Proceedings of the IEEE's* Special Issue on Advanced Power Generation Technologies, IEEE, New York, March, 1993.

N. Vosburgh, *Commercial Applications of Wind Power,* Van Nostrand, New York, 1985.

L. Willis, *Distribution Planning Reference Book,* Marcel Dekker, New York, 1997.

L. Willis and G. B. Rackliffe, *Introduction to Integrated Resource T&D Planning,* ABB Guidebooks, ABB Electric Systems Technology Institute, Raleigh, NC, 1994.

8

Distributed Generation and Storage

8.1 DISTRIBUTED POWER GENERATION

More than 95% of the electric power consumed on this planet is produced by traditional central-station generators of the type described in Chapter 5. Each one produces somewhere between 200 and 1,200 MW of power – enough for between 50,000 and 300,000 typical US homes. These generators are invariably quite large: a complete central-station generator might be the size of a 15-story office building. The entire plant facility might occupy 20 to 100 acres. But they achieve high efficiency – of fuel, of human and computer resources needed to control a site – due to their size. That efficiency has a price: the electric utility needs a delivery system – a transmission and distribution (T&D) system or wires and equipment to deliver power to all the homes and businesses scattered throughout its service territory.

Distributed generation (DG) takes an alternative approach with respect to size, and in large measure eliminates the T&D system. The DG concept entails using many small generators, of only .025 to 1 MW output (25 to 1,000 kW), all located close to the homes and businesses that will consume the power, rather than a few large generators. Often represented by proponents and DG manufacturers as a new, even revolutionary concept, distributed generation has actually been around for decades. The first electric systems consisted of small local generators. Over time, larger generators and T&D evolved as a way to improve efficiency. And today, millions of small local generators are installed at hospitals, police stations, radio stations, and office buildings throughout North America, as backup generators in case the utility power fails.

What *is* new, in the 21st century, is the technology and intent behind DG. Traditional "backup" generators were intended only to run only when the utility system was out due to storms, unexpected failures, etc. They were not durable (some would only run several hundred hours before needing an overhaul) and were inefficient (the power they produced would cost, in fuel alone, two to three times what electricity from the utility would have cost). Today many people believe that distributed generators are efficient enough and durable enough that they can be much more than backup generators. Some believe that DG units can replace utility systems altogether: everyone will have a small generator in his backyard or at his business location.

Hype Overshadows Promise

There has been a good deal of hype and more than a small amount of polarization within the power industry over DG and the promise it holds for the future. There are staunch proponents of DG who firmly believe in it and who are always trying to "sell" the concept of its widespread use, and many others who just as steadfastly maintain that DG is not and never will be viable, and that its use is nearly always a mistake. Something close to controversy exists because DG is potentially a disruptive technology: if everyone bought small generators and produced their own power, utilities, and the entire utility industry, would be out of a job. Blurring the facts is a good deal of political input to the process: DG is often perceived or at least represented by proponents as cleaner and more environmentally "friendly" than large central station generators and utility systems.

As is so often the case with a controversy, both proponents and opponents of DG make cases for or against their position using facts and analytical studies that are based on favorable conditions or interpretations that favor their viewpoint. But on balance, the authors are convinced that DG isn't and probably will never be efficient enough, economical enough, clean enough, or reliable enough to displace well-run, traditional utility power systems. On the other hand, DG has improved tremendously in the past two decades, to the point that it is competitive and beneficial in a much wider set of market niches than only its traditional backup-power role. It has a good future.

This first section will continue the discussion of DG by looking at some basic "rules" with regard to the laws of physics and how they impact generation design and costs, that will help put DG's advantages and disadvantages into clear perspective. The discussion will then turn to an examination of each of the major types of DG generator systems: piston, micro-turbine, and fuel cell, in Section 8.2. An allied technology, *distributed energy storage* (abbreviated as DS, for distributed storage), will be discussed in Section 8.3. DS may actually offer more value and promise for the future than DG, and when used in conjunction with well-designed, compatible DG, provides what utilities call "premium power quality" – exceptional reliability and voltage stability. Finally, Section 8.4 summarizes key points about DG's and DS's place in the power industry and their potential for the future.

Shifts in the Economies of Scale for Generator Design

Traditionally, electric utility systems used large, central station generators (as described in Chapter 6) because of the significant economy of scale that exists in electric generation technologies. Large generators produced power at less than one half the cost per kilowatt of small generators. The bigger the generator, the bigger that advantage it had over smaller units.

At the time of this writing (2005) there still is and apparently always will be a significant economy of scale favoring larger generators. This will be discussed below in "Physics Is Always on the Side of the Larger Unit." Traditionally (into the early 1990s) central-station generation + T&D had an overall cost advantage of about two to one or DG. That dropped to only about one and one half to one by 2000, enough to make a difference in some situations. Three reasons were behind that shift in economy of scale:

1) Innovation improved the efficiency of small gas turbines, combined-cycle, hydro, and fuel cells more than large ones, this having been the goal of much of the R&D for small generating technologies.

2) Improvements in materials, including new high-temperature metals, special lubricants, ceramics, and carbon-fiber, permit vastly stronger and less expensive *small* machinery to be built.

3) Traditionally, any generator needed roughly the same number of operators on duty whether it was large or small. Today, small and even some large units need no on-site operators, due to the use of computerization and "smart" control systems.

Distributed and Dispersed Generation Avoid T&D Costs

Proponents of DG point out that a small generator does not have to be less costly and more efficient than a large generator. It only needs to beat the economics of a large generator *and* the miles of wires required to move power from that large generator to the energy consumers. The DG units are already at those locations. Serving a community with small DG units located

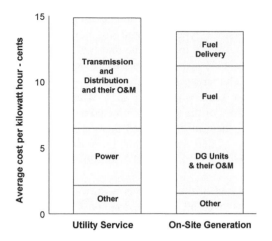

Figure 8.1 Cost of utility power (left) and distributed generation power (right) for a ski lodge in the Rocky Mountains from a 1996 analysis by one of the authors. The distributed generation alternative is nearly ten percent less costly than the utility's.

in each home's backyard or behind each business location would avoid any need for the utility T&D (transmission and distribution) system.

This is a valid argument, and one that means DG sometimes *is* the most economical alternative. The T&D system usually represents both a significant cost in initial capital, as well as in continuing operating costs. Figure 8.1 shows data from a case where the total cost for two alternatives for power at a ski lodge in the Rocky Mountains is compared on as even a basis as the authors could arrange. All costs are consistent and consistently treated.[1]   On the left is the cost of power if provided by a utility system: capital, O&M and energy (wholesale power) costs would average about 14.8¢/kilowatt hour. Note that T&D costs account for over half of the total. The ski lodge location is "off the grid" – quite far from existing lines and substations – and construction of a good deal of new line would be required.

On the right is shown the cost of power from DG, in this case a set of "high-tech" diesel generators. These are not the noisy, vibrating, and smoky

---

[1] Long-lifetime durable equipment like utility and DG units are typically evaluated over a 30-year period of ownership and use in order to compare lifetime costs. For more information including chapters on these cost analysis methods and numerous, much more detailed studies of costs, *see Distributed Power Generation – Evaluation and Planning,* H. L. Willis and W. G. Scott, Marcel Dekker, 2000.

diesel generators traditionally used for backup power, but a modern multi-cylinder, low-emissions type designed for continuous operation. Cost is 13.1¢/kilowatt hour, or 9% lower. Note that fuel delivery is a significant cost for the DG, nearly 40% – the fuel has to be delivered to the site by truck, once a month, which increases its cost.

This is a clear case where DG is a winner. Unfortunately for DG advocates, the owners of the ski lodge elected to take service from the power grid, because their local utility offered them a contract for service at a cost that averaged only 11.2¢/kilowatt hour, more than 15% better than the DG and fully one-quarter less than the 14.8¢/kilowatt hour cost computed for "utility power."

This brings up part of the controversy surrounding DG. Electric utilities average their costs over all their customers. Costs to serve easy-to-reach customers and hard-to-reach customers such as this ski lodge are all mixed

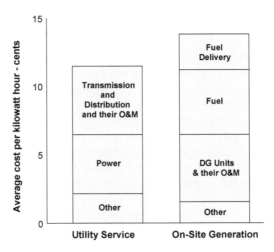

Figure 8.2 Cost of utility power (left) and distributed generation power (right) for *two* rather than one ski lodge. Compare to the single-lodge case in Figure 8.1. Not just utility generators, but T&D systems as well, have a substantial economy of scale: the transmission line needed to serve one ski lodge can serve two with very little increase in cost – its cost per kilowatt is cut very nearly in half. By contrast, DG's cost is scaleable but not much better: cost/kilowatt remains as in Figure 8.1's single-lodge cost. In this case DG is 20% more costly. Another increase in demand would drop the cost of utility power even more. In this and other cases, while DG looks good initially, in the face of extended load growth in a region, traditional electric utility system design looks better.

into one "rate-base" (set of costs) and allocated among all customers based on how much power they use. (This is a slight oversimplification of how utilities set their rates, but not a misrepresentation for purposes of this discussion.) This is how regulated utility rates are set. Thus, the local electric utility will "lose money" on service to this particular ski lodge, but it doesn't look at its costs or rates in that way.

Many DG advocates cry "foul" at such utility decisions. They argue that a change in regulations and rules is needed, that if DG were built in such situations, society as a whole, and thus "everyone in the larger sense," would save money. Although there are exceptions (Detroit Edison has looked seriously at using DG in such situations), most utilities and most power engineers oppose rules that would "favor" DG in such situations. One of the reasons is shown in Figure 8.2. Here, the same analysis of utility versus DG cost has been done, but for *two* ski lodges – one at this remote site and one just down the road from it. The utility system cost has dropped considerably because the utility only has to build one transmission line whether it is serving one, two, or many more other consumers in that area. The projected cost/kW has been cut nearly for T&D. But the DG cost has not dropped: a second lodge would need another full set of small generators, and another truckload of fuel delivered each month.

This is why most electric utilities oppose rules that would force them to use DG rather than let them decide as they want. DG is often cost-effective where small, very isolated loads *will remain isolated.* Utilities facing those situations (e.g., in central Australia) often serve most of the local load with DG. But looking to the future, a utility might invest in T&D for a now-sparsely populated area even if the first load growth there could be more economically served by DG, in order to invest in driving down long-term costs. In this particular example, the utility truly (and correctly as it turned out) expected the ski lodge to open up the region around it to further development. Within several years there were other ski resorts, and residential and commercial development in this area, all creating a much bigger electric load. "T&D" was definitely the right way to go.

Past Prospects for a Big Future

In the mid and late 1990s, proponents of DG and many others believed it had a big future as a viable and cost-competitive alternative to traditional utility systems. They believed it would become more economical, not just in isolated applications like that discussed above, but for the mass market, and so they forecast that DG would eventually displace much of the traditional central-station + T&D system infrastructure. Figure 8.3 shows why, with a third comparison of utility-to-DG costs that reflects the future as DG advocates saw it in the mid 1990s. Here, the utility's cost of service has been

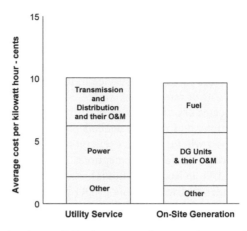

Figure 8.3 The situation as DG advocates and many others saw in the mid 1990s. This diagram compares the average cost of serving its customers for the utility from Figure 8.1 (rather than the cost to serve the isolated ski lodge with its need for a new transmission line) using the same 1996 cost basis, and DG using natural gas as the fuel, all computed with mid-90s prices for natural gas. DG beats the average cost of utility service. DG's future looked good.

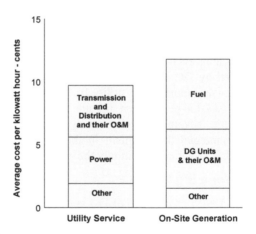

Figure 8.4 With fuel prices higher, wholesale power prices driven down by de-regulation and the competition it fosters, and recent improvements in T&D efficiency and cost reduction, the actual situation in the first decade of the 21st century is the opposite of Figure 8.3's vision. DG is at a 20% disadvantage.

recalculated for an average customer, rather than an isolated and more difficult to reach ski lodge. (This cost-to-serve-an-average-customer is basically the cost that defines its rates.) Shown at the right is a projected total cost for power from a DG unit running on natural gas. Both utility and DG cases use mid-90s fuel and other costs. The DG unit is 10% less costly.

Fuel cost escalation

Yet DG did not develop as expected, and probably will not. One must look only at fuel prices and how they interact with generator fuel efficiency to see why. The most efficient and cleanest DG units run on natural gas.[2] The data shown in Figure 8.3 are based on 1990s natural gas prices. Since then, natural gas prices have risen sharply, to somewhere between one and one half and two times those levels. Any one of the Figures 8.1 – 8.3 will show that DG costs are very sensitive to fuel cost: fuel represents a significant portion of its overall cost.

Utility system costs are also sensitive to fuel costs, but here is where the economy of scale benefits big generators at the expense of DG. As mentioned earlier, larger generators are more fuel-efficient than smaller units: they are affected less by fuel price escalation. Thus, when natural gas cost doubles, the net cost of power from a DG unit might increase by a third, whereas the wholesale cost of power from central station generation might increase by only a quarter. Further, utility system generation is a mixture of natural gas with other fuels such as nuclear and coal. The other fuels' costs have not risen as much as natural gas has, and DG cannot make use of them. Overall, other fuel increased in cost about half as much as did natural gas. Therefore, increases in natural gas prices hurt DG's price competitiveness against traditional types of utility systems.

Competitive power

A second reason DG has had a tough fight is de-regulation, and the competition it created at the wholesale power level (see Chapter 2). De-regulation forced the manufacturers of wholesale power (who own the huge central station generators now that the industry is de-regulated) to scrap older, less cost-effective generators in favor of newer equipment and to improve efficiency in all other areas. That led to noticeable price decreases for power, after one adjusts for increases in fuel costs, on the order of 10 – 20%.

---

[2] Actually, the cleanest DG units possible run on pure hydrogen. But hydrogen has to be manufactured from water by electrolysis using electricity or from natural gas or crude oil using chemistry: overall there isn't much difference one way or another.

T&D systems become more efficient

Finally, T&D utilities responded to de-regulation and 21st-century market pressures by becoming more efficient and lowering their costs, reductions which have in large part been passed on to their customers. The reasons for these improvements are intricately wound up in how the combination of regulators, utilities, and investors works (see others sections of this book). De-regulation and dis-aggregation forced energy delivery to "stand on its own" and run itself as a business rather than as (the traditional) ancillary support function to power generation. That, pressure from regulators for modernization, and a desire for higher profit and lower risk operation by executive management led many T&D utilities to downsize, to push for efficiency improvement, and cut overall costs during the period 1994-2004, in some cases by as much as 40%. Further, utilities are beginning to automate their T&D systems, further improving price/performance.[3]

The net impact of these changes is shown in Figure 8.4, the last comparison in this series. Here, the utility's power costs reflect a 50% increase in the price of natural gas and 20% increases in other fuels, against a 15% overall improvement in pricing due to the improvements above: overall power cost remains about the same. Utility T&D costs are shown as decreased by 15% – not as much as some, but representative of the average. By contrast, DG's cost remains about the same, except that fuel cost has increased by 50%, making it uncompetitive for the mass market.

Physics Favors Larger Generators

At the heart of all of this is one central fact: a small generator cannot be as fuel efficient as a large generator that has the same level of technology and quality of design. There are a number of reasons, but it is simple physics. The major barrier to high fuel efficiency is heat loss. All DG units (even fuel cells) create power in a way that begins with the oxidizing (burning, except in the case of fuel cells, which use a catalytic process) of a fossil fuel or pure hydrogen. In every case, the hotter this process can be made, and the bigger proportion of that heat that can be controlled and used, the more efficient the generator will be in producing electric power. But in all cases, the unit gives off heat to its surroundings, and with that heat goes a good bit of potential efficiency: it takes fuel to produce heat inside the unit and if some of that heat escapes, that is "wasted" fuel.

That waste of heat from a generator cannot be eliminated, but it can be

---

[3] And a quite realistic and fair comparison, too, for in order to serve in widespread use to meet modern, mass-market power requirements, DG systems would need a similar type and level of automation to that needed by T&D.

reduced. Designers can put insulation around the hottest parts of the generator, and cleverly design it to minimize wastage through the way power and air, fuel and exhaust flow through it: that is part of the design of any good generator unit. But such levels of innovation and attention to detail should be taken for granted in the design of any good generator, or any type.

But another way to reduce the amount of heat loss is to increase the size of the generator. Heat is lost through the outside surface area of the generator. Power is produced by its interior volume. Double the length, breadth, and height of a particular generator and it has eight times the volume. Its power output will be somewhere around eight times as much, at least if its designers do their job well. But its surface area, the area through which it loses heat to its surroundings, will have increased by only a factor of four. The ratio of production volume to power loss is cut by half, and potentially it is now more efficient. It really doesn't matter what type of generator one is talking about – reciprocating piston, steam or gas turbine, nuclear, or fuel cell. For all generation technologies, increasing the size of a particular design can lead to improved efficiencies, if the designers are savvy.

In the early 1990s, DG proponents raised expectations about DG fuel efficiency by claiming, quite correctly, that DG units then on the drawing boards would be more efficient than the existing base of large utility generators, because of the new technologies these DG units would employ. DG would outperform utility generation, its proponents said.

For the most part, what they said was true. But a flaw in the picture they drew, and a fact many people overlooked, was that future DG units wouldn't be competing against the contemporary base of utility generators. They'd be competing against new, large generators that would replace that older base of units. Those newer, big units would employ the same level of technology, the same automation, and the same design tricks as the new DG. And physics would be on the side of the bigger units.

Had fuel costs stayed in the range of $3.50 per MBTU or less, DG could still have been competitive. The difference in efficiency that size conveys is not much – often less than 10% – and one must bear in mind that that advantage has to cover the cost of the T&D system. But that efficiency advantage means a lot more when fuel is costly. And the bottom line is that natural gas prices have risen sharply, and are more likely to increase beyond what they are today, than to decrease to what they were in the 1990s.

CHP and Combined Cycle Generators: Using "Waste" Heat

There is a way that DG can be "unbeatable" in spite of this disadvantage, if it can put its "waste heat" to work. All fossil-fueled generators produce hot exhaust gases which contain a good deal of remaining energy that can, under some circumstances, be harnessed to provide additional value. Turbines and high-temperature fuel cells produce very hot exhaust gases. Piston-engine units produce a somewhat cooler exhaust but one that can still be used.

Co-Generation. One "trick" to boost the output of large generators, particularly gas turbines, is to use their exhaust heat to boil water, which produces steam, which then powers an auxiliary steam turbine in what is called *a combined cycle* turbine. A modified form of this idea, a *Cheng cycle* turbine, routes the steam back into the gas turbine itself, where it mingles with the fossil fuel gas. Combined cycle generators are often as much as one third again as fuel efficient as standard generators – 45% instead of 33% efficiency for example. However, they are more costly initially, their maintenance costs can be up to twice as much, and they have more constraints on rapid starting and operation.

Cooling-heating-power. Another option is to use a generator's exhaust heat to make hot-water or steam for industrial uses. It is possible to use the

Figure 8.5 A CHP system designed for a small apartment building applications. Typically located in the basement, it provides electricity, hot water, and heat and cooling to the building's occupants. Total energy use efficiency is over 75%.

exhaust heat in absorption-cooling machinery so that it actually provides cooling, hence the term cooling-heating-power (CHP). A CHP generator is a rather complicated machine, combining generator, air conditioner, and space and water heaters into a single package. Those various elements have to be sized to a particular application: so much generator, so much cooler, so much heater. But a good CHP application – one where the power, hot water and heating, and cooling outputs match the needs of the site – can have an overall fuel to end-use efficiency approaching 80%, meaning that 80% of the fuel is turned into something useful (see Figure 8.5).

This makes CHP-based DG so cost-effective that it is quite economical overall. In CHP, DG has an "unassailable" advantage over larger generators, because they cannot find useful applications in similar measures. To be valuable, the heating and cooling produced must be reasonably close to the consumer: one can transmit electricity hundreds of miles efficiently, over high-voltage lines, but not hot water, or cool air or water. DG, which is always near the home or business, is in the ideal spot in many cases.

DG's Future Role

What, then, is DG's future role if it has a limited or no future in the mainstream of utility power systems? It has several niches where it should do very well. First, DG will continue to fill its traditional niche as backup generation for power outages. Decreasing first costs for DG, and a growing need for reliable power in business and industry, mean a growing market for backup power for years to come. Additionally, when DG is bundled with distributed storage, a backup DG unit can provide very high levels of reliability *and* power quality, what is sometimes called premium power. This will be discussed later in this chapter. Second, there are situations more extreme than the ski-lodge case of Figure 8.1, where utility grid service is costly, and DG has a big cost advantage, a noticeable market segment.

Third, in some cases DG, including existing backup units, may be able to provide peaking power a few – perhaps several hundred – hours per year. During times of peak demand, the cost of wholesale power can double or triple, making the power from a DG unit economical by comparison. Utilities or private owners can run DG during such times to reduce the overall price of power. In some states in the US, and elsewhere, this requires considerably more licensing of the unit (it has to be certified as a generator, not just a backup unit). But it adds to the DG unit's economic value.

And finally, CHP is especially appealing to mid-size, residential building applications: apartment buildings, hotels, motels, nursing homes, hospitals, and such. The scale and character of energy needs in those buildings – with a 24/7 need for cooling and heat – matches DG CHP capabilities well.

8.2 TYPES OF DISTRIBUTED GENERATORS

Fuel Cells

Fuel cells are essentially fossil-fueled batteries, which, as long as they are provided with a flow of fuel and air, never run out of energy. They offer a completely different way of making electric power from fossil fuels, i.e., natural gas or gasified coal, than generators powered by turbines or piston-engines. Combustion, in the normal sense of the word, does not take place inside a fuel cell. There is no flame or burning. However, at a molecular level, oxidation, something like combustion, does occur as oxygen from the air is combined with hydrogen from the fossil fuel. In the presence of certain catalysts, hydrogen in the fuel and oxygen combine automatically (oxidize) without flame but with plenty of heat – it is the same chemistry as in combustion. Inside a fuel cell, this happens in a controlled chemical environment in a way that creates an electric flow that produces electricity.

Fuel cells were invented in the 1960s. Five major types exist, varying in the chemicals and catalysts that make them function. These are:

- Proton exchange membrane fuel cells (PEMFC)

- Alkaline fuel cells (AFC)

- Phosphoric acid fuel cells (PAFC)

- Molten carbonate fuel cells (MCFC)

- Solid oxide fuel cells (SOFC)

Table 8.1 gives salient data comparing these five types. Of these five, only one, AFC, is unsuitable for power generation purposes. The other four can run on air, taking oxygen out of the air as they need it. AFC requires pure oxygen, the other gases in air poison their chemical process. The rest of this chapter discusses, in turn, each of the four that offer DG potential.

In general, fuel cell types are listed in Table 8.1 from right to left in order of: increasing efficiency, increasing cost, and increasing "un-proveness." PEMFC units, on the left, are commercially viable at present, if a bit expensive; solid oxide units on the left are still in the laboratory. PEMFC are the least efficient of the types shown, but also the least inexpensive (for a fuel cell). SOFC units on the right are among the most efficient fuel-to-electric power device known, far more efficient than any other generator type.

Table 8.1 Characteristics of Major Fuel Cell Types

	PEMFC	AFC	PAFC	MCFC	SOFC
Electrolyte	Polymer	KOH & H_2O	H_3PO_4	$LiKaCo_3$	Stabilized
	membrane	Phosp. acid	Lithium carb.	Zirconia	
Typical construction	plastic, metal	plastic, metal	steel	titanium	ceramic
Need separate reformer?	Yes	Yes	Yes	No	No
Oxidant	Use air	Pure O_2	Use air	Use air	Use air
Internal Temp.	85°C	120°C	190°C	650°C	1000°C
Basic cell efficiency	30%+	32%+	≈40%	≈42%	≈45%
Typical application	car, spacecraft	car, other	DG	large DG	very large DG
Installed Cost, $/kW	$1,400	$2,700	$2,100	$2,600	$3,000

The basic fuel cell has no moving parts, making it potentially quite reliable, like a battery.[4] They are silent and produce little pollution even though they use fossil fuel. As a result, many people have claimed that they are the preferred generation source for the future. However, to be effective the units need complicated ancillary equipment: reformers to pre-process their fuel, pressurization pumps and pressure control systems, high temperature seals, and DC-AC conversion circuits. Some of this ancillary equipment is complicated and expensive in its own right – in fact, at current (2005) prices, it can cost about as much per kilowatt as the complete DG unit for some other types of units (like reciprocating piston generators). In a way, ancillary equipment needs are the fuel cell's downfall. While the fuel cell is simple and elegant in design, it is complicated in practical execution because it needs pumps, seals, and controllers that are not. In addition, several of the fuel cells types listed in Table 8.1 require rare materials that are very costly. (PEMFCs require platinum, already expensive), and whose widespread use would further drive up the cost of these rate materials.)

In the nearly 40 years since their development, fuel cells have been the subject of intense R&D to raise their efficiency, simplify their manufacture, increase their reliability, and, most of all, lower their cost. During that entire

---

[4] However, no machine is completely reliable. A fuel cell malfunction caused the explosion aboard the Apollo XIII spacecraft heading to the moon in 1970, which nearly killed the crew.

period, proponents of these very innovative and fascinating devices have maintained that success is just around the corner, that competitive commercial application will begin in a few years. But that has never quite occurred because other generation technologies have made just enough technical progress to stay out in front.

As a result of that research and development, modern fuel cells can produce electricity fairly reliably and silently, with little pollution, and tremendous fuel efficiencies – up to 50%, or half again as good as common utility turbine generators. Another advantage is that they can be manufactured in assembly-line fashion, in fairly small sizes – as small as 1 kW, and assembled into groups, if greater capacity is needed. Where a lot of power is required, hundreds of the exact same unit can be installed.

But despite these advantages, fuel cells are not yet viable generators for mainstream applications (e.g., competitive with utility system power price and reliability), for several reasons:

1) Like batteries, fuel cells produce DC power at very low (1 – 2 volt) levels. Therefore, at least a dozen or more must be linked in series to provide any reasonable voltage level, even for DC applications. For most household, business, or industrial uses, that DC power must be converted to alternating current (AC) to be of value. This is a straightforward task, but DC to AC conversion equipment increases the cost of the resulting generator set.

2) Most types of fuel cells slowly wear out. Over time, the catalysts inside them gradually become less effective. Fuel efficiency falls by as much as 10% over a period of several years. This can be corrected through disassembly and replacement of the catalyst, and is equivalent to the periodic maintenance that a steam or gas turbine requires, giving fuel cells no advantage in that regard, but no disadvantages either.

3) Fuel cells are quite complicated and quite costly to build, despite the simplicity of the concept upon which they work. First, the basic process only runs on pure hydrogen fuel, so most types need a *reformer*, a catalytic machine that strips the hydrogen out of natural gas, to be practical. Second, when operated at high temperatures and high pressures they need fuel and air pressurization pumps, and high-temperature/high-pressure seals. Special ceramics may have to be used for some internal parts because of the intense

heat, so hot it can melt any metal, and great amounts of thermal insulation are needed on the outside to vent the extremely hot exhaust gases safely. All these matters can be handled quite well, but that increases the cost of the fuel cell generators quite a bit.

Thus, taking into account the various capabilities and limitations, fuel cells really fall into two categories according to the type, design compromises, and cost:

(1) Those that operate at normal atmospheric pressure at high but not super-high temperatures, i.e., up to 400°C. They are relatively simple (no more complicated than the simplest reciprocating or turbine DG units), very quiet and fairly reliable, but not highly efficient. Their cost is higher than competing DG units but not greatly so.

These fuel cells fit special needs where dependable, quiet, continuous electric power is required at an isolated and seldom visited site. Since these cells are often available in small sizes, and are relatively non-polluting, they are used in outposts and isolated, environmentally sensitive sites where renewable power such as photovoltaic (see Chapter 7) is for some reason not viable.

(2) More expensive, high performance fuel cells, made of advanced materials and to designs that "push" technology more, with pumps and seals to permit operation at very high temperature and medium to high pressure. They tend to be complicated and to require considerable maintenance, although if maintained on schedule they can be fairly reliable. But they are among the most fuel-efficient electric power generators available, regardless of type of fuel. Cost is their only real downside: Both their initial cost and their maintenance and operating costs are so high that they are not really competitive except in niche markets.

Another category yet to be proven, but apparently quite viable, and even more effective, would be the proposed "combined cycle" fuel cells. These would begin with a very efficient ceramic, high-temperature fuel cell, and then use its hot exhaust gases to boil water to run an auxiliary steam turbine generator. Engineering calculations and a few experimental units reported results as of mid-1997, indicating that combined-cycle fuel cells may be the most efficient fossil-fuel generating units ever built. But they would also be among the most complicated and expensive electric generators ever built, combining all the complexity of an AC generator, with the very advanced DC technology of an exotic fuel cell.

Despite these disadvantages, fuel cells are competitive in some special situations, particularly for distributed resource applications at remote sites, where noise and vibration are objectionable. However, it is doubtful that the cost of fuel cells and their other performance characteristics will drop significantly to make them mainstream electric generators. While research will probably continue to refine them and lower their costs, competing technologies such as micro-turbines will undoubtedly also improve in a similar manner, making it unclear if or when fuel cells will ever emerge as a viable cost-effective DG technology.

It is interesting to note that fuel cells, like all other generation technologies, have an economy of scale. A large fuel cell generator can be designed to be more efficient than a small one. The degree of improvement with size for fuel cells is less than with some other generation types, but there is a definite economy that comes from bulk, if designed right. Partly for this reason, the first viable fuel cell generators might be large, almost central station size units, in the range of 50 MW, rather than smaller DG units. Such units would be slightly more fuel efficient than very small units. But the real efficiency gain would be in maintenance. Highly efficient commercial fuel cells will apparently be rather maintenance-intensive devices, and that maintenance will have a very great economy of scale: a large plant can justify the cost of one or two technicians on-site all the time to keep it running; a small plant cannot.

Micro-Gas Turbine Generator (MGTs)

MGTs are miniature gas turbines – small jet engines really – that are used to turn generators that produce electric power. Several important design innovations of the late 1990s made these types of units somewhat competitive when compared to some other types of power generation.

Like large gas turbine generators, MGTs run on fossil fuel that is burned to produce a very forceful stream of hot gas, which drives a turbine, which in turn spins an AC generator. MGTs are based on smaller types of jet engines, like those used in cruise missiles, and involve such innovations as "air bearings," i.e., a film of high-pressure air used for lubrication instead of oil, which eliminates metal to metal contact and wear, and completely automated control systems. They also include computerized designs that minimize exhaust emissions, and intelligent "self-diagnosis" computer controllers that automatically telephone for repair if needed. MGTs are manufactured in assembly-line fashion in units as small as 25 kW each. Physically, a 25 kW unit is about the size of a standard kitchen refrigerator.

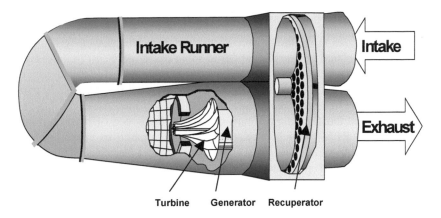

Figure 8.6 View inside a hypothetical micro-turbine generator.

Figure 8.6 shows a hypothetical micro-turbine generator, something close to an average of the various designs and features on the market. The unit shown is about five feet long and would be turned on end when in use (left side at the bottom), placed inside a rectangular metal case about the size of a refrigerator, with its intake and exhaust pointed out the top. It would produce about 25 kW of power.

The cut-out area shows this micro turbine's blade set, which has more in common with the turbo-charger in a large truck engine than with the turbine-wheel set in a jet turbine engine. The different physics at very small sizes dictates this change in turbine design. The generator is immediately behind and directly connected to the turbine. MGTs use a type of electric generator different from those of traditional central stations, a hybrid AC/DC generator, essentially a DC generator with built-in electronics to turn its output into pure AC power while still inside the generator.

The turbine's exhaust passes through a *recuperator*, a spinning porcelain disk that draws heat from the exhaust, then passes through the intake air before it enters the turbine, preheating the intake air. Preheating increases the MGT's fuel efficiency from about 18% to about 25%.

Fuel efficiency is not an MGT's strong suit, most being in the range of only 25% as compared to 40% for the best central station generator, and up to 50% for some very exotic types of fuel cell. But MGTs are reliable, quiet, and inexpensive. It is also important to some users that they can run on a wide variety of fuels – natural gas, diesel oil, kerosene, and even pulverized coal.

Internal Combustion Distributed Generators

Internal combustion engine powered generators produce power by using an internal combustion engine similar (or in some cases identical) to those used in automobiles and trucks to turn a generator (Figure 8.7). In the majority of cases (> 99%), the engine is a traditional piston engine, although Wankel and other types of engines have been used.

Internal combustion engines turn at roughly the speed of steam turbines (1,200–6,000 RPM), which is relatively slow compared to micro-turbines, (up to 100,000 RPM) and fast compared to wind turbines. This rather "normal" speed of generator rotation means that a traditional type of synchronous AC generator is used – the same type, although much smaller than, as those used on "real" (large central station) generators. The internal combustion engines powering the generator run on gasoline, alcohol, diesel, natural gas, or propane/methane. The choice of fuel is usually dictated by availability and emissions: natural gas is cleanest but requires availability of gas lines to the site.

Use of a synchronous generator is both an advantage and a disadvantage. The advantage is production of AC power directly, with no need for costly DC-AC converters that fuel cells and some micro-turbines use. The disadvantage is that – voltage and frequency – must be controlled mechanically (e.g., by varying engine speed, etc.) which is not quite as good as the electronic control achieved by DC-AC converters. On balance, however, since cost is the major barrier in front of DG use, this is an advantage overall.

Reciprocating piston, internal combustion engines are a proven, mature, but still improving method to provide power for a variety of needs including automobiles, water pumps, industrial equipment, and distributed generation systems. They have potential fuel economies as high as 45%, but their greatest advantages are a low-cost manufacturing base and simple maintenance needs. "First cost" – the cost to buy the unit and install it – is one of the biggest advantages of internal combustion powered DG units. Nothing else equals them in this regard, for they are just a simple AC generator combined with what is essentially an automobile engine.

Maintenance cost and consideration are another overwhelming advantage of this type of DG unit. Among all types of DG, only a reciprocating piston engine is so universally familiar that one can find someone to repair it virtually anywhere on the planet. In addition the synchronous AC generator is similar to the alternators and generators used in cars. For this reason, maintenance costs are low and availability is not a concern. This is definitely not the case with fuel cells or micro-turbines.

Disadvantages of piston-driven generators are a lack of good "waste"

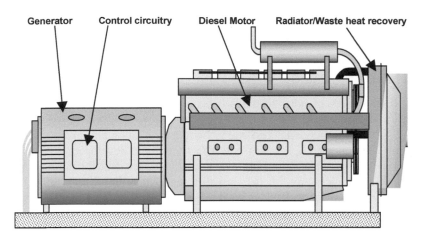

Generator   Control circuitry   Diesel Motor   Radiator/Waste heat recovery

Figure 8.7   A basic reciprocating piston internal combustion engine-driven generator. This unit uses a robust but cheap six cylinder diesel engine to drive a 480-volt, three-phase generator, producing 250 kW, enough power for 50 – 75 homes. About nine feet long, it would normally be housed in a rectangular metal case. Units just like this are the "backup power" for hundreds of thousands office buildings and small industrial plants throughout the US.

heat for co-generation compared to other DG types, marginal exhaust emissions, and considerable noise and vibration. In addition, they are noticeably heavier than other DG units for equivalent capability.

Proven by decades of use, the reciprocating piston, internal combustion engine is considered "old fashioned" by many, and it is certainly not as exotic as the fuel cell or even the micro-turbine. Its impending demise as the dominant small engine type has been forecasted by various proponents of newer designs for most of the last one hundred years. Its replacement has successively been predicted to be the turbine, various new internal combustion designs or "cycles" involving rotors, impellers, vibrating fluids, or other approaches for heat-to-motion conversion, and the fuel cell. Every such prognostication overestimated the pace of development and advantages of the technology predicted to take the pistons' place, and underestimated the improvement that would continue in piston engines. Despite 100 years of progress, there is apparently tremendous potential for further future improvement in fuel efficiency, maintenance costs, and emissions.

This type of generator currently accounts for 95+% of all distributed generation on the planet, a figure that has remained stable, or slightly increased, in the past decade. Pistons may be an older technology compared to some other DG types, but they are still a viable, competitive technology. They will likely remain the power source of choice for very small (< 250 kW) electric generators in the foreseeable future.

Other Distributed Generation Methods

A wide variety of other methods have been proposed, built, and in some cases abandoned for DG systems. There are many ways power can be generated, and there seems no end to human innovation and invention. As a result, one constantly hears of new DG types. The developers always tout their new concept as full of promise and advantages. But for the most part, these new devices aren't better, just different. Many innovators fall in love with their innovations. Backers want to get their investment back. Both try to create a "buzz" about the new device. But one should ask "Why is this an improvement?" Most aren't.[5] The three types of DG discussed earlier – fuel cells, micro gas turbines, and pistons – are the only ones that appear viable for mainstream DG applications. Numerous other machines are possible, and have been used, and have special applications. Among these are:

Stirling engine DG uses an engine with pistons and a crankshaft much like an internal combustion engine DG unit, but the fuel is burned outside the engine – *external combustion* – not inside. Stirling engines have some advantages with regard to controlling emissions, but face a slight disadvantage compared to internal combustion. Those units put the heat inside the machine, which is where one wants it to be from a loss standpoint, rather than outside. Stirling engines can be used in a renewable generator in which solar power, not fuel, is the external heat source.

Thermophotovoltaic cells are like the PV cells used in renewable solar generation (Chapter 7) but produce electricity from infrared "light" instead. When put close to a furnace burner they convert infra-red light given off by combustion directly into electricity. Portable propane heaters fitted in this way can run small radios, etc. This technology is not highly efficient and will probably not be as a major source for DG.

Hydrogen power. Hydrogen is a fuel, not a DG technology. It burns or oxidizes with low emissions and high efficiency and is therefore a desirable fuel for fuel cells, turbines, and both internal or external combustion engines. There are two barriers to the widespread use of hydrogen as a fuel: 1) developing a viable way to store a lot of hydrogen safely and cheaply, and 2) developing a manufacturing/delivery infrastructure that is as convenient and omnipresent as for gasoline and natural gas.

---

[5] In the past fifteen years the authors have seen only two inventions out of the many developed that seem to hold promise for significant impact: the Cheng cycle turbine discussed earlier, and the solar tower (see Chapter 7).

8.3 DISTRIBUTED POWER STORAGE

Distributed storage (DS) units do not create power. Instead, they store electric energy obtained at one time for use at a later time. Most use some form of battery to store the energy, although other types of mechanisms are sometimes used. Many of them also have a "smart" control system that permits them to control this energy to improve power quality or reliability of power availability at a site. Many readers will recognize the combination of battery storage and controller as an Uninterruptible Power Supply (UPS). Small in-house UPS systems are the most elementary, and certainly the most "distributed," types of DS units.

However, DS systems come in a wide range of sizes and purposes including some that are quite large. Figure 8.8 shows a BESS (battery energy storage system) installed in Alaska. This unit stores enough power to support an entire substation and its service area (equivalent to about 1,000 homes) for about 1/4 hour. Figure 8.9 shows a .275 kW (275 watt) UPS installed in the authors' home to provide backup and outage protection for the computer used to write this book. Although one is 100,000 times larger than the other, both are remarkably similar: each utilizes Ni-Cad battery storage combined with an electronic charger/inverter/control system in order to provide outage ride-through capability.

Distributed storage was originally conceived in the last half of the 20th century for *peak shaving*. The idea was that a home owner could install a unit at his home and "charge" it every night, when electricity was plentiful, and, for the utility, cheap. During the peak demand period of the day, usually late afternoon, energy would be drawn from the DS unit to reduce the amount of demand the home made on the electric system, thus "shaving" demand for electricity as seen by the utility. Similarly, units could be installed at businesses and industrial sites. The DS units would not only reduce peak demand for power, but also peak demand on transmission and distribution lines. The regulated utility would pass on the savings it saw in reduced energy and T&D costs to those customers who had the DS units.

There are situations where distributed storage is used for peak shaving. But most of its applications today serve another purpose entirely: reliability and power quality. The stored power in that homeowner's DS unit will permit him to "ride through" a utility power outage, something that typically happens once or twice a year to most utility customers in the US. In addition, if the control circuitry for the unit is designed properly, it works all the time to even out voltage, remove spikes and surges, and fill in dips in voltage, whatever their cause. For homeowners who are particularly concerned about power outages, and for business owners who have computers and other

Figure 8.8 Inside view of large battery energy storage system (BESS), essentially a "substation size" UPS. It is housed in a small warehouse located at a utility substation. This facility provides 27 MW of power for about 15 minutes, enough to ride through a brief outage or fill in during a period of needed energy stabilization.

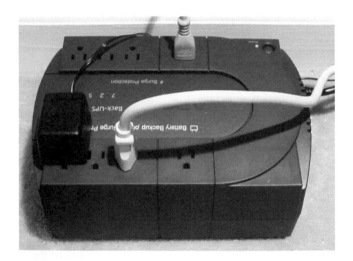

Figure 8.9 A UPS bought by the authors at an office supply store and used to provide backup power for a home computer. Although only 1/100,000 the size of the system in Figure 8.8, it is otherwise remarkably similar in design and purpose.

sensitive equipment at their location, DS units, often called UPS (uninterruptible power supplies), are quite valuable.

Distributed energy storage systems use a variety of methods to store energy – flywheels, compressed air, pumped hydro, super-capacitors, and various types of batteries. Significantly, *none* of these store electric energy in ready-to-use form. Flywheels store power as mechanical energy in a spinning mass, while compressed air and pumped hydro store it as a form of potential energy. Super-capacitors store energy as an electric charge, and batteries store it as a chemical charge imbalance that can instantly create a DC current, but none of these forms of electricity is readily useable for most purposes.

Thus, regardless of how a DS unit stores its energy, it needs a good deal of ancillary equipment to support that power storage medium. This includes equipment to convert incoming electric power into the form of stored energy (e.g., a battery charger or a flywheel spinner). It will need equipment to reverse the process, turning battery or flywheel or whatever form of energy the unit stores back into electric power. And it will require a controller to operate and to fine-tune the action of the storage medium. The controller and converters are usually built into one set of combined circuitry. Thus, a distributed storage unit can be viewed as composed of two major components:

1) The storage elements (battery, flywheel) which store energy and release it for use.

2) The charger/converter/control system, or "controller."

While many people focus mostly on the size and type of storage elements when specifying DS units, the controller is usually the more important aspect determining how much value the unit can provide to the energy consumer.

Controller Technology: Key to Reliability Benefits

The fact that any DS unit must have a basic energy controller built into it creates a unique opportunity that is the basis for most modern DS applications. Often it is only a matter of a few dollars – one additional circuit board and slightly upgraded circuit elements in the converter – to bolster that rudimentary energy controller with "smarts control" technology, turning it into a power quality controller.

"Smart control" means the ability to direct the stored power with great speed and precision, and to coordinate its use with factors external to or out of control of the DS unit, such as outages occurring on the utility system or faults or heavy appliance use within the consumer's home wiring. It means the stored energy is applied to solve power quality, energy consumption, or equipment operating problems.

Peak shaving

Distributed storage system, particularly those fitted with power quality control capability, can provide three types of benefit to utilities and energy consumers. As stated earlier, the first, and classic, concept for their use is *peak shaving* – "charging" the units at night when power is inexpensive and outputting that power during the daily peak when the cost of power is higher. In some cases this can reduce energy costs by one fifth. But most DS units lose about 10% of the energy that is stored for later use, so that net monetary savings are often limited – peak shaving just isn't worth it. Further, wear and tear on a unit, particularly one employing batteries, is almost directly proportional to the amount of its daily cycling. Cycling – charging and discharging – on a daily basis increases O&M costs substantially. As a result, peak shaving is among the rarest of uses for DS systems. On the other hand, DS units built only for peak shaving can be less expensive than those built for other purposes. Their control system only needs to be smart enough to "follow the lead" of the utility power, not to operate in isolation (as when the utility system is out of service).

Outage ride through

Outage "ride through" capability is by far the leading reason for DS purchase. A storm or unexpected equipment failure may have knocked out service from the utility, but a DS unit is a large UPS, permitting a plant/business/residence to continue its normal activity without interruption of electric service. The BESS unit in Figure 8.8 can provide 27 MW for 15 minutes, sufficient time to ramp up a backup generator or make other adjustments to the system serving the customers around it. The UPS shown in Figure 8.9 can provide power for a home computer for up to an hour.

A smooth transition for ride through

Power quality control at the instant of a utility outage, as well as on a day-to-day basis, is gaining recognition as what may be DS's greatest value. First, when a utility outage occurs, a DS system must be able to respond instantly to begin filling in with needed power. "Dumb" systems are often so slow to respond that a significant sag in voltage may occur for a second or more (long enough to shut down a digital computer). DS units fitted with good controllers react instantly and smoothly, taking over voltage with barely a ripple in the voltage that computers and other equipment sees.

Such capability is valuable. In many ways, a one- or two-second "blink" – just long enough to shut down computers, digital clocks, robotic looms, computer security systems, etc. – creates as much impact on a business or home as a one-hour outage. Therefore, proper handling of the transition as the lights go out is critical. A DS unit that can do this well is much more valuable than a "dumb" unit that might create a one to two second dip that is effectively like a short outage, before it can stabilize voltage.

8,760-hour power quality improvement

A DS or UPS unit's control circuitry can be designed so that it is always operating, even when the unit is not releasing or storing energy. Such power quality control circuitry is constantly ready to intervene with its stored energy to fill in voltage sags, prop up voltage during brownouts, trim surges, spikes and over-voltages, and stabilize transient events, as well as "ride through" complete outages. With a smart-enough and properly programmed controller, a DS unit can not only ride through outages, but stabilize voltage and mitigate system transients, acting as a type of "anti-lock brake" system to avoid cascading outages.

Distributed Storage versus Distributed Generation

One of the big roles for DG has been for backup power. But a standby generator has to start and come up to speed before it can produce power, and that takes at least a few moments. Remarkably, some automatically controlled piston engine DG units can do all that in less than seven seconds. Turbines and most fuel cells take a bit longer – up to several minutes. Regardless, the energy consumers must go through that period without power.

A DS unit can be used to augment a DG, so that it provides ride through for that brief period while it starts right after a power interruption. Many high-end backup DG sets contain a DS and a DG system in the same "box," with a unified controller that operates both while coordinating the action of storage and generator so that power transitions are smooth: the instant power supply from the utility is interrupted, the DS provides power, the DG begins to start, and within a few seconds, takes over as the power supply.

In addition, the use of a DS system with DG will improve power quality. Many people assume that distributed generation (DG) alone can also provide power quality advantages, but that is not really the case. While a distributed generator also has a source of energy (its generator) and a controller (to control the generator), the nature of both precludes the same level of power quality improvement as delivered by DS. A DG unit has to produce the power as needed and can react only as quickly as it can "ramp up" its output – fast by human standards but quite sluggish compared to the speed of many electrical events. By contrast, the DS unit has a reserve of stored energy it can release almost instantly, and more important, a capability to "suck in" power to damp over-voltages and stabilize transient power quality events. The very best "DG units" promise, and deliver, 8,760 hour power quality improvement and flawless outage ride through because they include a DS unit and its controller in the same box as the DG unit. A DG-DS combination is a very potent system, but it is the DS unit that provides the majority of the power quality improvement and the DG that provides long-term ride through capability.

Regardless, future developments in DS technology such as adaptive control (a DS system that "learns" how to program its power quality interaction by observing the system around it) and co-operative multi-unit control (a number of distributed units that can cooperate to stabilize system-level problems, not just solve local problems near each unit) will only increase the value and appeal of what may become the premier distributed resource.

8.4 SUMMARY

Distributed generators are small power generators located at or close to the energy consumer. They may be owned either by the utility or the consumer. Either way, they provide power by producing it close to the point of consumption, thereby avoiding the need for most of the T&D facilities that are required to route power from central station generators to energy consumers.

Despite a good deal of claims, small distributed generators will never be as efficient in terms of initial cost, fuel efficiency, or maintenance costs as large, central station generators: there is an economy of scale in generation that is somewhere between noticeable and overwhelming, depending on the specifics of technology and need in a particular situation. However, to be successful, small distributed generators need not be more efficient than large central station generators. They need only be more efficient and less costly than those generators and the T&D systems needed to distribute their power. To a great extent, the potential success of DG in displacing the traditional utility central-station-T&D-system paradigm rests on a single test: is the cost advantage rendered by economy of scale in generation greater than the cost of delivery of power over a T&D system. In other words, is the advantage of big generators over DG more than the cost of distributed power from the big generators to many small (DG-like) sites?

And for the foreseeable future, that is almost certain to be the case. The key element is fuel price: the advantage of big generators over small ones shows up mostly as better fuel efficiency. That matters less if fuel is really inexpensive, and more and more as fuel becomes more expensive. In a world where natural gas is $3 per MBTU, DG can compete well. But in a world of $5 prices or higher, it simply can't. Since natural gas prices are and look to remain high, DG does not appear to be a viable competitor to mainstream power utility systems in the near future.

One thing that DG advocates often tout is that DG units will improve in cost and efficiency over time. There is no doubt DG technology will improve, but so will that for large generators, and for T&D systems. The authors' detailed examination of the technological advances that benefit DG indicates that continued progress will also drive down the cost and improve the performance of central station generators by roughly the same levels: future utility investors will see no incentive to shift from large to small generators.

Therefore, it seems very unlikely that DG will become a "mainstream source of power" with small generators scattered "everywhere" displacing the traditional industry structure of a few large generators and an omnipresent T&D system.

But although DG is unlikely to displace the current power industry paradigm, it has a bright future. It fills several small but key niches in meeting the energy needs of isolated or special consumers. Its traditional role of backup power is one with growing market share, as more and more energy consumers see the need for increased reliability of electric supply. Further, when combined with distributed storage, it can provide premium levels of both electric supply reliability and power quality.

FOR FURTHER READING

G. Cook, *Photovoltaic Fundamentals,* DOE/SERI Report No. CH10093-117, 1993.

Institute of Electrical and Electronics Engineers, *Proceedings of the IEEE's* Special Issue on Advanced Power Generation Technologies, IEEE, New York, March, 1993.

N. Vosburgh, *Commercial Applications of Wind Power,* Van Nostrand, New York, 1985.

H. L. Willis, *Power Distribution Planning Reference Book – Second Edition,* Marcel Dekker, New York, 2004.

H. L. Willis and W. G. Scott, *Distributed Power Generation – Planning and Evaluation,* Marcel Dekker, New York, 2000.

H. L. Willis and G. B. Rackliffe, *Introduction to Integrated Resource T&D Planning,* ABB Guidebooks, ABB Electric Systems Technology Institute, Raleigh, NC, 1994.

9

Electric Utility Power Systems

9.1 INTRODUCTION

Electric power equipment such as generators, transformers, and circuit breakers is interconnected to form a *power system,* a group of electrical equipment whose characteristics, and connections one to the other, have been coordinated so that all parts function together in a smooth manner. A power system's purpose is to generate electric power and distribute it to consumers:

(1) It must do so *economically*, for cost is an important criterion in electric usage. Low cost electricity is very valuable to a nation, but high cost electricity is far less useful.

(2) It must do so *reliably*, because electric power is very close to a necessity in developed societies. Power availability on the order of 99.99% is expected in many parts of the world.

(3) Most important, it must deliver power *safely*. Electric power is a form of energy, and if left unchecked, like any energy source, can hurt people and destroy property.

(4) It must function well within a de-regulated industry. Generally, this means limitations on the operation and pricing rather than on the design, but it is a consideration to keep in mind.

Generation, Transmission, and Distribution

Traditionally, it has proven easier and much more economical to produce electric power in "giant economy size" generating stations, each providing enough power for a small town, rather than in many smaller locations scattered throughout a region, and closer to the electric consumers.[1] This *central station generation* concept of producing power from a small number of very large generators has dominated the design of electric power systems all over the world for the last 100 years. It has resulted in vast power systems that are built around a few dozen big generating stations, with gigantic transmission lines linking them.

Typically, electric utilities and power engineers distinguish equipment in a power system as belonging to either generation, the *production* of power, or to transmission and distribution (T&D), the *transportation* of power. Generation and T&D are disparate engineering disciplines, requiring different skills and training. In traditional vertically-integrated electric utilities, generation and T&D were separate departments. In a de-regulated electric industry, those who own and operate generation don't own or operate T&D facilities, and vice versa. Chapters 7 and 8 discussed electric power generation. This chapter looks at electric power *systems*, particularly at the transmission and distribution, explaining the types of equipment used, their structure and functions, and the way power is moved from generating plants to consumers.

9.2 T&D SYSTEM EQUIPMENT

A typical power system is composed of many different types of equipment, some of which, like *service transformers*, are used by the tens of thousands. Most equipment has two important ratings. Its *capacity* rates its ability to carry electric power. Equipment capacity is rated in amps, i.e., this conductor can carry 460 amps, or in watts, i.e., this transformer can carry 25 kilowatts. Equipment is also rated by *voltage class*. Some equipment is good only for service at below 15,000 volts, and thus is rated 15 kV, while other equipment may be rated at 75 kV, or 400 kV, as the case requires. When equipment is asked to carry more current, or power, than its rating, it overheats, and fails. When equipment is subjected to voltages beyond its design limit, it *flashes over*, usually failing quickly, in a manner that precludes subsequent repair. There are really only two major types of equipment that deliver power:

- Transmission and distribution lines, which move power from one location to another

---

[1] The reasons for the economy of scale and some of its implications are discussed at length in Chapter 8, Section 8.1, particularly in the sub-section titled "Physics Favors Larger Generators."

- Transformers, which change the voltage level of the power.

A power system is built from thousands of units of these two building blocks. Added to the lines and transformers are three categories of ancillary equipment:

- Protective apparatus to provide safety and "fail safe" operation

- Voltage regulation devices to keep voltage as constant as possible, even as the load changes

- Monitoring and control gear to measure equipment and system performance and feed this information to control systems, so that the operators know what is happening.

What Is Transmission and What Is Distribution Equipment?

Definitions of transmission and distribution vary greatly among different countries, companies, and power systems. Generally, three types of distinction between the two are made:

1. *By voltage class:* Transmission is anything operating at normal voltages above 34.5 kV; distribution is anything below that.

2. *By function:* Distribution includes all utilization voltage equipment, plus all lines that feed power to service.

3. *By configuration:* Transmission includes a network; distribution is all the radial equipment in the system.

Generally, all three definitions apply simultaneously, since in most power systems, any transmission above 34.5 kV is configured as a *network*, and does not feed service transformers directly, while all distribution is *radial*, built of only 34.5 kV or below, and does feed service transformers. Substations are often included in one or the other category, or sometimes as separate level. Equipment listed below can be had in transmission or distribution levels:

Buswork. This is the set of structures designed to move electricity between pieces of equipment inside a substation. A *bus* is usually a large pipe, or a very big wire, ten to fifty feet long and horizontal to the ground, used as a common connecting point. Buses allow many more than two pieces of equipment to be connected. Rather than join a circuit breaker directly to a transformer, each is connected to the bus. Monitoring and other equipment are attached to the bus, to measure the performance of the system, rather than onto the transformer or breaker directly.

Cables. These are electric lines built for installation underground. They consist of between one and three conductors (often each made of many strands of aluminum or copper), insulating material, and a metal sheath, which provides protection and grounding. Figure 9.1 shows several cables in a duct bank.

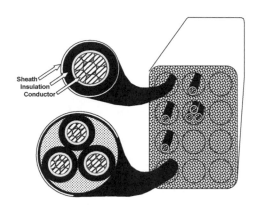

Figure 9.1 Distribution cable consists of conductor, either solid or stranded, and wrapped with one or more types of insulation, and a sheath. Shown here are cross-sections of typical cable, contained in a 3 × 4 concrete duct bank. From *Power Distribution Planning Reference Book – Second Edition,* by H. L. Willis (Marcel Dekker, 2004).

Capacitors. A type of voltage regulation equipment, which by correcting the power factor can improve voltage under many heavy loads. Power factor is a measure of how well voltage and current in an alternating system are in sync. In a perfect system, voltage and current would alternately cycle in conjunction with one another, reaching a peak, then reaching a minimum, at precisely the same times. But in distribution systems, particularly under heavy load conditions, current and voltage fall out of phase. Both continue to alternate 60 times a second, but during each cycle, voltage may reach its peak slightly ahead of current, due to a slight lag of current behind voltage, which has the effect of decreasing voltage level. This reduces the effectiveness of the power flow. Capacitors restore the timing of voltage and current, increasing the useful power from the line.

Circuit breakers. These are switches that can disconnect the power flow. The difference between a normal switch and a circuit breaker is that the latter can operate very fast, quicker than the blink of an eye, and can break the flow of huge amounts of current, which occur when there is a *fault,* or short circuit.

High voltage circuit breakers are marvelous devices, intricately designed using advanced concepts of physics and engineering. At high voltage, particularly near generating plants, the current's flow during a short circuit can be very difficult to stop. The electricity flows in tens of thousands of amps, almost a form of liquid fire, but with a force that can vaporize metal, and create magnetic fields that can crush metal. Circuit breakers use very clever designs to ensure quick, safe interruptions of these flows.

Conductor. The wire used in overhead distribution and transmission lines. Electrical conductors are available in various capacity ranges, generally corresponding to the metal cross-section. It is usually made of aluminum strands, each about 3/16 inch in diameter, wound around several steel strands. The aluminum conducts electricity well, while the steel provides high strength. Typically, the conductor is between 5/8 inch and 1½ inches in diameter, and weighs *between one and eight tons per mile!* Other things being equal, a thicker, heavier wire carries more power.

Conductors can be all-steel, which is quite rare but sometimes used where winter wind and ice loadings are severe. An all-steel conductor is used only where strength is really critical because while steel is stronger than most other metals, it does not conduct electricity nearly as well. Gold is the best metal to use for conductivity, but is prohibitively expensive. Copper is the next best, but only slightly better than aluminum, which is less expensive, and much lighter, a decided advantage because it allows for less expensive poles and towers. A popular type of conductor is ACSR (Aluminum Clad, Steel Reinforced), made of aluminum strands, each about 3/16 inch in diameter, wound around the outside of an inner core of several steel strands. The aluminum conducts electricity very well, while the steel provides high strength. A very modern type of conductor just coming into use is carbon core conductor, which uses aluminum strands wrapped around a core of carbon fiber. It is more than three times as expensive as other conductors, but the carbon fiber is quite strong, and does not stretch when it gets really hot. As a result it can be loaded to a higher current than similar weight of ACSR wire.

Interestingly, only the aluminum in the ACSR or carbon-center wire carries electric current. Due to a phenomenon called the *skin effect*, caused by the interaction of magnetic fields with the electric current, the power flows mostly on the *outside* of any conductor, which in ACSR, is just aluminum.

Distribution lines. These are wooden or metal poles supporting metal conductor, through which the electric current flows. Figure 9.2 shows a typical line.

Ducts. These are underground concrete, plastic, or metal piping, through which electrical cable is run. Figure 9.1 shows a duct bank.

Line switches. Sets of moveable connectors, they can open and close electric flow between two feeders. Either manual (hand cranked) or motorized, they are heavy, and if covered with ice are difficult to operate. Unlike circuit breakers, they do not function quickly, but take several seconds, and cannot open

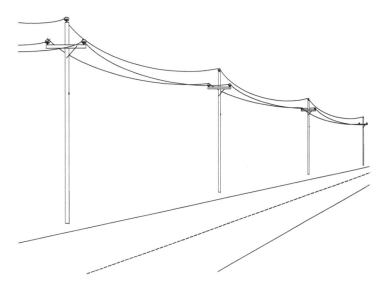

Figure 9.2 A distribution line running alongside a road. Poles are typically 100 to 200 feet apart and hold the conductor about 30 to 40 feet off the ground.

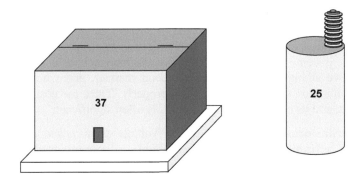

Figure 9.3 Service transformers convert primary voltage incoming power to utilization levels. Left, a 37 kVA (37,000 volt-amp) pad-mounted transformer, about three feet wide, for converting 7,200 volts to 120/240 volts. Cables enter and leave it from underneath. Right, a 25 kVA pole-mounted transformer for 34.5 kV to 120/240 volt service.

(interrupt) the high current flows that occur during short circuits. Line switches are used because occasionally it is good to vary the connection of line segments within a power delivery system, particularly in the distribution feeders. Switches are placed at strategic locations so that the connection between two parts of the feeder system can be opened or closed.

Overhead equipment. This includes transformers, conductor, switches, regulators, and associated apparatus and supplies built for the above-ground portions of the power system.

Pad-mounted equipment. Consisting of transformers, switches and other electrical equipment designed for use with underground lines, it is built to site just above ground on concrete pads (see Figure 9.3). This equipment is located inside locking metal or plastic boxes, to keep it protected from the weather, and from contact with animals and humans.

Poles. The structures that hold distribution lines in the air, Figure 9.4 shows several different types of distribution poles.

Protective Equipment. When electrical equipment fails, such as when a line is knocked to the ground during a storm, its normal function is interrupted. Protective equipment is designed to detect these conditions and isolate what's damaged, even if this means interrupting the flow of power to some customers. This minimizes further damage to equipment as well as the chance of fire or injury due to contact of electric power with people and property. *Circuit breakers, sectionalizers, fused disconnects, control relays,* and *sensing equipment* are all part of the protection system. Their job is to sense unusual conditions and instantly isolate damaged or suspect equipment, and also to try restoring service, automatically and safely, if it is interrupted. Most protective equipment is complex, containing sensitive electro-mechanical parts, many of which move at high speeds in a split-second manner, and which depend on precise calibration and assembly to work properly. Thus, the cost of protective equipment, and its control and maintenance, is often significant.

Service transformers. These lower the voltage from *primary distribution voltage* to *utilization voltage* and are available in pole-mounted versions for overhead systems, and pad-mounted and vault-type units for underground systems.

Substation. It is the entire set of equipment used to control and route power taken from one voltage level onto another. *Distribution substations* route power from the transmission or substation level onto the distribution feeders. Figure 9.5 shows a substation and the major equipment inside. *Transmission substations* convert and control high voltage power. Substations typically require from ¼ to 10 acres of land (see Figure 9.5).

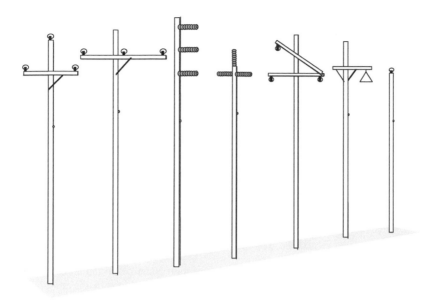

Figure 9.4. Different types of distribution line pole assemblies. The one on the far right is for a single-phase lateral circuit, the other six are all for three-phase lines.

Towers. Lattice-like metal structures that hold transmission lines above ground.

Transformers. These are the fundamental device in alternating current power systems and are the reason they work well. A transformer takes power at one voltage and converts it to another. For example, a transformer with a *turns ratio* of ten will take power at 15,000 volts on its low side, and convert it to power at 150,000 volts on the high side. Current is inversely transformed. If there are 1,000 amps at 15,000 volts incoming, the transformer will output 100 amps at 150,000 volts. The amount of power, i.e., the product of voltage and current, remains the same. Transformers function equally well in either direction. They can take power at 150,000 volts and convert it to power at 15,000 volts. Figure 9.6 shows a substation transformer. Figure 9.3 showed two small, neighborhood service transformers.

Transmission lines. Extending between wooden or metal poles, or lattice-like steel towers, they consist of metal conductors, through which the electric current flows as shown in Figure 9.7, or cables buried in underground duct pipes.

Figure 9.5 A nearly completed distribution substation under construction. Equipment in the foreground includes the distribution buswork and switching. In the middle is the transformer, and in the background to the left are the subtransmission switching and buswork. This particular substation is a modular design, built at a factory in several sections and trucked to the site, then assembled. Such substations require far less work than the traditional approach, which was to assemble all the components at the site.

Underground equipment. Consisting of transformers, conductor, switches, regulators, and associated parts and supplies, it is built for portions of the power system built below ground.

Voltage regulators. These devices can control the voltage in small increments, boosting it should it start to drop during periods of high demand, or lowering it if it becomes too high. They are often installed at the substation, at the point where power is first routed onto a feeder, in order to ensure that the voltage supply to the feeder remains at the correct level. *Line regulators* are often installed one to two miles from the substation in order to control voltage at intermediate points in the distribution system. Whether located at a substation, or on a feeder line, voltage regulators are intricate mechanisms with many moving mechanical parts inside. They are relatively expensive, create electrical losses, and require considerable periodic maintenance because of their many moving parts. Hence, they are avoided where not absolutely necessary.

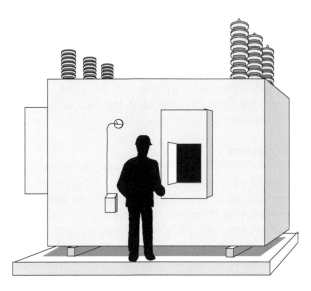

Figure 9.6 A small three-phase distribution substation power transformer. This one is roughly nine feet high and weighs about twenty tons. It can convert 15 MW of power, enough for about 4,000 homes, from 115 kV to 12.47 kV.

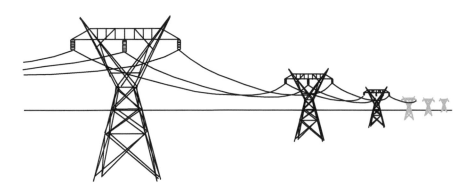

Figure 9.7 A 345 kV transmission line built on steel lattice towers, which are fabricated of steel beams arranged for minimum weight yet great strength, to hold the conductors above ground and resist expected wind loadings. The towers in this picture are about 150 feet high and 800 feet apart.

9.3 T&D SYSTEM LAYOUT

A large power system consists of a dozen or more generating stations, in concert with a complicated transmission and distribution system. The T&D system comprises thousands, perhaps millions, of units of equipment scattered throughout the service territory, operating together to achieve uninterrupted power delivery to customers.

Multi-Voltage Systems

T&D systems consist of several very different *levels*, each with a far different operating voltage than the others. While voltage at any point in a power system is kept fairly constant, that at various locations may be very different by design. Typically, power systems are built with between three and five distinctive voltage levels, e.g., 345 kV, 138 kV, 34.5 kV, 13 kV, and 120/240 volts. Each level has a separate set of equipment, interconnected and operating at the same voltage. But these levels are interconnected only at certain special points, by transformers, that permit different voltage levels to be joined.

Voltage costs less to use than current

The reason for these various voltage levels is that high voltage is generally less expensive to handle and more efficient to use than high current, but less convenient for small amounts of power. Since power equals voltage times current, power engineers can get considerable power when they need it by increasing either voltage or current, or both. However, moving high levels of current, more than several hundred amps, means that *a lot* of metal must be used in the conductor. High voltage means only that the wires have to be kept far apart, i.e., by several feet or yards. The cost of high-current systems would be prohibitive, in both money and use of natural resources, because of the great amount of aluminum or copper required. For example, if all the metal in a typical automobile were converted into electric wires, such as those used in an average transmission line, it would serve only for about ½ mile of line. Use of higher voltages means high capacity lines do not need to use a great deal of metal or weight, reducing cost substantially.

Thus, when a lot of power is needed, the choice is most often to use higher voltage, and keep current levels in a range well below one thousand amps. For technical reasons, power is usually produced by generators that operate at about 20,000 volts. However, transmission lines work most efficiently at around 200,000 volts, allowing power to be moved in larger amounts, much farther, at far less cost, when voltage is in the neighborhood of a quarter million volts. Thus, the power flowing out of each generator is first run through a *step-up*

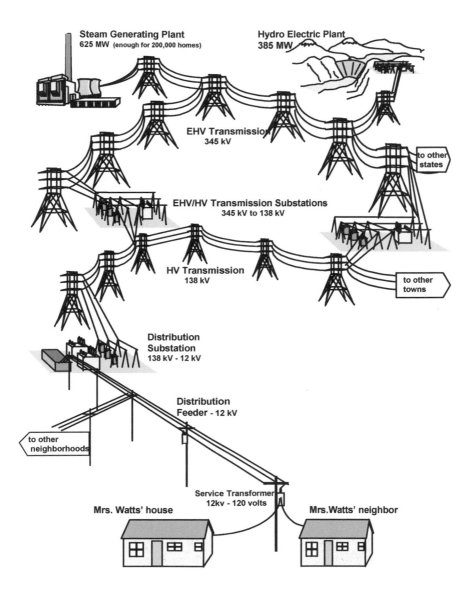

Figure 9.8 A power system consists of several levels: generation, extra high voltage (EHV) transmission, high voltage (HV) transmission, distribution, and utilization.

transformer that boosts its voltage from about 20 kV, thousands of volts, to 200 kV. At this high voltage it is directed onto the transmission network, where it will now flow to whatever part of the region requires it. (See Figure 9.8.)

Different voltage levels for different purposes

A transmission line moves a lot of power, perhaps an entire city's worth. Once that power has been brought to the vicinity of the city, it has to be split into various neighborhood-sized portions, and distributed to customers. Transmission lines have too high a voltage for this purpose. While they are efficient in moving great quantities of power long distances, at this point the utility must move only medium-sized amounts of power perhaps a dozen miles. As a result, voltage is reduced to about 10,000 volts, split into smaller amounts by routing it onto different distribution feeders, and each feeder is run through neighborhoods on distribution lines so that power reaches within a few hundred feet of all consumers. Distribution lines typically use wooden poles, or are buried underground inside concrete ducts (pipes).

Finally, no homeowner or small business needs power at 10,000 volts. Power is reduced in voltage again, right before delivery to the customer, to 120/240 volts in the US, 230 to 250 volts in Europe and many other countries, and around 105 volts in Japan. At this *utilization voltage,* it is routed into homes and businesses.

Thus, a power system (Figure 9.8) generally consists of at least three levels of power equipment:

- *Transmission,* operating at around 200,000 volts for moving city-sized amounts of power long distances

- *Distribution,* operating at around 10,000 volts and moving neighborhood-sized amounts of power several miles

- *Utilization,* where the power is consumed, and never moved more than a few hundred feet.

Often, a large power system will have two levels of transmission: extra high voltage (EHV), at 345, 500, or even 765 kV, and high voltage (HV), at 230, 138, 115, or 69 kV. Some power systems use what is called a subtransmission voltage level, in the range of 34 to 69 kV, to move bulk power more conveniently around large cities and over rural areas, before being put on the distribution system.

Power is split into smaller portions as it gets closer to the user

Electric power is generated and transported from generation plants to customer over this multi-level system, flowing down from high to low voltage.

Everyone's power starts out as part of a large bulk quantity and ends up delivered in home-sized amounts at their locations. For the past 100 years, this hierarchical system has been an effective way to move and distribute power. *The key element in this structure is the "reduce voltage and split" function*, a dividing of the power flow together with a simultaneous reduction in voltage. Usually, this happens between three and five times as power makes its way from generator to customer.

For example, the 5 kW used by a particular customer, Mrs. Watts at 123 Edison Street in Ampsville, might be produced at a 750 MW power plant more than 300 miles to the north of her city. Her power is moved as part of a 750 MW *block of power* from plant to city on an EHV (345 kV) transmission line, to a *switching substation*. (See Figure 9.8.)

Here, the voltage is *lowered* to high voltage (138 kV) by running it through a 345 to 138 kV *transformer*, and immediately afterwards, the 750 MW block is split into five separate flows in the switching substation, each of these five parts being roughly 150 MW, and each routed onto a different high voltage transmission line.

Now part of a smaller block of power, Mrs. Watts' electricity is routed to her side of Ampsville on a 138 kV transmission line that snakes about 20 miles through the northern part of the city, ultimately connecting to another switching substation. This 138 kV transmission line feeds power to several *distribution substations* along its route, among them the substation that serves a number of neighborhoods, including Mrs. Watts'. (Transmission lines whose sole or major function is to feed power to distribution substations are often referred to as "subtransmission" lines.) Here, her power is run through a 138 kV/12.47kV *distribution substation transformer*.

As it emerges from the substation transformer at 12.47 kV, the primary distribution voltage, the 40 MW are split into six parts, each about 7 MW, and routed onto six different distribution feeders. Mrs. Watts' power flows along one particular feeder for two miles, until it gets to within a few hundred feet of her home. Here, a much smaller amount of power, 50 kVA, sufficient for perhaps ten homes, is routed to a *service transformer*, one of several hundred scattered up and down the length of the feeder.

While Mrs. Watts' power flows through the service transformer, it is reduced to 120/240 volts. As it emerges, it is routed onto the secondary system, operating at 120/240 volts (250/416 volts in Europe and many other countries). The secondary wiring splits the 50 kVA into small blocks of power, each about 5 kVA, and routes one of these to Mrs. Watts' home along a *secondary conductor* to her *service drops*, the wires leading directly to her house.

Within her house, the electric power is further split into various paths and routed to different rooms and appliances, on individual circuits, which each have a separate fuse or breaker in her *electric panel box*.

The Various Layers of a Power System

Due to the various voltage levels, as well as the different types of equipment used, a power system can be viewed as composed of several hierarchical *layers of equipment (Table 9.1)*. Each consists of many units of similar equipment, doing roughly the same job, but scattered throughout the utility system. For example, the transmission grid covers the entire power system, moving bulk amounts of power from where available to where needed. The transmission system can do its job wherever needed throughout the utility service territory.

Similarly, while feeders, alone, cannot do the entire job of the power system, as a layer, they cover the entire utility service area, in the sense that there is always one within a few hundred feet of any consumer location. Likewise, there is always a substation within the required distance away from any consumer, and so forth. Figure 9.9 illustrates the layering effect of this structure. Every layer is necessary. If any one layer failed completely, the entire power system would cease to function.

Power flows "down" through these layers, from that of generation to that of consumer. As it moves from the generation plants, the power travels through the transmission layer, to the subtransmission layer, to the substation layer, through the primary feeder layer, and onto the secondary service layer, where it finally reaches the customer. Each layer takes power from the next higher layer in the system and delivers it to the next lower one. These layers share several characteristics:

- Each is fed power by the layer above it, i.e., the next higher level is electrically closer to the generation, which is where the power begins. Power reaches distribution layer by flowing first through the substation layer, and power reaches that level through the transmission, etc.

- Both the nominal voltage level and the average capacity of individual equipment units *drop* as one moves from generation layer to customer layer. Transmission lines operate at voltages between 69 kV and 1,100 kV, capacities between 50 and 2,000 MW. By contrast, distribution feeders operate between 2.2 kV and 35.5 kV, with capacities between 2 and 35 MW.

- Each layer has many more pieces of equipment in it than the one above. A system with several hundred thousand customers might have 50 transmission lines, 100 substations, 600 feeders, and 40,000 service transformers.

Overall, the net capacity of each layer (number of units times average size) *increases* as one moves toward the customer. A power system might have 4,500 MVA of substation capacity, but 6,200 MVA of feeder capacity, and 9,000 MVA of service transformer capacity installed. This arrangement of having

Table 9.1 Equipment Layer Statistics for a Medium-Sized Electric System

System Layer	Voltage kV	Number of Units	Avg. Cap. MW	Total Cap MW
Generation	produced at ≈ 20kV	5	300	1,500
Transmission	345, 138	13	125	1,625
Subtransmission	138, 69	27	65	1,755
Substations	139/23.9, 69/13.8	45	44	1,980
Feeders	23.9, 13.8	227	11	2,497
Service Trans.	.12, .24	60,000	.05	3,000
Secondary/Service	.12, .24	250,000	.014	3,500
Customer	.12	250,000	.005	1,250

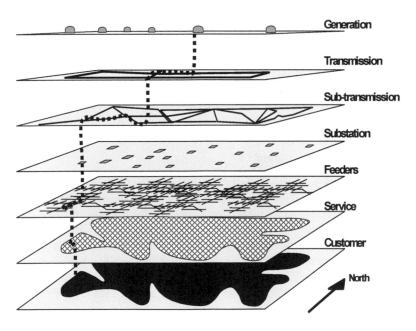

Figure 9.9 A power system is composed of service layers, each performing its function over the entire service area. Power flows down from the top (generation) to the bottom (customer) layers. Power follows a route through all layers on its way to any customer (dotted line).

greater capacity at every level is deliberate and required, both for reasons of reliability and accommodation, and load coincidence. The fact is that all customers do not have their peak demand for electricity at the same time.

Generally, reliability of electric availability drops as one moves closer to the customer. A majority of service interruptions are the result of failure (from either aging or severe weather damage) of transformers, connectors, or conductors very close to the customer. This is due simply to the tremendous number of such components in a power system, and their exposure. There are *tens of millions* of things that can go wrong in the service layer, the components layer closest to the customer, but only *hundreds of thousands* of components that can fail at the transmission level.

The transmission layer

The transmission system is a network of three-phase lines operating at voltages between 115 kV and 765 kV. Capacity of each line is between 50 MVA and 2,000 MVA. The term *network* means that there is more than one electrical path between any two points in the system, as shown in Figure 9.10. Networks are laid out in this manner for reasons of reliability and operating flow. If any one element (line) fails, there is an alternate route to avoid power flow interruption.

Besides moving power, the largest portions of the transmission system, namely its major power delivery lines, are partly designed for stability of operation. The transmission grid provides a strong electrical tie between generators, so that each can stay synchronized with the system and with the other generators. Much of the equipment and cost in the transmission system is for these stability reasons, not solely or even mainly for moving power. (Chapter 16 will discuss this in more detail).

The subtransmission level

Some utilities and power pools call the lower voltage parts of their transmission system *subtransmission*. Its lines deliver power to distribution substations along their routes. A typical subtransmission line may feed power to three or more substations. Normally, subtransmission lines range in capacity from 30 MVA to 250 MVA, operating at voltages from 34.5 kV to 230 kV. With occasional exceptions, subtransmission lines are part of a network grid, where there is more than one route between any two points. Usually, at least two subtransmission routes flow into any one distribution substation, so that feed can be maintained if one fails.

The distribution substation level

Substations are the meeting point between the transmission grid and the distribution feeder system. They are locations where a fundamental change takes place within most T&D systems. The transmission and subtransmission systems

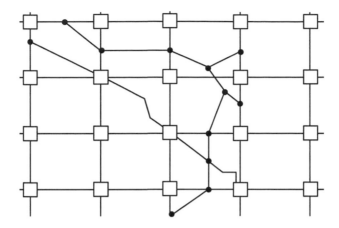

Figure 9.10 A network is an electrical system with more than one path between any two points. If properly designed, it provides electrical service even if any one element fails. With the large amounts of power involved, transmission is almost always built as a network: If any transmission line (solid lines) fails, there is another path for power flow. To design such a network to always have a second or even third path for power flow is difficult and expensive, but worthwhile because power used by so many consumers is involved. Here, there are at least two pathways between any two substations (squares).

above the substation level usually form a network, with more than one power flow path between any two parts, thereby ensuring outstanding reliability. But from the substation to the customer, a network configuration would be prohibitively expensive. It is simply not justified based on the smaller number of customers whose power is now involved. Therefore, most distribution systems are *radial, with only one path* from substation to each customer. Often (see Figure 9.8), many customers share parts of a common, but single path, built to be as reliable as possible

A substation usually occupies an acre or more of land. Its equipment consists of high and low voltage *racks* and *buses* for the power flow, *circuit breakers* for both the transmission and distribution level, *metering equipment*, and the *control house,* where the relaying, measurement, and control equipment is located. But the most important apparatus, what gives this substation its capacity rating, are the substation *power transformers*, which convert the incoming power from transmission voltage levels to the lower primary voltage used for distribution.

Individual substation transformers vary in capacity, from less than 10 MVA to as much as 150 MVA. They are often fitted with *tap-changing mechanisms* and control equipment to vary their *windings ratio*, so that they maintain the distribution voltage within a very narrow range, regardless of larger fluctuations on the transmission side. The transmission voltage can swing by as much as 5%, but the distribution voltage provided on the low side of the transformer stays within a narrow band, perhaps only ± 0.5%.

Very often, a substation will have more than one transformer. Two is common, four is not unusual, and occasionally as many as ten are used. Multiple transformers increase reliability. Most transformers can handle much more than their rated load for a short period, e.g., up to 200% of rating for two hours. Thus, a substation with two or more transformers can serve all of the customer load (demand) even if one transformer is out of service briefly for repairs. Equipped with from one to six transformers, substations range in size, or capacity, from as little as 5 MVA for a small, single-transformer substation serving a sparsely populated rural area, to more than 400 MVA for a large six-transformer station, serving a very dense area within a big city.

Substations consist of more than just the different types of electrical equipment and their costs. The land, or site, has to be purchased and prepared, i.e., it must be excavated, a grounding mat (wires running under the substation to protect against an inadvertent flow during emergencies) must be laid down, and foundations and control ducting for equipment must be installed. Transmission towers to terminate incoming transmission, and feeder getaways, ducts or lines to bring power out to the distribution system, must be added.

The distribution feeder level

Feeders, either overhead distribution lines mounted on wooden poles, or underground buried, or ducted cable sets, route the power from the substation throughout the neighborhoods for several miles around it. The most common *primary distribution voltage* in North America is 12.47 kV, although from 4.2 kV to 34.5 kV is widely used. In Europe, 11 kV is popular. There are primary distribution voltages as low as 1.1 kV and as high as 66 kV operating in portions of the world. Some distribution systems use several primary voltages, for example 23.9 kV, 13.8 kV, and 4.16 kV, in different areas of the system.

A typical feeder distributes power to between 500 and 5,000 customers, delivering between 2 MVA and 30 MVA, depending on the conductor size, the layout, and the distribution voltage level. The layout is called a *dendritic configuration*, i.e., repeated branching into smaller branches, as the feeder moves out from the substation toward the customers (Figure 9.11). In combination, all the feeders in a utility constitute the *feeder system*. An average substation has about five feeders, but the count can vary between one and 40.

The main, three-phase trunk of a feeder is called the *primary trunk* and may divide into several main branches. These terminate at open points where the feeder meets the ends of other feeders, points at which a *normally open switch* serves as an emergency tie between two feeders.

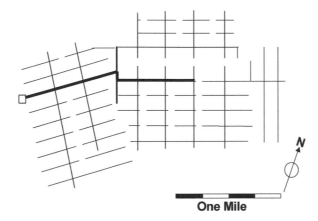

One Mile

Figure 9.11 Map of a typical feeder, in this case 12.47 kV as used in power systems throughout much of the suburban United States. Line width indicates size of feeder segment (capacity to carry power). This particular feeder, in Texas, provides power to consumers over an area of about three square miles, serving 1,085 homes and businesses, and carries a peak load of 4.57 MW. Its larger portions follow streets, and smaller portions run down the middle of blocks along the backyard property line of houses.

The primary trunks and switchable segments are usually built using three phases. The largest size of distribution conductor is about 500-600 MCM, but conductor over 1,000 MCM is not uncommon, and there are feeders with 2,000 MCM conductor, used for reasons other than just maximum capacity (e.g., contingency switching needs). Often a feeder has excess capacity to provide back-up for other feeders during emergencies.

The majority of distribution feeders worldwide and within the United States are overhead construction, i.e., wooden pole with wooden crossarm or post insulator. Only in dense urban areas, or in situations where esthetics matter, can the higher price of underground construction be justified, because it costs from three to ten times what overhead does. The primary feeder is built from insulated cable, pulled through concrete ducts that are first buried in the ground.

Often, however, the first several hundred yards of an overhead primary feeder are built underground even if the system is overhead. This underground portion is used as the *feeder getaway*. Particularly at large substations, the underground getaway is dictated by practical necessity, as well as by reliability and esthetics.

At a large substation, ten or 12 three-phase, overhead feeders leaving the substation mean from 40 to 48 wires hanging in mid-air around the site, with each feeder needing the proper spacings for electrical insulation, safety, and maintenance. In a tight location, there is simply not enough overhead space for so many feeders. Even if there is, the resulting tangle of wires looks unsightly, and, perhaps more important, is potentially unreliable: One broken wire falling in the wrong place can disable a lot of power delivery capability.

The solution is the underground feeder getaway, usually consisting of several hundred yards of buried, ducted cable that takes the feeder out to a *riser pole*, where it is routed above ground and connected to overhead wires. Very often, this initial underground link sets the capacity limit for the entire feeder, i.e., the underground cable *ampacity* is the limiting factor for the feeder's power transmission.

Feeder switching

Each feeder is divided, by normally closed switches, into several switchable elements. During emergencies, segments can be re-switched to isolate damaged sections and route power around outaged equipment to customers who would otherwise have to remain out of service until repairs were made.

The feeder consists of all primary voltage level segments between the substations and an open point (switch). Any part of the distribution level voltage lines, three-phase, two-phase, or single-phase, that is switchable is considered part of the primary feeder.

The Lateral Level

Laterals, short stubs of primary-voltage line that branch off the distribution feeder, are the final primary voltage part of the power's journey from the substation to the utility's customers. The vast majority of a utility's customers are fed via these lateral circuits. Laterals are relatively short (usually less than ½ mile long) lines, operating at the primary distribution voltage (e.g., 12.5 kV), that branch laterally (hence the term) from the main feeder trunk.

Typically, the primary three-phase feeder circuit may leave the substation and run for several miles through neighborhoods nearby, usually on easements alongside a major road or street, a railroad right of way, or similar "public use" land. Some utilities will design feeders that have a few major, three-phase branches that connect off the trunk to run down other main streets, but others do not − a main or trunk feeder is a single route of three-phase primary feeder that winds along major streets in the area a few miles from the substation. Either way, a series of laterals taps off the primary feeder as it passes through the community. These laterals branch off the feeder frequently (every block or so along the street it follows) to route power down side streets to the vicinity of homes and businesses in that area.

Usually, laterals do not have branches (in other words, no other primary-voltage lines "lateral" off them), and most laterals in American-type systems are only of a single phase. All three phases are used in a lateral only if a large amount of power is required, or if three-phase service must be provided to some of the customers the lateral feeds. Normally, single-phase laterals along a feeder trunk are arranged to tap alternatively different phases, so as to balance loading as closely as possible: the lateral running down this street uses phase A, the next one along the feeder uses phase B, the next phase C, then back to phase A, etc. In this way loading on the trunk feeder's phases is kept roughly equal.

Laterals deliver as little as 10 kVA for a small single-phase lateral to perhaps 2 MVA for a large three-phase lateral. Generally, even the largest laterals use a small conductor (on the order of ¼ to 3/8 inch diameter) as compared to the conductor size used on the trunk feeder (on the order of ½ to 1 inch in diameter there). When a lateral has to deliver a great deal of power, all three phases are normally used, still with a relatively small conductor for each, rather than using a single-phase with a large conductor. Power flow, loadings, and voltage are maintained in a more balanced state if the power demands of a large lateral are distributed over all three phases.

Service transformers

Service transformers (see Figure 9.3) convert primary voltage to service voltage (that which the customers will use), which in the US is nominally 120/240 volts. An important result of the laws of nature, as applied by mankind, is that while this voltage is ideally suited for typical household and business use, it can move power efficiently only about 100 – 200 feet. Beyond that, equipment costs for wire, electrical losses, and voltage drop all become a bit extreme. Therefore, a utility will use many service transformers arranged along the feeder and laterals so that there is at least one within 100 to 200 feet of every utility customer.

Most service transformers will be relatively small as regards their capacity, at least compared to other equipment in a power system, because each will serve only a few energy consumers, those within a short distance around it. A typical service transformer is about 25 kVA (capable of serving five homes) whereas a typical transformer at the utility substation might be 25 MVA – one thousand times that capacity.

On typical distribution feeder, which might consist of six miles of primary trunk off of which branch 60 half-mile-long laterals (for a total of 36 miles of primary circuit), there would be about 500 service transformers, or one every 380 feet. This would mean that if the laterals and these transformers were artfully arranged, none of the customers on this feeder should be more than 380/2 = 190 feet from the transformer closest to them.

The secondary and service level

Secondary voltage circuits, sometimes called service voltage circuits, or utilization voltage circuits, lead from each service transformer to the customers it serves. The term "secondary" was coined in the few decades of the electric era, when power systems consisted of only two voltages: the "primary" voltage that the utility used, which was typically several thousands of volts, and the lower "secondary" voltage provided to customers and suitable for home appliances, which was on the order of 110 volts.[2] Service and utilization voltage are more modern terms. All are in wide use although usually one is heavily preferred at any one utility company. The authors, paying homage to the century-plus tradition of the power industry, use secondary except when specifically addressing voltage issues related to the customers' equipment or needs.

A utility's secondary voltage circuits are usually arranged so that each service transformer serves a small set of customers in the area right around it (see Figures 9.12 and 9.13). The lines leading to a home or business are called its service drops, and lead from the secondary circuit, or the service transformer if it is close nearby, to the utility meter box on the building.

Although every secondary circuit and service drop is quite short (typically less than two hundred feet) cumulatively there is more total of length secondary-voltage line than anything else in the utility system. The feeder with 36 miles of primary circuit and 500 service transformers would typically serve about 2,000 customers (4 per transformer), with an average of about 150 feet of service line needed to reach each customer. The total would be about 300,000 feet, or over 55 miles of secondary level circuit.

Design of overhead lateral and service circuits

Primary-voltage laterals are built as either overhead construction (on wooden poles) or underground (using buried electrical cable) as meets the local requirements. The associated service lines are usually of similar type: buried underground if the laterals are, and overhead if the laterals are overhead.

Overhead construction is usually something like the configuration shown in Figure 9.12, which shows a typical single-phase 12.47 kV overhead lateral and

---

[2] Early power utilities targeted 110 volts as the voltage delivered to customers, probably because early power engineers and utilities thought something close to 100 volts would be most suitable for small household and business appliances, and allowed 10 volts for the voltage drop through household wiring. Regardless, over more than a century, the 110-volt target crept upward as utility engineers, desiring to do a better job, gradually "gave the customer more." Today, industry practices (and many engineering standards and codes) have fixed on 120 volts as nominal, and most electric equipment for home use is designed so it will work with anything between 110 and 125 volts. Many utilities deliver 125 volts (near the maximum permitted by standards) to their customers, in order that appliances in the home will very close to 120 volts even allowing for the typical range of three to five volts drop as the power makes its way through the wiring in a modern home. (Out of idle curiosity, the authors plugged a voltage meter into the socket serving the computer being used to create this footnote, and measured 121 volts.)

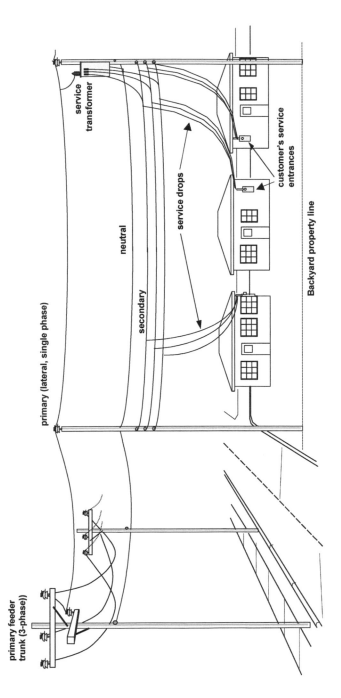

Figure 9.13 An overhead primary lateral and the service level equipment associated with it.   See text for details.

the service level equipment that is used to route power from it all the way to the electric consumers. The single-phase lateral branches off the three-phase feeder that runs along a major street (left side of the figure). The lateral runs down the property line in the middle of the block of homes it services. Service transformers spaced at every other house or so reduce voltage from primary voltage to service voltage (120/240 volts) and route it onto secondary circuits and hence to service drops leading to each nearby home. For simplicity's sake the equipment here is shown without lightning arrestors or the fused cutouts which many utilities install to improve reliability.

Wooden poles are used to support the lateral's single primary voltage conductor, which is usually the highest wire in the construction, as shown here. A number of feet below that will be the neutral, a similarly sized wire. The neutral is placed below the primary for two reasons. First, it operates at a lower voltage (theoretically, very close to zero, but it is never safe to touch the "neutral" because in some circumstances it can have enough voltage on it to harm a person).

Second, if the primary conductor breaks, perhaps because a tree falls on it, the primary conductor is most likely to hit the neutral wire first as it falls to the ground. This contact would cause a primary voltage fault (short circuit) which would instantly trigger a fuse or breaker operation to de-energize the circuit. This protects the low voltage service lines immediately below the neutral, and anything on the ground further below it, from contact with an energized high-voltage conductor (because at the point where the falling conductor would make contact with them, it would be de-energized).

Design of underground lateral and service circuits

Unlike underground transmission lines, which are almost always installed in expensive concrete or metal duct banks, single-phase laterals are sometimes *direct buried;* the cable is put inside a plastic sheath that looks and feels much like a vacuum cleaner hose, a trench is dug, and the sheathed cable simply dropped into it and buried. Figure 9.13 shows an underground lateral and associated equipment, serving the same customers as were by the overhead in lateral in Figure 9.12. Like the overhead example, this is a single-phase lateral that branches off the three-phase feeder that runs along a major street. The lateral is made of underground cable which combines both conductor and neutral along with insulation (see Figure 9.1). This cable begins in the air on the primary feeder "terminal pole" (so named because it terminates the overhead route of the circuit). From there it runs down the pole to the ground through a protective metal pipe. Now under ground and in a metal or plastic sheath, it crosses the street and runs down the property line in the middle of the block of homes it serves (or it might run along the sidewalk near the street). Every so often, it rises up under a pad-mounted service transformer connecting to its high

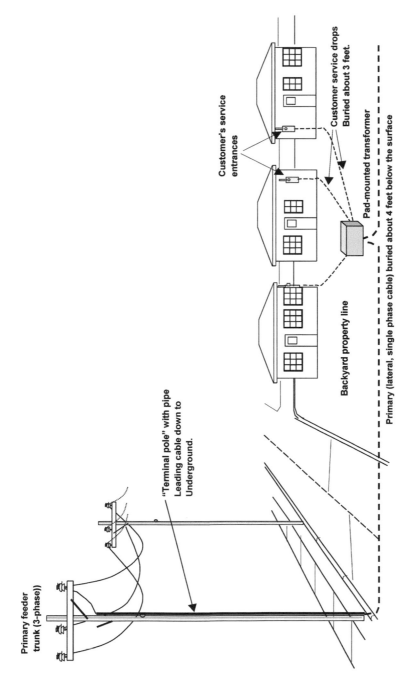

Figure 9.13 An underground primary lateral and the service level equipment associated with it. See text for details.

voltage side, thus providing power to it. Underground service drops take the power, now at 120/240 volts, from the low side of that transformer to each nearby home. Often called URD (underground residential distribution), this type of construction is in some cases no more expensive than overhead, although it tends to take longer to fix when something goes wrong (particularly if the cable fails) than to repair overhead lines. For simplicity's sake the equipment here is shown without the lightning arrestors or the fused cutouts which many utilities install to improve reliability.

Overhead and underground service transformers

In overhead construction, service transformers are mounted on the poles, between the levels of the primary conductor and the secondary wires. Most overhead service transformers look like a large gray soda can three to four feet high (right side, Figure 9.3). Most are single-phase units, and available from numerous manufacturers in capacities between 5 kVA and 166 kVA. When the utility needs to provide three-phase service, there will most often be three separate phase conductors on the overhead lateral (hence it looks much like the feeder trunk on the left of Figure 9.12, and the utility will use two or three, rather than one, pole-mounted service transformer, mounted side by side on the pole (which is usually thicker than a standard pole so it has additional strength to handle the weight of all three) and interconnected together, in what is called an *overhead transformer bank.*[3]

Underground distribution most often uses pad-mounted transformers (left side, Figure 9.3), each a strong, locked protective steel box with a transformer and its fusing/switching equipment inside, mounted on a concrete foundation (pad) on the ground. The primary-voltage cable comes up from underneath and lower voltage cables route out of it to the customers, all buried and safe from human contact or interference.

Single-phase pad-mounted transformers are typically between about 15 kVA and 166 kVA in capacity, and have physical dimensions on the order of two by two by three feet to three by three by five feet. Unlike the case with overhead systems, in underground systems three-phase service is almost always provided by using a three-phase pad-mounted service transformer rather than a group of two or three single-phase units. The three-phase pad mount transformer is a single steel box with a three-phase transformer inside it along with three sets of primary phase terminations and switches, etc. Banks of

---

[3] There are, and some utilities use, three-phase overhead transformers, but many don't. They install separate single-phase transformers – either two or three. (There is a "design trick" which can be used to apply only two transformers to still provide three-phase power in some cases, if the amount of power needed is not too much and a few other criteria are met.)

single-phase units are not used in underground construction, as they are in overhead designs.[4]   Three-phase pad-mounted units are available up to about 2,500 kVA capacity and are correspondingly larger in physical dimensions than the lower capacity single-phase units.

The vast majority – probably close to 98% – of underground system transformers are pad mounted.   In some cases, utilities will use a *vault transformer.*   Here, something akin to a pad-mounted unit is installed in an open concrete box, called a vault.   The vault is large enough that there is workspace all around the transformer, for any maintenance needed. It is buried with its open top flush with the ground level.   A heavy, locked protective steel grate, strong enough to walk on or even support light vehicles, serves as the top of the vault.   One often sees these steel grates in the sidewalks of downtown areas of major cities. Below many is a vault with a power transformer.

Some manufacturers offer, and a few utilities use, *direct-buried transformers,* which are installed completely underground, along with the power cables, so that nothing of the electric system shows above ground.   Direct-buried transformers are not widely used for several reasons. First, a series of designs tried in the 1960s had poor durability, giving direct-buried transformers something of a reputation for being unreliable. Those early failures were mostly due to the second reason: Some types of soil act as heat insulation, keeping the heat inside the transformers where it builds up to high temperature and causes pre-mature aging of the insulation, leading to pre-mature failure (by contrast, air around a pad-mounted unit cools it fairly well). This heating problem can be handled through proper design, but that increases the cost.   Third, a search to find and an effort to dig (perhaps creating a mess in someone's prize front lawn) is required if and when the utility needs to check them or do maintenance. Finally, and perhaps most important, even when the heating is not a big issue, direct-buried transformers cost a bit more than pad-mounted units of equivalent capacity, both to buy the unit and to install it.

Urban Service Systems: Nearly Always All Underground

In heavily congested and really built-up areas such as the downtown areas of major cities, an entirely different type of underground distribution system is used. It is difficult to generalize about it in detail because the utilities serving major cities have evolved significantly different designs to meet their unique needs: the downtown system designs in New York City and in Chicago, for

---

[4] Technically, a bank of three single-phase pad-mounted units could be used to provide three-phase service, by connecting three nearby single-phase units together with the proper cables, etc. No doubt, somewhere in the world, this has been done. But the authors are aware of no utility in North America (or anywhere else for that matter) where this is standard practice for underground design.

example, are significantly different in quite a few respects.

In most heavily built up urban areas, *all* of the electric system is located underground, including perhaps the substations and the transmission lines leading to them. Certainly, all primary trunk feeder circuits and laterals will be underground, usually running under the center of the streets inside concrete ducts or pipes installed specifically for that purpose. The transformers will be pad-mounted or vault type, or designs suitable for "inside duty" and located in the basements of major buildings. These units are often called a "dry type" transformer because they contain no oil, as do transformers designed to be "outside." This means they cost a bit more, and that they cannot take short-term overloads as well, but that even if they suffer the very rare catastrophic failure, they are less likely to cause a fire.

Another difference in distribution in dense urban areas is that it is exclusively three-phase, simply because none of the customers want or use single-phase power. A large office tower can require many MW of power, too much to provide at single phase, and typically will have a number of needs for three-phase power anyway (e.g., the large motors for its cooling compressors). Thus, there are no single-phase laterals and no single-phase transformers. For technical reasons, the feeder trunks, all exclusively three-phase, have no branches. They are often designed as *loop circuits,* each connected to the substation at both ends, and winding up and down a number of streets in the area around the substation, in and out of all large buildings, where they energize the local dry-type service transformers with three-phase power. For more detail on underground urban distribution design, see the *Power Distribution Planning Reference Book – Second Edition,* pages 456-465.

9.4 SUMMARY

A power system must generate and distribute electric power to its consumers economically, reliably, and safely. Traditionally, the concept was to produce power via *central station generation,* consisting of a few large generating stations linked with gigantic transmission lines, instead of having smaller generating stations scattered throughout a region, and closer to the electric consumers.

Two major types of equipment deliver power: transmission and distribution lines to move power from one location to another, and transformers to change the voltage. Ancillary equipment consists of: protective apparatus for fail-safe operation; voltage regulation devices to maintain voltage as constantly as possible; monitoring and control machinery to measure system performance and relay the data to apprise operators of what is happening. The following are general axioms about power systems:

(1) Voltage costs less to use than current. Usually, when providing a certain amount of power (voltage times current), it is more

economical to use high voltage and less current than the other way around.

(2) Different voltage levels are used for different purposes. High voltage is more appropriate for moving large amounts of power long distances. Low voltage is needed by appliances and home wiring.

(3) Power is split into smaller portions, at lower voltage, as it moves from generating plants to neighborhoods and then on closer to the customer.

(4) Power systems are composed of several hierarchical layers of equipment, each one having more pieces in it than the layer above, and thus greater capacity as one moves toward the customer.

FOR FURTHER READING

James J. Burke, *Power Distribution Engineering: Fundamentals and Applications,* Marcel Dekker, New York, 1994.

Enrique Santacana, editor, *Electrical Transmission and Distribution Reference Book,* ABB Electric Systems Technology Institute, Raleigh, NC 1997.

H. L. Willis, *Power Distribution Planning Reference Book – Second Edition,* Marcel Dekker, New York, 2004.

10

Regulation and De-Regulation

10.1 INTRODUCTION

This chapter provides an overview of the power system regulation and de-regulation, a "big picture" overview with concentration of the important concepts and differences between regulation and de-regulation, and the "why" behind choices governments and society make about whether and what type of de-regulated structure they choose for their electric energy infrastructure. It is a good chapter for a reader interested in understanding the "big-picture," policy and societal issues to read if he or she only has time to read one chapter. Later chapters will delve into the details that show some of these key aspects, such as open access transmission, are implemented. Here, focus is on what open access and other de-regulation/regulation issues mean, why each might be chosen as a part of a de-regulated structure, and how each interacts with the other aspects.

Neither regulation nor de-regulation is better than the other. Each brings advantages and dis-advantages that, under difference circumstances and from different perspectives, make one more appealing than the other. But for reasons that are discussed here, and in more detail later, in the 1990s many governments around the world decided to change from the traditional regulated paradigm, which permitted no competition among electric suppliers, to some form of management that did. Basically, that is what de-regulation is all about.

This chapter begins with an overview of the concepts behind the regulated, or traditional electric utility industry in Section 10.2, what regulation means, how it was applied to the power industry, and how electric companies and customers interact in that environment. Section 10.3 then discusses the reasons that brought forth de-regulation: What advantages it was perceived to bring under certain circumstances. Section 10.4 summarizes what de-regulation entails and the re-structuring with which governments implement it. Section 10.5 discusses the changes in industry priorities that it will cause.

10.2 WHY WERE ELECTRIC UTILITIES ORIGINALLY REGULATED?

Strictly speaking, regulation means that the government has set down laws and rules that put limits on and define how a particular industry or company can operate. Nearly all industries in all nations are regulated to some extent, even if it is only with laws that constrain them to do business under "fair" or fully disclosed business practices, or to operate their facilities within recommended safety guidelines. Very competitive businesses, such as auto manufacturing, airlines, and banking, are all heavily regulated in this sense, with myriad government requirements defining what they must, can, and cannot do, and what and to whom and when they must report their activities.

But regulation as used here refers to an even much more rigorous set of rules that have structured the power industry for over a hundred years, and include:

Monopoly franchise: The government grants one company the sole right to sell electricity to consumers in a certain area, its *franchise territory.* Within this territory, no other company can sell electric power.

Obligation to serve: This local power company must provide for the needs of all electric consumers in the region.

Guaranteed rate of return: The government guarantees the utility that its regulated rates will provide it with a "reasonable" profit margin above its costs, if it plays by all the rules.

Prescribed operating and business practices: The government may put stringent limitations on how the local power company functions. These could be requirements on how it builds its system (*"All power lines will be put underground"*), or the way it does its planning (*"All new capital projects will be submitted for approval before construction commences"*), to strict definitions on how it finances its operation (*"No individual will own more than 5% of the company stock"*).

Least-cost operation: The government will define how the utility computes costs and sets its prices. Usually, it requires that the utility operate in a "lowest cost" manner, and defines specific ways that it should and should not finance its operations.

Regulation of electric utilities is not the only way a government can control a power industry within its jurisdiction. As described in Chapter 1, another popular way is either to own and operate the power company directly, as a government or municipal utility, or to set up an agency or administration that provides some or all of the electric power industry functions, thereby having a hand in fashioning what the local power industry is, and how it operates.

But whatever advantages ownership of the electric utility might bring to a government, it brings one major disadvantage: The government, itself, now has to pay for the electric system. If it grants a monopoly franchise, someone else's money pays for the electric power system and its operation. If regulation is done correctly, the government gets what it wants, electric power available to all its citizens at a reasonable cost, and the investing company gets what it wants, profit from its investment.

Each of these five major concepts is discussed in more detail below. The reader should remember that there are exceptions to nearly every rule: With so many utilities, in so many countries, for over more than a century, there has been an exception to nearly every rule and generalization that could be made. What is given below describes the *mainstream majority* of the industry.

Regulatory Jurisdictions

There are many regulatory organizations that have some authority over electric utilities, and as a result, almost every electric utility company will find itself subject to several levels of overlapping government regulation, including federal, state, county, and municipal authorities. In the United States, the Federal Energy Regulatory Commission (FERC) regulates interstate transportation of energy, including electricity, gas, and other power sources. It has set up standard reporting requirements, which nearly all utilities in the United States adhere to, including those that may not, strictly speaking, need to comply. These reporting requirements dictate, among other things, the type of accounting to be used. In addition, the Nuclear Regulatory Commission (NRC) has very specific definitions about the operation of any nuclear plant and requires adherence to a wide range of procedures by any utility owning and operating one.

Each state has at least one, and in several cases, two or more, commissions or agencies that regulate public utilities. Usually these are called the Public Service Commission (PSC) or Public Utilities Commission (PUC), but may include unusual names like Commerce Commission or Energy Agency.

Finally, each municipality has a department or agency that interacts with the electric utility in its area. Usually, within the United States, the relation between these authorities works in the following manner:

FERC sets standards and regulations on all aspects of interstate trade, which generally means power pools, wholesale wheeling, and the operation of large, multi-state utilities. It regulates inter-regional transportation and commerce in energy, makes certain that electric power on the national level is subject to common operating standards and practices, and often sets policies of operation to assure that utilities comply with national interests.

State agencies set policies for all regulated utilities in their states, usually including overall rate schedules, definitions of how costs are to be computed, how decisions on spending are to be made, and disputes resolved. These state regulations may include guidelines or set limits for the various municipal regulatory authorities and their franchise agreements.[1]   In some cases, county, rather than municipal franchises, have been granted, and franchise regulation rests with the county, not municipal, governments.

Rural electric membership cooperatives are generally regulated, within the United States, by the Federal Government's Rural Electric Authority (REA), which dictates standards, design guidelines, financing, reporting, and numerous other aspects of their operations. In addition, rural and municipal utilities served by the Tennessee Valley Authority (TVA) or Bonneville Power Administration (BPA) must adhere to guidelines and operating frameworks set down by those government agencies. Beyond that, some rural utilities, usually those outside of TVA, BPA, and similar jurisdictions, are subject to either mandatory or voluntary compliance with regulation by their respective state regulatory authorities.

As a result, many utilities in the US find themselves regulated by at least three entities, and a similar hierarchical overlapping structure exists for utilities in other parts of the world, too. A utility may have local franchise agreements with towns and cities, which it must obey: These grant the exclusive municipal franchise, but have certain requirements for the utility. There is a state public utility commission, commerce commission, or some other regional entity that exercises authority over practices, rates, and so forth. Usually, this body standardizes operating practices, service standards, and perhaps rates (prices) and sets guidelines on how much authority the municipal jurisdictions can exercise. But despite these overlapping authorities, regulation is a fairly simple concept, based on the five principles stated earlier. Someone will require these of each utility, as well as reporting to prove adherence to such rules.

---

[1] In some states the Public Utility Commission (PUC) authority extends over investor-owned utilities, municipals, and some REAs operating in the state. In most, however, depending on the law, its authority covers only investor-owned utilities, and municipal utilities are not regulated by the state.

Monopoly Franchises

A monopoly franchise gives one particular company exclusive rights to sell electricity in a certain geographic area, the franchise territory, which often constitutes the governing or regulatory body's entire area of jurisdiction. For example, a municipal government will grant a franchise covering all areas within its city limits to a single company. Dealing with two or more local power companies, each serving a part of the city, or overlapping in some manner, would make life complicated for the government and confusing for the consumer.

The monopoly granted varies greatly depending on the government, the culture, and the time, e.g., regulation was quite different in the early 20th century vs. the late 20th century. Usually, the franchise is *a total monopoly* on retail sales of electric power. In the territory, only this company can sell electric power to end-users. At one time it also included a monopoly on commercial production of electric power: Only this one company can own commercial power production equipment located in this territory.[2]

The monopoly franchise nearly always includes ownership of "public" transmission and distribution facilities. The utility's franchise does not prevent private individuals or companies from owning private electric systems on their own land. For example, every homeowner owns the "system" or wires running through his home, and every industrial plant has a private power system to deliver power within its grounds. However, only the local franchisee can build and operate T&D equipment on rights of way and easements throughout the franchise territory – the government prohibits anyone else from doing so.

A privately owned power system is an option for some consumers

The franchise covers only public and commercial power and power systems. Usually, unless other laws and regulations not directly connected to the franchise prescribe otherwise, any private individual or business can own and operate a private power system. This is certainly true throughout the United States. A homeowner can buy and run a private generator, providing for personal power needs, although few do so because of the cost and hassle involved. But many large manufacturing plants own several sizable generators, and have a staff to maintain and operate them, providing all or most of the power the plant requires (see Co-generation in Chapter 6). The plant may also own extensive T&D equipment, to distribute this power to all the pumps,

---

[2] By contrast, if this is not the case, another utility could operate a generating plant located inside the franchise territory, but because it did not have the local franchise, could not sell the power there. It would have to move its power, once produced, over transmission lines to its own territory.

heaters, rollers, stamping equipment, offices, warehouses, and whatever else needs electricity at its site.[3]

But there would be two things the local utility's franchise precludes the private industrial plant from doing with its electric system:

First, it can not extend its T&D facilities outside its property, as, for example, if it expanded from its original site to an extension across the street and wanted to run wires from its generators to its facilities there. Strictly enforced, a traditional regulated electric franchise allows *only* the local utility to move power across public roads and land. The plant would have to build generators and separate electric systems on opposite sites of the road, not connected by a line passing over or under the street.[4]

Second, this industrial plant is not allowed to convey any portion of the power it produces to any other company or individual. Strictly speaking, this means that it cannot run a line over the fence between its own and a neighboring company's plant, and sell the excess power its generators might be able to produce to that company.[5] Even though the fence runs along the boundary between two private industrial sites, the monopoly franchise prohibits this.

Further, the franchise proscribes the plant's reselling power it buys from the utility: It cannot buy a large amount of power from the utility, perhaps gaining a quantity discount, and then resell a portion of that to its neighbor.

A franchise guarantees utilities it will have customers

Originally, one big reason governments would grant an exclusive franchise for electric service was to attract investors, because electric utility systems cost *a lot* of money. No businessman would invest in a system designed to serve all the customers, as opposed to only a portion deemed profitable, unless he were given assurances that his costs would be covered with at least a small profit. A second reason was that 19th and early 20th century technology made it virtually impossible for several different generating and energy sales companies to operate over the same power grid. No one wanted to build two or more

---

[3] Within the United States many industrial sites have private power systems rivaling all but the largest public utilities. For example, Dow Chemical's plant near Freeport, Texas has roughly 2,000 MW of power generation, and encompasses 345 kV high voltage transmission lines.

[4] This has been the subject of numerous court cases and settlements around the world, with nearly every type of result, including settlements allowing limited construction of private lines, to compromises whereby the local utility owns and operates the line for a set fee. But traditionally and usually, *any* extension off private property is prohibited.

[5] This, too, has been the subject of numerous court and regulatory cases involving all kinds of situations and results, often including very subtle nuances of interpretation, e.g., what if the two industrial plants are different divisions of the same company? Or what if they are separate companies but owned by the same holding company?

overlapping grids, given the higher costs and greater esthetic and environmental impact they would make. One power system would do, and it had to be owned by *the* electric company in a region.[6]

Franchises tend to institutionalize the utility

Monopoly franchise agreements usually cover a certain period of time, e.g., three, five, or ten years, and are set up for renewal and renegotiation of their terms at these periodic intervals. Conceivably, at these times the government could select a company, other than the current franchise-holder, to be the local electric utility. Usually the government finds it ineffective to do anything but negotiate, sometimes fractiously, with the utility to change policies it does not like. The reality is that only one company can be institutionalized as the local utility, and that in all but rare cases this decision is final.

The reason is that utility systems are among the most capital-investment intensive of businesses. They require expensive equipment, careful engineering, and considerable amounts of valuable land, all items that create a very high "up-front cost." As a result, after the franchise holder has operated for a while as the local utility, it has built up a considerable amount of transmission and distribution facilities that are scattered throughout the public environment, *property it owns.* Similarly, it has hired and trained a staff expert in the operation of these particular facilities.

Thus, if the government wanted to shift the franchise to another company, it would have to find another one willing to buy those assets at a "fair" price from the present franchise holder, and to hire and train a similar group of people to run them. Probably, the new company would just hire all the employees of the former franchise holder. The purchase price for the new system would be high, but could, and would be, determined under the regulatory umbrella of financing rules and regulations established for the utility. Under the new franchise agreement granted, sufficient income to cover those expenses would be assured. But even in the best of circumstances, such a sale would be very complicated and expensive to administer. It would involve a lengthy and perhaps contentious evaluation of what the existing system included, and what it was worth. It also might be quite difficult to find some other company willing to take on the purchase of the assets, particularly if it asked, "Will you do this to me when my franchise runs out?"

For this reason, when a franchise is to be renewed or renegotiated, seldom is there any serious consideration given either to finding a new utility company, or to the government's taking over the local electric utility. Instead, the utility and

---

[6] Modern computers and data communications technology make it quick, feasible, and economical, for many companies to use a common set of T&D lines, just as they have enabled similar "multi-company use" of common telephone lines.

the government must sit down and work out the new agreement, knowing that both will be dealing with one another in the foreseeable future. Over time, although they might squabble about rates and controversial projects, they become symbiotic.

Obligation to Serve

In accepting the monopoly franchise within its service territory, the local power company gives up one cherished prerogative of a de-regulated company, the right to say "no" and walk away from unprofitable business deals or projects. Within limits set down by the franchise agreement, the utility is required to provide electric service in any amount needed, to anyone who wants it and is willing to pay its standard regulated rates anywhere in the franchise territory. Usually, there are limits on all of these aspects of any particular transaction. The franchise agreement will spell out what is considered "normal" or typical service, usually with a set of different classes of service (residential, small commercial, large industrial, etc). Such definitions will encompass the majority of homeowner and business needs.

But anyone who wants large amounts of power (i.e., 1,000 MW), or wishes it delivered to a difficult to reach spot (the top of a mountain), or has other needs that fall outside the "normal" definitions of the utility's franchise agreement, may have to negotiate with the utility. Such an electric consumer may also have to wait a reasonable amount of time while the utility puts facilities in place, and may be required to contribute cash up front to pay for part of the special facilities to service them. It may have to pay a price determined by a negotiation process under the framework of the utility's regulated operation.

Obligation to serve is included in almost all franchise agreements for two reasons. First, to guarantee that all customers are offered service in a non-discriminatory way. Second, to assure that the grid is eventually extended to all places where it is needed. Within the United States at the beginning of the 21st century, this last reason is hardly a serious matter, because the power system reaches almost everywhere (there are some isolated homes and businesses in northern Nevada and elsewhere in the western US that are "off the grid."). But this was a serious concern in the early years of the 20th century: Regulation provided a framework that required utilities to make the grid ubiquitous.

Monopoly Franchise from the Customer's Perspective

Under regulation, electric power consumers have only two ways to obtain the electric energy they need:

(1) They can provide for their own power needs *privately*, buying, building, or operating any type of generator and power system that does not violate local rules on safety, pollution, etc.

(2) They can buy electric power from the *holder of the local monopoly franchise*, which has both a standard set of rates for power, and a standard set of terms and conditions, prescribed by law, for doing business and providing the service.

There is a narrow gray area of permitted competition that was, under the traditional regulated structure, limited somewhat by law but more by economies of scale. A third party could offer to install and operate a "private" generation/power system at a consumer's home or business site. For example, an energy service company could build, maintain and operate a generator at a customer's home or business as a service. Under traditional regulatory frameworks, there were usually various regulations that severely limited this practice. But frankly, there were so many possible loopholes in such limitations that some form of this competition would have occurred in the last 100 years, if anyone really wanted to do it.

The major reason that "third party energy service companies" offering "self generation" did *not* develop under the traditional regulated power industry structure is that there was insufficient economic incentive. Given the economies of scale involved in the production and reliability of electric power (see Chapters 5 and 6), small "private" power systems could not compete with a large "public" utility. There was simply no potential business that offered economy to the customer and profit to the third party company.[7]

Thus, the consumer has to buy from the local power company, choosing from among a set of standard services and price schedules. Under traditional regulation, these are usually offered with little choice or options available in the quality, type, or conditions of service. On the other hand, the local power company cannot say, *"No, we're sorry but you're not in a profitable location, and we just don't have any more power to sell at the moment."* The utility is obliged to provide the power when asked, and at a "low" cost – low in the sense that it must operate in a reasonably priced manner acceptable to the government, charging its customers only enough to recover that cost, plus a small increment of profit.

Guaranteed Rate of Return, and Regulated Prices (Rates)

A cornerstone of regulated-monopoly operation of electric utilities, or just about any other industry, is regulated, *least-cost prices*. The government authorities overseeing the utility define a *rate schedule* of prices the utility must charge. These make certain that the utility, which has a local monopoly, does not charge too much. The prices are set so that they cover the utility's costs, and provide a

---

[7] Under de-regulation, this changes: The energy service company now has "open access" to the power grid, where it might provide economic advantage.

reasonable profit. The amount permitted is generally more than the utility's stockholders could make from a "risk free" investment, such as government bonds or certificates of deposit. After all, there must be an incentive for investing in the utility rather than in something else.

The allowed rate of return for a regulated utility is, however, much less than the profit that might be made from investment in a non-regulated company, like an electronics firm. Still, while some of those firms make 40% profit, there are usually several that go belly-up.

The concept behind these prices, *cost recovery* and a *regulated rate of return*, means that prices will be set by the government so that the utility is certain to recover all its costs and its permitted profit. It can't lose, except if conditions change drastically in unforeseen ways, or it makes real blunders in its decisions. Nevertheless, the utility's risk is quite minute compared to that of most businesses, and under normal circumstances it is "guaranteed" to make money.

Generally, the allowed profit margin is based upon investment, not revenues. The guaranteed profit is not a function of the utility's gross revenues, as in, "You get to keep 5% of every dollar taken in." Instead, the guaranteed profit is a certain level of return on investment. The utility may be promised a return on investment of 13%, meaning that it will earn 13 cents on every dollar it invests. The government is giving the monopoly franchise to the utility in order to convince it, and its stockholders, to invest in utility facilities and equipment needed by the public, and so the utility will cover the other costs necessary to run the electric system in the government's jurisdictional territory. The incentive the government offers is a very safe return on this investment.

Least-Cost Operation

A strict canon of regulated monopoly pricing is *least-cost* operation of the utility. Since a monopoly franchise holder is in a position to deliberately "run the cost up" in its operations so that it makes more money (because it makes a percentage on all expenses), a good deal of the regulatory procedure governing its activities is designed to limit its ability to do so. For example, new furniture for a manager's office is a capital investment. A regulated utility that went unchecked in its operations could therefore increase its profits by buying new furniture for all its managers.

Suppose a utility is "assured" a 13% rate of return. Perhaps it can borrow money at only 8% interest. Then it can make a million dollars (13% - 5%) profit just by borrowing $20 million to refurnish all its offices. In fact, it could borrow $40 million and make two million dollars of profit by spending twice as much, buying expensive solid-wood office furniture along with original oil paintings and sculpture for decoration.[8]

Only qualified expenses go in the rate base

Government oversight of the electric utility's business and operating practices stops abusive spending of the type outlined in the paragraph above in two ways:

The first way that cost is controlled is by requiring that every expense be certified through a process of meeting certain guidelines, before it can be put in the *rate base:* The set of investments upon which the utility's rates are based. New furniture might qualify to be included in the rate base, if the need is real *("We were opening a new operations center as an expansion of our facilities, and had no old furniture available."),* if it is of a justifiable type *("We specified furniture meeting the recommended guidelines set by the American Office Workers Association."),* and the utility can show it spent no more than necessary. *("We solicited bids publicly, and selected one of six bidders, whose combination of price, payment terms, and warranty we judged to be the lowest overall combination of cost.")*

It is very doubtful if solid-mahogany furniture and original artwork would ever qualify for inclusion in an electric utility's rate base, although if such is considered "normal practice" (say for top executive offices) it would be permitted.[9] A regulated utility is still free to buy such luxuries any way it wants, but it must do so at its own risk: It cannot put non-essential items into the monetary base upon which its rates are computed and for which it is guaranteed a profit. If it went ahead with such a purchase, the cost would have to come out of its profits, and would be viewed as a non-regulated investment, which if it yielded a profit would be the utility's own good fortune and outside of its earnings from utility operations. *("Wow, who would have guessed this painting we bought from that guy named Picasso, for only $15,000 in 1951, would be worth $25 million today!")* Investment in art or any other item that does not qualify for the rate base is entirely at the risk of the utility and its stockholders, as are all the investments of any unregulated company. The regulatory process will not guarantee any profit.[10]

---

[8] A savvy utility would buy only works by local artists, through a program of supporting local artists and galleries. Reasonable amounts of such "local community involvement" can be put in the rate base and, in fact, are often expected of a local utility.

[9] The utility would argue, rightly in most cases, that it is competing for top executive talent and that it must match the perks that its executives could get if they decided to work in a drug or auto manufacturing company, if it wants good top management.

[10] To avoid abuse, most regulatory processes require fairly well defined separation of the regulated and unregulated expenses in a utility company. Many utilities would not be allowed to buy such art, even as unregulated investments. Their stockholders and executives would be free to set up a separate company to do so (e.g., "Big Electric Art Investments, Inc."), but it would have to be a legally separate business entity.

Least-cost evaluation of expenses

The second way that cost is controlled in the regulated monopoly framework is through requirements for the utility's evaluation of expenses in its planning and decision-making. Assuming new office furniture is justifiable, the utility must follow procedures to ensure that it buys the least-cost furniture, and is able to document that it did so, if asked by the regulatory authority.

Least-cost does not mean the utility must always take the "cheapest" alternative. Instead, it means that it must seek the most economical way of doing the job at hand, furnishing offices, in this case. The specific interpretations of least cost vary slightly from one regulatory agreement to another, but usually the term means decisions are made so that the utility needs as little revenue as possible to cover its overall costs. In other words, given two options that solve a problem or fulfill a need, the utility should select the one that will require it to ask the least amount of money from its customers.

For example, suppose a new substation is being built to serve demand growth on the outskirts of a major city. It that will have a load of 8 MW next year, and 15 MW within a few years, followed by slow growth thereafter. Table 10.1 shows six alternative transformers the utility can consider. The smallest and least expensive transformer listed in Table 10.1, while having the lowest initial cost, $189,000, creates relatively high electrical losses, because it will be loaded close to or above its nominal rating at times. These losses will consume $111,873 of power annually, a cost the utility will have to pass on to its customers. This transformer's 10 MW capacity also means it cannot meet the long-term needs of the area, and will have to be replaced in only seven years,

Table 10.1 Options Available in a Utility Spending Decision

Planning Option	Transformer Type	Initial Cost	Losses $/yr.	O&M $/yr.	Replacement Years	Cost	Annual Revenue
1	10 MVA HBD-reb.	$189,000	$111,873	$8,786	7	$215,000	$145,501
2	10 MVA standard	$234,000	$91,000	$8,553	9	$215,000	$127,211
3	12 MVA standard	$267,000	$80,000	$10,202	9	$200,000	$119,291
4	15 MVA standard	$287,000	$59,800	$10,350	23	$205,000	$101,001
5	20 MVA standard	$365,884	$57,000	$15,675	never		$103,775
6	20 MVA low loss	$490,000	$45,855	$17,955	never		$105,460

requiring a newer, larger transformer – another expense. Thus, over the long haul, this "inexpensive" unit will prove very expensive. In fact, of the six options shown, it is the most expensive in the long run, requiring revenues of over $145,000 per year in the long run.

Buying a large-capacity, low-loss transformer (Option 6) at a higher initial cost of $490,000 provides a lower overall cost and smaller revenue requirement because savings in long-term losses and replacement cost more than make up for higher initial price. But there is an even better option that splits the two approaches, option number four, where a standard 15 MW transformer will be used but replaced after 23 years.

The key concept at operation in this selection, and used similarly in almost all other spending decisions by a regulated utility, is that initial and long-term operating costs are carefully balanced against one another to obtain the lowest overall requirement. Oversight of a utility by the regulatory authority generally requires a decision-making process that takes this, or a similar approach. Planning, engineering, and operating practices and reporting throughout the utility are all required to correspond to these rules. While the specific details about how costs are computed and the decision-making implemented vary from one regulatory authority to another, several key concepts don't. They are a part of almost all regulatory frameworks for monopoly franchise electric utilities:

> *Comprehensive costing:* The evaluation includes all costs of ownership and operation, not just the initial expense.

> *Long-term view:* Cost evaluation is done over either the lifetime of a typical bond payment, or the equipment, whichever is shorter. This is usually 30 years.

> *Revenue requirements minimization:* The utility must act in good faith to try to minimize the total revenues that must be collected from the customers. [11]

Costs based on these principles, for items that are considered within the venue and necessary for the electric utility, are permitted.

---

[11] This last is different than just seeking least-overall cost. One option open to the utility in the example cited here might be not to build a new substation at all, and thus, not buy a new transformer: The new subdivision would be served from an existing substation far away, perhaps overloading it and shortening the life of its equipment, and creating high power delivery losses. But the costs of high losses and shorter lifetimes of existing equipment would be weighed against the cost of building the new substation: The utility has to borrow money to build it, and that interest has to be added to the rate base. Overall, it might work out to be the least costly alternative.

Rates and Rate Cases

The regulatory process always includes some method through which the government regulatory authority and the utility agree on the schedule of rates to be charged. Usually, at periodic intervals, or whenever conditions change sufficiently that the utility or the regulatory authority deems it necessary, the utility will submit a *rate case*. This is a detailed analysis of its expected cost of operation over the next few years, along with the proposed set of rates it will charge, an analysis of the profit these rates will produce, and arguments and justification about why that profit level is reasonable. *("We expect to have a total of $500 million invested in our facilities. Operating costs will be $600 million per year. The attached set of rates is expected to raise $665 million, giving us a 13% profit on our investment, which we think is fair given that we could invest in government bonds from the Democratic Republic of Outer Mordovia and make 11%.")*

Actual rate cases are much more complicated affairs than the brief example that ends the paragraph above. Rate cases can often run to tens of thousands of pages of reporting and backup material. Costs are broken down by type of expense in numerous ways (capital vs. operating, equipment vs. labor, overhead vs. capacity vs. energy-related and many other distinctions, all made simultaneously so that costs are segmented into a variety of tiny cubbyholes of detail), and by attribution to classes (some costs can be tied solely to residential service, others only to the utility's efforts to meet the needs of its commercial customers, while others cover all customer types) and causes. *("This cost is due to our need to serve the peak demand, while that one is due to requirements to provide reliability of service, and that third cost to safety requirements set by the federal government.")*

Projected revenues under the new rate schedule are based on a detailed forecast of future customer needs as well as on their reaction to the new prices. The difference between projected cost and revenues, the projected profit, makes assumptions about unusual expenses. *("We expect equipment and labor costs to slowly increase as they have for the last ten years. We expect one major hurricane requiring $80 million in contingency equipment repairs, because we have one every three years," etc.)*

Typically, the regulatory authority and the utility negotiate the rate case's details. The regulatory authority acts as a kind of consumer advocate, intent on obtaining the lowest possible rates for the utility's customers. The utility wants to justify the highest rates, so that it can make a good profit.

The regulatory authority and the utility seldom argue about the actual rates schedule, *per se*. The rates are, in fact, defined by the assumptions, forecast, etc. Usually, then, argument and negotiation center on the assumptions in the forecasts that lead to the rates. For example, the utility has an incentive to base its case on the highest possible forecast of further growth in customer demand.

A high forecast justifies building new electrical facilities, every dollar of whose cost will provide the utility with some additional profit. Even if the new facilities are not needed, because the load growth was not as great as forecast, the utility will still earn money on them, because they qualified for the rate base.

The utility also has an incentive to forecast the highest possible values for future equipment, fuel, and labor costs in its rate case. The higher projected costs will justify higher rates, and if by some chance actual costs are less than projected, the utility will probably get to keep the difference as extra profit. If by contrast, future costs are more than the utility forecast, the difference comes out of its profits.

Usually, the regulatory commission spends a good deal of time poring over the utility's cost and sales (load forecast) projections, and similar operating forecasts. It may not agree that these projections are reasonable – it may feel that lower costs are more likely and thus the utility's rates can be lower.

Often, the total amount of money the utility will collect is not the issue – the regulatory commission or agency may agree with the utility's case on the total amount of revenue it should have. Instead, the controversial issue often is the relative rates among customer classes, i.e., who pays what share of the total revenue requirement? Disputes often erupt over whether commercial customers are paying their "fair share" versus the rates charged residential users. The commission may also have to resolve disputes from customer advocacy groups that believe large industrial customers, who have negotiated special contracts, e.g., to deliver 1,000 MW to the top of the mountain, have gotten favorable discounts from the utility, at the expense of smaller customers.

Rate cases can become quite adversarial, with numerous parties involved, each staking out their position, and with a good deal of money, often millions, riding on the outcome of the regulatory decision. In most cases, the regulatory agency is not trying to batter down the utility to the lowest possible rates, but merely trying to balance the utility's needs against society's in general. For one thing, regulatory authorities and commissions recognize that a financially healthy utility is important to the local economy and the community as a whole.

Certification of Need or Request for Capital Construction

Usually, regulated electric utilities must seek permission from their regulatory commission for any major construction project, again through a process carefully laid out by law and in their franchise agreements. For example, if a utility wants to build a new substation or transmission line, it must apply for permission, which might be called a "Certificate of Need" or a "Capital Construction Authorization," or something similar. This gives it the right to proceed with the project, in the name of meeting public needs, and to put the cost in its rate base.

Mandated Operating
and Business Practices

Regulatory frameworks at the municipal, state, and federal level all have very exact rules on how an electric utility company can function in areas beyond financing, setting rates, and making buying decisions. These rules cover the full range of operating considerations, from safety to financing, from electric equipment to hazardous materials usage. They vary widely, including very special, local additions.

For example, the U.S. government, through FERC, requires certain types of reporting of costs, revenues, and operational expenses, which largely define how major utilities must do the accounting and reporting of their expenses. Since they have to report this to FERC, "FERC-type" accounting and reporting has become standard throughout many parts of the U.S. utility industry.

The U.S. Rural Electrification Administration (REA) has very specific requirements for how utilities under its authority engineer and plan their systems. It stipulates certain types of load projections, plans, cost evaluation, and decision-making. It requires certain types of design and equipment utilization. For utilities served by the Tennessee Valley Authority (TVA), the rates they could charge their customers were traditionally set by TVA, in a large rate matrix that defined what and how they had to design and how to apply their rates.

State utility commissions have their own requirements. For example, many demand that utilities under their jurisdiction plan use a method called *integrated resource planning,* which means the utilities must consider conservation and energy efficiency, not just power system construction, among the options when they plan their system expansion. Other states insist that the utility operate distribution equipment at the lowest practical voltage levels possible under schemes referred to as conservation voltage reduction (CVR), to lower societal energy usage. These are only two examples of myriad requirements to "do it our way." The utility complies because it is a public company and the regulators speak for the customers, putting the cost of such measures in the rate base.

Often, cities and towns will add requirements to their franchise agreements and constrain the utility to only certain types of construction or operating practices in the city. This may require that for esthetic reasons the utility build only underground lines and facilities in certain areas, and locate all large equipment, like substations, at least 200 yards from a major intersection or street. The utility may be required to co-operate in certain ways with local sewer, water, and city planning officials in all of its plans. Occasionally, a utility will be required to maintain offices in certain outlying towns, even if they are not economically justifiable, as part of a community participation policy. Such requirements are not economic, but political.

Land Needed for Facilities, Rights-of-Way, and Easements

Rights-of-way are land the utility owns as property, for its power transmission lines. It also has to buy land for substations and other similar facilities. *Easements* are public facilities – usually along streets, roads, and alleyways – that are designated for utility use and in which it can locate distribution lines, and some types of smaller transmission lines. Typically, when a utility needs a new site, or new right-of-way for its lines, its planners will identify the alternatives available and work out the cost and benefits of each. It will then try to obtain the necessary easements or land. Easements are generally no problem, but rights-of-way and key sites for major installations can be. An electric utility, acting in the public interest under its franchise agreements, has the right in some cases to exercise *eminent domain*: It can take, at fair value, land needed from the owner, whether the person wants to give it up or not. Such cases can be contentious, the legal fees so expensive, the risk of losing at a late date when other aspects of a plan are locked in so great, and the publicity from such cases so bad, that it is rare for a utility to use eminent domain. Instead, some compromise is usually worked out.

Summary of Regulated Utility Structure and Functions

Regulated electric utilities provided the industry with stable growth and good service for more than a century. Table 10.2 lists the salient aspects of the regulated electric utility.

Table 10.2 Key Characteristics of the Regulated Electric Industry

Monopoly franchise	Only the local electric utility can produce, move, or sell commercial electric power within its service territory.
Obligation to serve	The utility must provide service to all electric consumers in its service territory, not just those that would be profitable.
Regulatory oversight	The utility's business and operating practices must conform to guidelines and rules set down by government regulators.
Least-cost operation	The utility must operate in a manner that minimizes its overall revenue requirements (amount it must bill its customer base).
Regulated rates	The utility's rates (prices) are set in accordance with government regulatory rules and guidelines.
Assured rate of return	The utility is assured a "fair" return on its investment, if it conforms to the regulatory guidelines and practices.

10.3 WHY DE-REGULATE? THE GOOD
AND BAD OF UTILITY REGULATION

Basically, the drive for electric industry de-regulation began because governments valued the advantages of competition among energy suppliers, and wide choice for electric consumers, more than they did the continuing benefits of utility regulation.

The Original Need for Regulation

Both governments and business favored utility regulation during the early history of the industry. Often, they vehemently disagreed on the details of what and how regulation would be implemented, but from the beginning, both groups recognized that it was necessary. From the perspective of the businessmen running the early utilities, regulation brought several important benefits.

1. *It legitimized* the electric utility business. Government franchises and regulation clearly implied to a possibly skeptical public that civic leaders thought electricity was a "good thing."

2. *It gave utilities recognition* and limited support from the local government, in approving ROW and easements, and by generally co-operating with them as they expanded their embryonic companies.

3. *It assured a return on investment,* regulated as that might be.

4. *It established a local monopoly.* Early utility leaders could focus on building up their systems and the quality of them, without having to worry about competitors undercutting prices to gain market share, etc.

Municipal leaders wanted regulation, too. It assured them of universal electric service – under regulation the businessmen running these utilities were obliged to provide service to everyone. Monopoly franchises simplified the buying process, too, which was important for early consumers. Electricity was new and confusing enough, without the added burden of having to deal with the conflicting claims, standards, and offerings of different power companies.

Regulation originally reduced risk, as it was
perceived by both business and government

Beyond the reasons given above, and perhaps most important for both government and business, regulation offered an acceptable, *risk-free way to finance the creation of an electric industry.* At the dawn of the electric era, government leaders were not about to invest large amounts of public capital in a new and untried technology, no matter how attractive it appeared. Regulation took care of that – the businessmen would risk their money, not the government's. True, the government was guaranteeing them a fair return on their investment, but only through regulated rates: If electric technology didn't work, or if this new energy source didn't sell in the marketplace, then the businessmen lost money, not the public.

For their part, early "electric businessmen" like Westinghouse, Edison, Brush, and their contemporaries, knew their technology was sound, but they were reluctant to borrow and invest the huge sums needed to build an electric system in an environment where a competitor with deeper pockets might deliberately choose to lose money to take their customers away from them. What they wanted was a local monopoly, a stable market, and an assured return on their investment to minimize their risk. Thus, regulation provided both sides with the risk minimization they needed.

> *Without utility regulation and government sponsorship or backing of electric utilities, a universal electric system, reaching all homes and businesses, and the infrastructure to support it, would never have been built.*

The Driving Forces in Favor of De-Regulation

Many changes in technology, business, energy usage, and politics led to the worldwide trend toward electric industry de-regulation. But several seem particularly important.

Background factor: The long-term technology trend of lower economies of scale in power generation

One force that contributed to the de-regulation of electric power was the change in generation economies of scale that occurred throughout the 1980's (see Chapters 5 and 6). Until about 1980-1985, truly "cost-competitive" generation could be achieved only by building truly monstrous coal, natural gas, or nuclear power plants. It was difficult enough for a monopoly franchise holder to finance such power behemoths. No one could envision how competing companies, each with only a segment of the market, could afford to offer the lowest possible prices. But the situation changed beginning in the 1980s. Technical progress led to more efficient small turbines and generators. Smaller generators could nearly match the efficiency of the enormous units, particularly if they were run on natural gas, rather than coal. Simultaneously, the price of natural gas declined.

Thus, in many instances, it was possible to build new power plants that could provide energy at a lower price than what utility customers were paying for that coming from the existing old, giant power plants. At this point, a great many industrial and commercial users of electricity began to build and operate their own plants to produce power at a price lower than what they could buy it from the utilities, while others demanded to know why they could not "shop around" and change suppliers to get the lowest price power possible.

Many people do not count this as a contributing factor to the push for de-regulation, but the authors believe it had a significant effect. Large industrial users of power, seeing that they could buy and run their own generators at a lower cost, naturally asked why an even better choice was not available to them: "Why can't we hire a third party (i.e., another company) to do this for us so we get the advantages of the power cost and can continue to concentrate on only our core business?" De-regulation occurred in the U.S. largely because large industrial and commercial users of power lobbied the federal government for just this right.

Reason 1: The need for regulation changed

But far more fundamental than any one of the many reasons often mentioned for change was the fact that overall the basic need for regulation of electric utilities, *when all issues were taken together,* had largely abated long before the end of the 20th century. *First,* the original need for regulation, which was to encourage relatively risk-free investment for electric infrastructure development, passed into unimportance decades earlier. *Second,* the electric generators, transmission grids, and other infrastructures had been built: There was virtually no place in the United States, or in Europe, where electricity is needed but not available from "the grid." *Third,* the omnipresent electric system that had been created had been "paid for" decades earlier. While utilities continued to borrow money so that they can add to and re-new their systems, that incremental investment did not represent the same level of risk as creating the grid did in the late Victorian era.

Reason 2: Privatization

In many of the countries where electric utility de-regulation first occurred (e.g., Argentina, England), the government was also privatizing the industry (see Chapter 12). *Privatization* means the government sells its state-owned electric utility business to private investors. Usually the motive, as in England, was the government's firm conviction that private industry could do a better job of running the power industry.[12] The push for privatization, and the accompanying political perspective, nearly always led to favoring de-regulation.

De-regulation does not have to be a part of privatization efforts: A government can sell its utility system to several regional investor-owned

---

[12] Undeniably, the motive behind privatization in some developing nations was simply to raise cash for the government, which finds in the national electric system an asset it can sell in return for a significant amount of hard international currency.

companies, offering each a monopoly franchise in its area. But de-regulation is coincident with privatization in most *national* arenas, being necessary in order to attract good investment. Asked to put up large amounts of money to buy the power systems, business wanted to know they could make good money if they did the job well. Thus, de-regulation to "free up" the rules nearly always accompanies privatization.

Reason 3: Cost was expected to drop

Competition breeds innovation, efficiency, and lower costs. In 1990, the date most observers assign to the beginning of the world-wide push toward electric utility de-regulation, the electric utility industry had not seen cost competition among electric utilities in nearly a century. As a result, while electric rates had declined somewhat, they had not dropped as much as where there had been competition, for example, among equipment manufacturers who designed and built equipment for utility systems. Figure 10.1 compares the cost of the average kW of power sold by six large US utilities, to the cost of a standard substation transformer, over the period 1930 to 2000. Electric rate reductions did not keep pace with the decreases in equipment prices. The reasons why are manifold and complicated, to the point that there is no agreement on why, if, or what could be done about it. Many people believe that competition will bring about significant decreases in cost. Others vehemently disagree. One thing that seems certain is that customer value will improve, whether price, itself, goes down or not, as discussed later in this section.

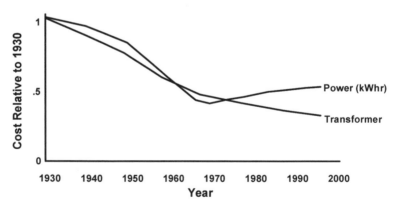

Figure 10.1 Relative price of a kWh (average of six utilities), compared to 1930, over the period 1930 to 1998, and relative lifetime ownership cost of a standard 24 MVA, 138/12.47 kV transformer offered by the industry's largest two manufacturers.

Reason 4: Regulation provided dis-incentives for innovation

The regulatory process and the lack of competition gave electric utilities no incentive to improve on yesterday's performance or to take risks on new ideas that might increase customer value. In fact, they provided several dis-incentives: If a new idea succeeded in cutting costs, the utility still made only its regulated rate of return on investment; if it didn't work, the utility would usually have to "eat" a good deal of the cost of the failed attempt, as imprudent expenses. Thus, the system always asked, "Why rock the boat?"

Furthermore, why would a regulated utility want to use new ideas to lower its costs under a regulated rate of return framework? That just cuts the cost base upon which the utility receives a guaranteed rate of return. In other words, it potentially reduces the utility's level of profits.

Thus, despite the technological progress in electrical, electronic, and computer technologies that occurred in the half century following World War II, progress which wrought spectacular changes in most other industries, in 1990, just before de-regulation began, *electric utilities were still providing their customers with the same product and the same type of service that they had provided fifty years earlier.*[13]

And there are many valuable improvements that could have been made. People who argue that there are no needed or useful improvements possible in electric service comparable to the improvements made in the computer, auto, or home entertainment industries suffer from an inability to envision what is possible. There are literally dozens of examples affecting generation, transmission, distribution, and customer levels of the power system. Only two will be presented here.

Example 1: Automatic Power Delivery Tracking and Trouble Response. By the late 1990s, when the first edition of this book was written, the unregulated package delivery company, Federal Express, had in-truck computers, satellite communication, and advanced information management systems that could track individual package deliveries in real time, anywhere in the United States, all to assure reliable on-time delivery for transactions that averaged only about $10 each. Customers concerned about a particular package could (and still can) telephone or access a Federal Express computerized tracking system and find out, automatically the where, when, and why of their package's delivery.

---

[13] Reliability was incrementally better, and price in real terms was slightly lower mostly due to economy of scale: Homeowners and businesses both bought much more power in 1990 than in 1940, allowing a utility to spread its costs over many more kWh. However, compared to any other industry, costs did not drop as much, or services or performance rise as much, as would have been expected.

At that time, no electric utility in the United States had a system anywhere nearly as comprehensive to track the power it delivered to their customers, even though it already had a wire leading to every home and business, the technology was available, and its average weekly transaction with each customer was far more than the $10 Federal Express charged per package. At most electric utilities, they did not know a customer had problems until a telephone "trouble call" was answered and processed by a human being, then radioed from a central dispatch center to a driver in a service truck, in exactly the same way it was in the 1940s. Yet, systems that were basically the same used in package delivery tracking could have cut power outage times significantly and provided *much* better service. A homeowner does not even have to telephone, and in fact does not even have to be at home, or aware of the problem, for the utility to already be taking action to fix it.

Under the regulated industry structure, utilities had little incentive, based either on improved profits or bigger market share, to try these and many other similar technological improvements. Only in the late 1990s, as de-regulation started to take effect, did utilities start to implement such systems. At the top of their lists of reasons to make investments in this technology is "needed for de-regulated operation and competition."[14]   Today, many utilities have such systems in place and are working on systems that go even further, to track and coordinate the efforts of their forces in the field to optimize customer service quality.

Example 2: Digital Transmission Relaying. The example cited above is far from an isolated instance, as one further transmission level example will illustrate. As a whole, the electric utility industry was very slow to adopt electronic (smart) relays. *Relays* are small, and until the advent of electronics, mostly mechanical, devices that monitor and control the safety, lightning protection, and other contingency and reserve systems in the utility transmission grid. Traditional "electro-mechanical" designs were developed in the 1930s through the 1950s, and are broadly similar to precision wind-up watches, with tiny springs, gears, tension bars, latches, cams, and innumerable miniature moving parts. Millions of these are still in service, and they are marvelous devices, just as the wonderful watches of the 1930s and 1950s were. But they need very careful calibration to work well, are sensitive to dust, vibration, temperature change, and wear, and require frequent and expensive maintenance by very skilled technicians.

By comparison, electronic relays, first commercially developed in the early

---

[14] It is worth noting that, even if the distribution level remains regulated as discussed in Chapter 2, and in Section 10.5, the partial de-regulation of the industry and the focus put on delivery as it now stands alone will spur regulated local distribution companies to adopt such performance improving measures.

1980s, are less expensive and more reliable. Beyond that, they operate much faster and much smarter, to the extent that they provide noticeable improvements in transmission system performance. The reasons why are highly technical, but electronic relays can be thought of as an anti-lock braking system for a power transmission grid. Mechanical relays, no matter how well designed and calibrated, are simply too slow in comparison to the speed with which electricity "can go wrong" in some situations. As a result, they cannot act quickly enough, to put "the brakes on" some types of trouble, e.g., some types of equipment failures and short circuits. Traditional transmission system designs had to avoid any conditions where that might be a issue, by using conservative designs, by accepting the fact that there might have to be occasional blackouts, and by occasionally "de-rating" transmission lines by up to 50% of their capacity. (This is somewhat analogous to a person who owns a truck with poor brakes, who has no choice but to drive slowly and not carry big loads, in order to avoid situations where the truck would not be able stop safely.) Transmission systems equipped with electronic relays can often be "uprated" in electric power-carrying capacity at almost no additional cost, while still providing improved reliability and service quality.

But despite these great advantages, at the time the first edition of this book was written (late 1990s) some electric utilities still had not employed electronic relay systems. The fact is they have had no incentive to improve, because what worked last year will work just as well next year as far as their regulatory priorities are concerned.   Today, after roughly a decade of de-regulation, few utilities would consider using anything but modern electronic relays in any new system, and many have replaced almost all their old relays with digital units.

Reason 5: Competition will improve customer focus

Another theme of de-regulation is that it promotes *customer focus* on the part of suppliers, and increases *customer choice*. Although monopoly-franchise utilities have an obligation to serve all customers, that does not promote the same type of focused, pro-active attention to customer needs that competition does. A monopoly-franchise utility listens to its customers when they explain their needs, and then responds. A competitive electric service company *anticipates* their needs and responds in advance of their articulating them.

Competition and customer focus mean choice, not just low cost

People who ask only, "Will the cost go down if we de-regulate?" miss some of the point about de-regulation and competition in the electric industry. More important than lowering the cost is increasing the customer value. To some customers, lowering cost might be the best way to increase the value of electric service. But to others, additional or premium services might provide greater value, even at a higher cost.

The de-regulation of the long-distance telephone industry is a good case in point. Opponents of de-regulation will argue that long-distance calling costs did *not* go down after de-regulation of that industry. That's not really true, but then they didn't plunge, either – the drop in cost was initially modest although a steady trend has resulted in much lower costs today. But the real point is that de-regulation *provided more choice*. Customers who wanted low price could pick calling plans offered by economy long-distance companies, that did under-cut the rates available prior to de-regulation. But customers who wanted higher quality, more features, extra services, or greater flexibility could choose to pay more, and get more.

What is also undeniable is that the telephone industry experienced a technological revolution following de-regulation, with much of the expertise it gained devoted to providing ways of offering innumerable customizable "calling plans" and a host of new calling services. These are widely and inexpensively available, i.e., call waiting, call forwarding, caller ID, cellular phones, portable phones, and a host of choices in type, color, and features for one's telephones. Almost certainly, had the telephone industry stayed as it was in the early part of the 20th century, these choices would not have been available, at least not as soon or as widely and inexpensively as they were under de-regulation.

Table 10.3 Reasons Why De-Regulation Was Appealing in the Late 20th Century

Regulation not necessary	The primary original reason for regulation, to foster the development of a universal electric system infra-structure, had been achieved.
Electricity prices may drop	The price charged for electricity is expected to drop from one to three cents per kilowatt hour due to innovation and competition.
Customer focus will improve	Good as service quality and customer response were under regulation, competition is expected to give much wider customer choice and more attention to improved service.
Will encourage innovation	A competitive electric industry will provide rewards to risk takers and encourage the use of new technologies and business approaches.
Augments privatization	In countries where the government wishes to sell state-owned utilities, de-regulation enhances the value of their asset as perceived by potential buyers.

It would be grossly unfair to accuse only electric utilities for an unwillingness to take risks and their lack of technological progress and intense customer focus under regulation. They were simply responding to a system of incentives and rules set down by government and society, and they behaved no better or worse than companies in many other regulated industries like telecom and air and rail transportation. The problem was the regulatory system itself. It provided growth and stability when needed, but stability can mean stagnation, and that was ultimately the result for the utility industry.

Thus, by the late 20th century, what needed to be "fixed" was the regulatory framework, and hence, de-regulation became a popular option. Table 10.3 summarizes the reasons why de-regulation became so appealing to so many governments and electric consumers. Often, these were not the reasons cited for de-regulation, however. Most often, price and customer focus were issues heralded as the sought-after improvements. In addition, privatization and simplification of bureaucracy were goals in some nations (see the next section, and Chapter 12).

10.4 GOALS FOR AND EFFECTS OF DE-REGULATION

Chapter 2 provided an overview of how de-regulation works, why it was done, and what the goals were for policy makers that drive the change. This section will look at some of the goals and features of de-regulation in more detail.

Create Competition

In many countries throughout the world, governments have both de-regulated and re-structured their nation's electric power industry. This has been a gradual but steady trend beginning in the late 20th century and continuing into the 21st. For the most part, the wholesale level – manufacture of power (generation) and transportation of bulk power (transmission) – has been "de-regulated." This usually leaves the transmission system as a type of regulated monopoly (i.e., there is only one), its operation heavily defined by its role of enabling intense and open competition at the generation level. This was discussed in some length in Chapter 2. Competition at the generation level is an established norm, and works well, in many countries around the world.

Originally, most regulators intended to move toward retail de-regulation – a framework in which consumers would have a choice among competing providers of electric energy. Many countries went this far, including the UK. However, retail competition is not nearly as common as wholesale-level de-regulation, and may never become widespread in the United States. Regardless, this section discusses the basic concepts of de-regulation and the functions used to implement it at both the wholesale and retail levels.

Unbundling Energy from Delivery: A Summary

A key concept of electric industry de-regulation in nearly every nation that has implemented it is that no one entity should have a monopoly on either the production or the sale of electricity as far as any buyer is concerned: there should be competition in the sale of the power that every single person buys. This definition may seem unusual, in that it stresses the "every single buyer" issue. But that was a key goal, that *every* buyer, not just some or most, would see a competitive marketplace of suppliers vying for his business. And that goal, through a circuitous set of circumstances, led to the requirement that a lot of the focus on de-regulation be on the power delivery grid, not just on the power production and sales.

In many de-regulated industries, just assuring that there is competition in the production of a product will assure that all buyers see a competitive situation. For example, any buyer who wants to avail themselves of "competition" in the supply of cornflakes can go to a grocery store that has competing corn flakes offerings on its shelves. If the range of offerings is not good there, they can go to another. Assuring that competitive corn flakes manufacturers exist pretty much assures anyone and everyone will benefit from the competition. It's the same way in many, many industries.

But this is not the case with electricity. The root cause of this is that electricity cannot be stored – it can't be put on a shelf. It must be consumed at the moment of production, which means buyer and seller must be connected in real time. And that means that a particular buyer might be in a portion of the grid where only one local generator can really serve him, and thus effectively "at the mercy of" that seller. That can happen, in fact it is sure to happen, if the grid is operated in certain ways. The way to avoid that is to: A) make certain that no one owns both generators and the power delivery system, and B) that whoever operates the delivery system operates it in a way that assures nothing like that happens.

It is worth noting that no one, anywhere in the world, seriously proposed trying to create competition with respect to *delivery* – there were no proposals to allow competing T&D companies as there were to have competing generation companies. Rather, de-regulation schemes all envisioned changes in the rules governing how the single, local, regulated delivery system is to be owned, used, and operated, so that it supports a competitive generation market. And everywhere, de-regulation settled on the same solution: de-regulate the generation industry, permitting and in fact encouraging multiple, competing producers of power, but keep power delivery as a single, regulated system in any region, just change the rules for operation of the grid, so that a key goal of delivery system operation is to run it in a way that never permits any one generator to have a "local monopoly" on being able to sell power in an area.

Thus, just about everywhere de-regulation was applied, an early goal was to re-structure the local electric industry so that production (and perhaps retail sale) of power was competitive, while delivery was still a regulated, monopoly franchise business but operated so that it supported wholesale and retail competition.

This meant *unbundling* of the traditional vertically-integrated electric utility structure. Traditional utilities had owned generation, and T&D, and retail sales. Business unbundling separates generation (power production and sales) from operation and ownership of the delivery system. (This was accomplished by dis-aggregation of utilities, see Chapter 2.) It separated *electric energy*, which is a commodity under de-regulation, from *power delivery*, which remained a regulated monopoly franchise.

The changes in ownership, operation, and management of the power grid ended up being the changes that were the hardest to get right, and that took the longest to implement (in truth, they have not all shaken out even at the time of this writing). By comparison, creating competition in power production and getting that to work well proved quick and relatively painless. The background and details on this subject will be discussed at length later.

A big change: Conceptually and functionally

Unbundling had a big impact on the industry, as might be expected, because it required not only a host of ownership, organization, and functional changes, but in addition it led to a change in perspective that led, at least in the authors' view, to some significant if unanticipated changes in power industry management, most of them for the good.

The traditional vertically-integrated utility paradigm did not distinguish between the costs and value of power production and delivery (see Chapter 2). It treated power as a service that was measured and priced only at the point of delivery (the customer meter), and paid for by the customer in a single bill that did not distinguish the cost or value of the power from the cost of value of its transport. Most regulated electric utilities never calculated separate costs for the power and delivery, and never really thought about the value, cost, or pricing of delivery as a separate service. Electricity was a service to customers that simply required a very "long" machine that stretched from generator to customer meter. Utilities had never looked at the costs of transmission or distribution in detail or studied those functions as a separate business. Those functions had always been lumped into "the cost of doing business" of the vertical utility.

"Unintended" functional and conceptual impacts

Unbundling and functional separation of production from delivery forced everyone in the power industry to begin actually thinking of power delivery as a stand-alone function and business. That was new. And it led to great improvement in the business efficiency of power delivery, something the authors view as an unexpected consequence to some extent, or de-regulation.

The people given the task of running the newly dis-aggregated transmission and distribution companies had to develop ways to measure their costs, determine the structure of both their companies and their pricing systems, and operate their now dis-aggregated power delivery businesses in some sort of "cover our costs and make a profit" manner. As they focused on how to make their businesses successful, they began comparing (benchmarking) one transmission company against another, one distribution company against its peers in the industry, and power delivery companies in general against companies outside the industry. They made two discoveries. First, some power delivery utilities were doing a much better job (lower cost, fewer people, better results) than others. Second, it was possible for even the best of these companies to further improve on its cost/performance ratios by applying both traditional and new technical and business tools to determine where and how to improve its efficiency.

As a result, from 1995 to 2005, the business performance (results per dollar) for T&D has steadily improved across the entire industry. Some of the results have been significant – one utility cut spending by over 20% and improved *all* quantitative measures of customer and employee satisfaction and service quality. Performance/price will likely improve more over the next decade.[15] The nearly industry-wide move to asset-management based operations, to activity-based cost (ABC) tracking, and to the extensive use of field-force and system automation are all due to this new focus on making the delivery function and *efficient* stand alone business.

Open Access, De-Regulation, and Competition

In examining the forces that changed, and are still changing, the power industry, it helps to separate cause from effect, and to recognize that three simultaneous changes were and still are at work. While related, each is a unique cause and effect of the overall change in the distribution landscape:

> **De-Regulation**. A better term might be "re-regulation": The rules have changed, but there will always be rules. Again, traditionally governments

---

[15] The reader should not assume that traditional vertical utilities were intentionally sloppy or meant to be inefficient. They adjusted and improved their operation as they saw it and as they were required to look at it by their government regulators. They did not have the tools to measure costs as modern utilities did, nor any incentive to develop them. There is no doubt that the entire industry had settled in to a very comfortable paradigm of vertical management and costing that simply failed to see many of the inefficiencies that 21st-century, de-regulated paradigm has driven out of the system. But that is as much the fault of the regulators and regulatory system as the utilities.

granted each electric utility a vertically-integrated franchise monopoly for commercial electric production, delivery, and sales in its service territory, in return for which the utility had to operate in a regulated environment with cost-based pricing of its product. Under de-regulation, governments are changing the utility environment and the regulations to foster competition. Price for power is no longer cost-based, but market based (in some sense). There are still rules (many of them).

Competition. The fundamental goal of de-regulation was and remains to foster competition among energy producers, i.e., generators, and in many cases also among energy sellers, i.e., retail companies. Competition is more important and fundamental a goal for de-regulation than assuring pricing is market based. It is more correct to regard market-based pricing as the mechanism that de-regulation authorities see as the best mechanism to promote healthy competition.

As discussed earlier, no one has seriously proposed competition in the movement (transmission and distribution) of power, because that is impractical. But competition among energy producers exists in many countries, where the power industries have been de-regulated, including the United States. Pricing in nearly all cases is market-based, although the rules vary greatly.

Open Access.   Under open access, all qualified parties, not just the delivery system owner, have comparable rights to use the power system to move power from one location to another. This is necessary to assure that competition is "fair." Open access is not the sole way to promote fair competition. Competitive local franchise bidding (New Zealand) and "monopsonistic" pools, as in the UK during the first few years of de-regulation, are two other ways. But it seems to be part of the most effective way regulators see to promote a type of competition that is apparently stable under "de-regulation" – one that sends price signals (gives business incentives to the right people) to "fix" problems or deficiencies in the power supply or grid as they develop, thus assuring a viable long-term power industry.

De-regulation creates competition, through the combination of permitting multiple ownership and competitive (market-based) pricing through mandated *open access*. When looking at various countries, or industries, for analogies to what might happen in the electric industry, it is important to identify which of the above three changes was the root cause of any particular alterations in that industry, in order to extrapolate them properly to the electric industry.

For example, the airline industry is often held up as an example of a de-regulated industry, where both success (Southwest Airlines) and failure (Pan American Airways) followed as a direct result of de-regulation. Airlines *are* de-regulated. They *do* have competition. But they *do not* have complete open-access: All airlines are not permitted to fly on all routes. Certain airlines are granted routes, and only limited competition exists on each: American and United own "joint custody" of the Raleigh-San Jose transmission route, etc. Thus, changes that occurred in the airline industry are mostly due to de-regulation and competition, but not to open access. Maintaining a distinction about which of the three causes leads to any particular result is especially important when studying distribution systems, because distribution may not be entirely open access, nor will it be entirely de-regulated.

The need to limit any one generator's market power

Effective competition requires that no one supplier be so big, or have an advantage, that would permit them to dictate prices in some way. To avoid this, government policy makers insist that no company have *market power* – the ability to dictate prices all of the time or under certain conditions (e.g., perhaps by owning all the peaking generation, so it can dictate prices during peak conditions). Around the world, governments that have driven power industry de-regulation have ordered that any really large generation owner – e.g., the only one in a particular region or one that had market power as determined by certain economic and financial tests – break apart its holdings (dis-aggregate them) into several separate and completely independent companies which subsequently have to compete against one another.

The need for open access

As alluded to earlier, joint ownership of generation and transmission facilities (or more properly, joint *control* of both) in a region can create situations where a supplier has market power, regardless of its size or the number of other suppliers. If a supplier happens to control the only transmission line into a city or region, it could conceivably restrict the use of that line by its rivals, meaning only it could sell power there, driving up the price it can command. In the real world, there are few situations that are quite that extreme, but there are many where a generation-transmission owner could influence the situation enough to create a much higher price in a region than if unfettered competition were taking place. To assure that nothing like that does take place, the various generation companies vying for business in a region all need to have *open access* to the grid, with none having an advantage.

The concept of open access T&D is quite similar to how our society's road systems have operated for several centuries. Any person qualified to use the road system can; it is shared by all on what is essentially a "first come, first served" basis. For example, competing package-delivery companies like Federal

Express, United Parcel Service, DHL, and others all deliver packages over the same shared road system. If one company somehow gained ownership or even limited control of a particular road, that would give it a big advantage in the area accessed by that road.

The solution that guarantees "fair" competition in both package delivery and electric power is to assure that all players have equal access to the transportation grid they need – the road system or power grid as the case may be.

One solution the power industry developed, widely used in the US, is for an independent system operator (ISO) to operate the grid in a particular region. The transmission lines might actually still be owned by companies that own generation, but they have no control over them (they are paid for the use of their lines, though). The independent delivery system operator has two duties under de-regulation:

1. To operate the regional transmission grid well, at high reliability and as economically as possible

2. To make sure that competing power sellers and all the buyers have equal access (opportunity) to use the wires.

The power industry traditionally worked at only the first of these goals: reliability and economy have always been priorities. The second is new, a function of electric industry de-regulation and the chief responsibility for which independent operators were created.

Transmission operation has proven most difficult to "get right" under de-regulation

This aspect of power industry de-regulation – the operation of the power grid co-existent with competitive generation – has proved the most difficult aspect of de-regulated policy to get right. The generation level itself proved surprisingly amendable to de-regulation (at least, compared to the dire predictions of naysayers). Competition in power supply is working and working well in most jurisdictions around the world.

But transmission ownership and operation, and open access, have frustrated the industry since the beginning of de-regulation. At the time of this writing, fully ten years after implementation of de-regulation was begun, many regional grid policy-makers are still wrestling with issues that go beyond fine-tuning of rules, to even what type of pricing and operating rule set they will have (e.g., ERCOT). One cannot claim that de-regulated operation of the power grid has been a failure. It has worked, just not well enough to meet all requirements people believe need to be satisfied. There will continue to be major changes in structure and rules for some time.

10.5 COMPARING FOUR APPROACHES TO REGULATION AND DE-REGULATION

Four possible "paradigms" – sets of values and guiding principles – will be examined here to illustrate the spectrum of choices a society has with respect to running its power industry. In its turn, each of the four approaches discussed below entails a greater degree of competition than the one preceding it. The point to be learned here is that any of these approaches can be made to keep the lights on. Presumably, a society choosing one over the other three (or over combinations or variations on them) values the changes that particular framework brings over those that the others would provide.

1. Franchise Monopolies

In this traditional, regulated-utility scenario there is no competition. A particular company is granted a monopoly to be the only company allowed to produce electricity and deliver it to local distribution companies and/or final consumers in a region. Sometimes, a separate delivery company exists, with a monopoly on retail delivery, which is part of, or works with, the monopoly power producer, or may be separate. But there is no competition at either the generation or transmission level. Customers must deal with one, and only one, supplier.

This century-old arrangement provided low risk for the original investors in the power grid and encouraged development of large-scale transmission systems and large power plants. It is doubtful if the industry would have gotten off to as good a start or reached as stable a structure within only a few decades without it. And there is considerable belief around the world (with some justification based on results) that an economy-of-scale – a critical mass that permits the local power industry to become a viable business – can be more quickly established if one owner-driver is in control. Thus, monopoly franchise continues to be an approach to managing power industry operation in many developing countries.

Arguments to continue this approach center on the fact that "the lights stayed on" for a century under the monopoly franchise structure around the world, and that this approach achieves massive economies of scale which should provide the lowest possible cost.

2. Purchasing Agency That Buys from Competitive Generators

This is an interim type of de-regulation, in which the power producers (generators) are de-regulated, fragmented into multiple companies so that none has no market power, and permitted (forced) to compete. But the transmission grid is not de-regulated, because there is only one buyer: a "purchasing agent" who represents all the consumers in an area, and controls the grid, too. The agent "shops around" among the various independent power producers, thereby prompting competition for his business. This arrangement is often called a "single-buyer paradigm," or a *monopsony*. It avoids the need for transmission open access operation, because there is only one potential user of the grid – the buyer agency – and it has total control of the grid. Regardless, energy consumers still see a monopoly supplier, as they did traditionally. No doubt they will be told they are getting the benefits of competition, and they may be.

In most proposals for this type of industry structure, the purchasing agent is usually envisioned as the local distribution utility, which buys power from competitive independent power producers who bid for its business, and buys long-term wheeling contracts over the local power pool. An important part of this arrangement is that the contracts between producer and buyer are almost always envisioned as *very long-te*rm (five years or more).

This framework is almost identical to how many municipal utilities did (and still do) business, but just implemented on a much wider (i.e., national) scale. In the mid 20th century, many a city electric department, having no generation of its own, would have long-term contracts with several large, nearby generator companies (usually, nearby investor-owned utilities) utilities. Traditionally, there was no formal mechanism established for the municipal utility to shop for power quickly and efficiently, as there is today, but they managed through individual initiative. Today, even if regional de-regulation rules provide a way to buy power dependable on a short-run basis, many municipal utilities still prefer to establish long-term contracts.

This approach is often seen as an interim framework – step one on the way from traditional regulation (covered above) to complete wholesale operation as covered next; the government de-regulates generation first, acting as the grid agent while changes at the generation level are finalized, then after a period, the grid is de-regulated in a second step. This was often done because governments realized that acting as the agent/grid operator was a "no-win" prospect for them. The purchasing agent/grid operator never satisfies anyone: buyers complain about high prices they must pay and suppliers complain about low prices they must accept, and the agent gets blamed for all reliability and blackout events. This alone prompted some governments to quickly move one from what is a viable and workable approach to de-regulation of the generation level.

Arguments in favor of this approach are that it gains most advantages of competition while providing a foundation for long-term financing of large, i.e., efficient power plants. By committing a plant's output on perhaps a ten-year contract, the competitive power producer is guaranteed revenues as long as the plant runs, even if later competitors appear with lower prices. With a guaranteed market, the generators can finance by sustaining a high proportion of debt, which reduces the prices they have to charge. Disadvantages are that this approach passes much of the market and technology risk on to the customers, and it limits the benefits of competition somewhat by protecting the plant owners from the effects of technical change and market force that full competition would bring.

3. Wholesale Competition through Open Transmission Systems

This is the situation as currently exists in most of the United States and in many other parts of the world. In many jurisdictions this is the final planned state for de-regulation; in others it is envisioned as an interim step which will eventually lead to full wholesale and retail competition (described in the final part of this section). Here, there are multiple wholesale sellers as there were above, but also multiple independent and competing buyers, with none permitted to have market power.[16]

There is open access at the transmission level and some mechanism, i.e., a wholesale power exchange, that supports efficient bidding and transactions of short-term power contracts of bulk power. Local distribution/retail companies (LDCs) act as purchasing agents for their customers, buying electricity on the wholesale market, picking from among competitively bidding producers and thereby, presumably, getting the best price and conditions they can. However, there is not one central buying agent: while each LDC is the sole distributor of power in its area, there are many in a region trying to find the best bargain among the sellers, creating a rich "competitive" market for power purchases and assuring (when the rules are done right) more competition than in the monopsonistic approach covered above.

These distribution companies still retain a monopoly franchise over the retail consumers in their area, who see no real day-to-day difference from (1) and (2) above except that perhaps wholesale competition drives down the net cost of the power they use. But they still must deal with one local power company whether they like it or not (not necessarily a bad situation in the eyes of many consumers, apparently, based on responses in some regions to (4), below).

This framework is close to the British system operating right after privatization in 1990, and is the model that is envisioned by many in the United States. However, in both Britain and in places in the United States, policy makers are moving past this.

4. Wholesale And Retail Competition

Under this approach there is open access at both the transmission level (open access transmission, often called *wholesale wheeling*), and at the distribution level (open access, or *retail wheeling*). As a result, at least in theory, any supplier can transact business with any consumer. The big change here from the previous three approaches is that retail customers now see a choice of suppliers – retail sales of power is competitive, and separate from distribution.

---

[16] A very large buyer can have similar market power to a very large seller, basically being so much of the market that they effectively can dictate prices – "I control most of the demand, so sell at the price I demand or I'll buy from someone else."

This fourth system, *retail wheeling*, is what countries like Britain, Norway, Chile, Colombia, Australia, and the United States had expected to move to in their original de-regulation plans. Some (the UK) did. Others (many states in the US) have not and may not for some time.

Theoretically, in this framework all final consumers are brought into the market. Retail competition increases transaction costs by necessitating more complex trade arrangements and metering, something that can be handled by computers and proper processes, but that nonetheless concerns many people. For small users, the cost of tracking such transactions may outweigh the benefits of retail competition. Another worry is that the responsibility for poor service may be hard to assign, when the local distribution company is not the retailer as well. Opponents of retail wheeling attack the concept on this ground. Proponents point out that if service is poor, competition will bring forth people willing to fix it for a profit, leading to improvement of service. This framework can be made to work and work well. Where it does, a minority of residential and small customers like it but the majority seems not to care too much – often few switch to another supplier even though they have lots of options. However, larger power users – commercial and industrial – see good benefits from retail competition and weigh heavily into the political process in support of it.

Dis-Aggregation of the Traditional Vertically Integrated Utility

As noted above, unbundling of power generation from delivery, and of delivery from retail sales, is a theme in the more advanced schemes for de-regulation. Electric utilities are expected to split apart into "unbundled companies" in a similar manner, too, with each utility re-aligning itself into several separate companies that respectively focus on each part of the new industry, i.e., power, delivery, retailing. This is called *disaggregation* and was covered in Chapter 2.

Perceptions and Incentives under De-regulation

As mentioned earlier, the people promoting de-regulation and the governments effecting it believe that its benefits far outweigh its costs. In many cases it is difficult to distinguish clearly whether a change wrought by de-regulation will be good or bad – most often it will be a bit of both. But no doubt, de-regulation will change the *goals* electric companies work toward, the *incentives* that motivate investment, and affect in many ways "the way things work."

No Striving for System Synergism

A vertically integrated electric system could be planned, engineered, and operated as a single entity, with all its parts – generation, transmission, distribution, and customer equipment – selected, arranged, and interconnected, with all the equipment designed to work together efficiently. As an example, transmission systems were always designed to allow "economic dispatch" of the generators – they were configured so that the utility's set of generators could operate in their most economical mode all the time.

Under de-regulation, the various parts of the system are now separate entities with separate focuses. They have no mechanism to plan and operate their interaction in such a finely coordinated, synergistic manner. For example, there is no incentive for the transmission system owner/operator to plan the system in conjunction with the generation dispatch.

While there is some truth is this argument, it is not clear that this is necessarily bad. Traditional utility transmission systems might have been built to work well with their local generation, but that often meant they were inappropriate for the regional grid uses they often found themselves handling, i.e., the transmission grid could efficiently deal with its owner's power, but when asked to support power from other utilities, it was a dud. De-regulation asked the transmission owner/operator to take *a bigger perspective*. It is not clear that this will lead to obvious improvements, but it is unclear that it won't either.

In truth, there was little less vertical synergism in traditional electric utilities than might be expected. Transmission and Distribution were two very separate entities – designed by different groups, often at different locations, with little communication between them, and operated by different dispatch operations centers. Similarly, the retail sales department (usually called Marketing or Customer Service) in most vertically integrated regulated utilities seldom worked closely with the Distribution department. Thus, there is not a lot to be lost, although there may not be much to be gained, either.

Shorter-Term Focus

Traditional least-cost utility planning focused on obtaining economy over the long-run. Decisions were often made on the basis of 30-year evaluations of benefit. In a competitive industry, investment is generally done only if the payback is less than five years, not thirty. De-regulation and competition do force a shorter term focus on all construction, business decisions, and other aspects of electric power.

But again, this is not necessarily bad, just different. Many of the problems the electric utility industry faced in the 1980s and 1990s were because it took such a long-term view. The "thirty year present worth minimization" concept of always seeking the lowest cost long-term solution became a dogma that could drive utilities to the brink of bankruptcy. For example, the apparently good, long-term economics of nuclear power plants led many utilities to invest billions of dollars in long-term debt in the 1970s. Consequently, many nearly went broke, when subsequent co-generation costs dropped significantly.

Stranded Assets

Most electric utilities borrowed incredible sums to built large generators that would last decades. The loans also lasted decades, during which technology changed, making those generators, still being depreciated, obsolete. That is why there are stranded assets – generators not yet paid for, which cannot compete on even terms with newer units. Similarly, many T&D systems are full of old equipment because the utility took the long view and bought large, long-lived transformers and breakers, with lifetimes of up to 60 years, because it appeared the best bargain over the very long run.

To put this in perspective, imagine that in 1975 a young family justified buying a large Mercedes, instead of a small Ford, on the basis that while it was more expensive, it was a much better car, with room for a growing family, and that it would last much longer than the Ford. Viewed from a certain perspective, it *is* the low-cost alternative, e.g., less than buying a series of "cheap" Fords over the next few decades. But to be an affordable alternative, i.e., to keep the monthly payments below those of a series of Fords, this family had to finance the Mercedes over 30 years, justifying that term on the fact that the car will last that long or longer. Twenty years later, they own a very workable, reliable, but twenty-year-old car, one that requires frequent and expensive maintenance, gets poor fuel economy compared to modern cars, and has none of the safety and convenience features (air bags, MP3 player) in modern cars. In addition, although not foreseen, they do not have a large family, and thus do not need a large car, period. But they still owe *ten more years of payments* on the vehicle!

Their biggest mistake was to plan for and finance the expensive car over its *physical*, rather than *technological* lifetime (and to plan for only one scenario – a large family). In a similar manner, many of the "good long-term decisions" made under the traditional regulated industry structure look non-optimal from later perspectives, because these purchases were financed over the time the equipment would last, rather than within the time frame before it became obsolete.

Then, too, the long financing and planning periods used by regulated utilities assumed stability of conditions during that time. And over the long range, conditions change in ways not originally anticipated. Returning to the car purchase analogy, maybe the young family buying the big Mercedes sedan never had children, and thus never needed a big car to hold a big family. Or perhaps they had many children, and thus really needed a mini-van. Uncertainty is nearly always much greater than anticipated, simply because one can never imagine all the factors that one doesn't know.

The net result is that many utilities in the United States still owe great amounts of debt on older generators and transmission systems, serviceable equipment that they cannot afford to replace because they still owe a great deal on it, but which has been made partly obsolete by technological progress. This equipment has lower heat rates and higher electrical losses (the equivalent of poor fuel economy) and requires more maintenance than modern equipment. It is also not upgradeable to automation and other advanced features, in the same way an old car cannot be retrofitted with air bags.

Thus, while many people rue the passing of an era when power systems were "designed for the long haul," there is reason to expect that the more pragmatic, shorter-term, "business case" perspective of a competitive market might reveal noticeable benefits, if not outright savings.

It is important to realize that these often very costly failings of the traditional regulated power industry were not completely the fault of the electric utilities. Their every decision was approved, in fact most were *mandated by*, their regulatory commissions. The utilities argue, not without some justification, that they are being treated unfairly. They were told to make such purchases, and now, burdened by high debts and with partially obsolete equipment (stranded assets), they are being told to compete with newer players who have no such liabilities.

10.6 INCREASED SERVICES FROM AND FINANCIAL PRESSURES ON LDCs

In most industries, partial de-regulation moves the still-regulated parts of the industry to behave a bit more like competitive companies. This is/will be the case with LDCs (still regulated) in the electric industry.

For example, when the telephone industry was partially de-regulated in the 1980s, only long-distance service was "opened" to competition. In a broad way, telephone de-regulation was similar to electric industry de-regulation as it has developed in the US. Transmission (long distance) lines were opened to use by companies that competed for sale of services it conveyed. However, local distribution was left as regulated monopoly franchises. Local telephone companies (LTCs) were in charge of both local distribution (wires) and retail sales, the equivalent of LDCs in the electric industry.

In spite of this, LTCs moved a step closer to competition than had been the case prior to telephone industry de-regulation. Within a decade of the de-regulation of the long-distance part of their industry, these local companies offered multiple retail plans to fit customer needs. They gradually brought forth optional services (caller ID, blocking, forwarding, answering services, call waiting). They became more aggressive in driving down price than had been evidenced prior to de-regulation. Even though they were still regulated, they acted in some ways like competitive companies, moving beyond what regulation required of them.

This is already evident in the electric industry and will become much more so over the next decade. Even if de-regulation in the US proceeds no further than the wholesale generation and transmission levels, the LDCs that remain will all tend to act a bit more "competitive" than in the past and the retail sector will benefit. Several large LDCs already offer several grades of service quality (premium service, interruptible at a discount, etc.). Almost all are working hard to improve service quality, in a tireless and never ending round of benchmarking themselves against "competitors" – other regulated LDCs – for top performance ratings in the industry.

Equally important – perhaps more so to executives and occupants of the board rooms – regulated LDCs have taken on a more "business-like" approach to financing and management than the traditional regulated utility ever did. Many LDCs do not borrow money to the limit that their regulated structure, with its ability to "dump" any cost that is justified into the rate-base, will permit. Instead, they operate in a way that keeps debt-equity ratios and other financial metrics nearer to those that any de-regulated company would target.

Investor Psychology

A contributing reason for the increasing focus on customer service is that unbundling put the spotlight on delivery performance in a way impossible before de-regulation. Delivery is now a stand-alone function. What was once buried deep in the mix of a vertical utility is now out in the open, where its performance, cost, and competency are easy to judge. But this cannot explain all of the change, particularly not the great focus on financial measures: many utilities are benchmarking themselves against other LDCs with financial metrics as well as for customer service metrics. Why?

The major reason seems to be investor psychology. Although investor-owned LDCs are regulated, they are now part of a de-regulated industry. In any industry, investors (and the stock analysts who shape their opinions) tend to see all players, regulated and de-regulated, in the industry as somewhat related and to have expectations for both that are somewhat similar. Thus, Wall Street sees LDCs alongside the completely de-regulated generation companies and expects a certain parity in attitude if not performance and risk: they expect to see a certain similarity in financial management and business approach. LDCs that expect their stock value to be high must respond.

Indirectly, this explains much of the basis for customer service quality. Investors cannot look at an LDC's market share, product distinction, or price premiums as an indicator of where it has sound management: its market share is 100% (it is a monopoly franchise), the basic service provided by all LDCs is identical (distribution and retail sales), and price is set by regulation. Customer service quality emerges as about the only sound way that investors can determine if they are putting their money in a company whose management knows what it is doing. Thus, many of the effects of retail competition occur even if the retail level is not completely de-regulated.

FOR FURTHER READING

T. W. Berrie, *Electricity Economics and Planning,* Peter Peregrinus Ltd., London, 1993.

P. C. Christensen, *Retail Wheeling–A Guide for End-Users,* PennWell Books, Tulsa, OK, 1996.

H. L. Willis, *Spatial Electric Load Forecasting,* Marcel Dekker, New York, NY, 1996

H. L. Willis and G. B. Rackliffe, *Introduction to Integrated Resource T&D Planning,* ABB Guidebooks, ABB Electric Systems Technology Institute, Raleigh, NC, 1994.

11

De-Regulation At The Wholesale Power Level

11.1 GENERATION AND TRANSMISSION IN A DE-REGULATED INDUSTRY

How does the power industry work under de-regulation? What do power producers actually do? Who sees that the vast electric grid operates smoothly and reliably? Where do electric customers go to buy electricity, and how does it get from that place and producer to their location? Under the best of circumstances, the answers to these questions are confusing, because the operation of a de-regulated electric industry is more complex than that of a vertically integrated electric utility.

Adding to the confusion is the fact that there are several different ways that a government can decide to order de-regulated operations to work. All are intricate, but each to a different degree and in a different way. This chapter and the next will cover the main concepts of these various structures at the generation (this chapter) and transmission system (Chapter 12) levels, and attempt to show how each works, and, more important, indicate the common threads that run through all the approaches.

Wholesale and Retail

Two "levels" of competition are permitted, and in fact encouraged, in a de-regulated electric power industry. At what can be termed the *wholesale level*, Gencos produce and sell bulk quantities of electric power. Even a small,

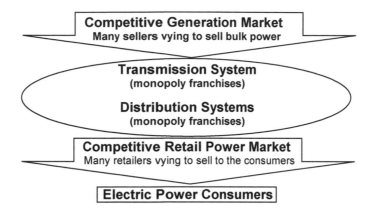

Figure 11.1   The completely de-regulated electric power industry. Power from a competitive wholesale market (top) flows through the transmission and distribution systems (monopoly franchises operated in an open access manner) to a competitive retail market (bottom). Often retailers are the buyers at the wholesale market.

traditional generating plant produces power for 10,000 or more homes. Power is typically sold in *bulk quantities* to other companies, or very large industrial customers, through some de-regulated power market mechanism,   to be discussed later. These bulk quantities of power are moved over the Transco's transmission system(s), which can be likened to a railroad system   for electricity: It can efficiently move substantial amounts a long distance, but it reaches only one or two points in each community, after which "local delivery" must be arranged.

Locally, retail delivery is accomplished by *retailers*, who compete for the business of the consumers in the area by offering low price, good service, and unique features. These are the companies buying power at the wholesale level and arranging for transport to each community where they do business, so that they have power to divide up and sell to individuals locally. Their power is shipped over the local distribution system to their customers' locations. Local distribution companies (Discos) own and operate the distribution system as a monopoly franchise.[1]

Thus, a re-structured, completely competitive electric industry is a sandwich

---

[1] In some nations and states, retail competition is not permitted. Consumers must buy from the local Disco, which maintains a monopoly franchise both on the "wires" and on retail sales and thus fulfills the Resco functions, too. In such cases the Disco buys power on the wholesale market, and competition exists only "at the top" of the industry.

of competition above and below an open access delivery system(s), as shown in Figure 11.1. This structure can be conveniently divided into *wholesale* (generation and transmission) and *retail* (distribution and customer sales) levels.

This section highlights some of the most important concepts in de-regulation re-structuring at the wholesale level. Section 11.2 considers the generation level marketplace, and the various systems used to assure competition and equity in wholesale power. Section 11.3 looks at an important issue in de-regulated wholesale markets: Do buyers bid a maximum price they are willing to pay for power, or do they submit only the amount of power they need, basically saying they will pay whatever price the market settles upon? Section 11.4 discusses some of the details of what is being sold: Energy, capacity, or both. Section 11.5 looks at pricing and price volatility.

New Power and Delivery "Hub" Entities

In order for a de-regulated power industry to work well, and in addition to the Gencos, Transcos, Discos, and Rescos discussed in Chapter 8, two additional entities or functions must be created by the industry's re-structuring:

> *Power market:* There must be some way for power producers to sell their power, and for buyers to buy power: Business control.

> *System operation:* The transmission system can move power from sellers' production points to the buyers' locations, but it must be kept under electrical control.

Both of these functions must be accomplished in one form or another in every de-regulated electric power industry. Both require objectivity and some type(s) of "fairness" or equality of operation. Thus, none of the competitive companies involved (Gencos, Rescos) can possibly fulfill either of these goals. System operation can be accomplished by the Transcos and Discos, under some types of de-regulated structure, but the power market is a concept that was completely alien to the power industry prior to de-regulation. For this reason de-regulation usually requires that one or more new entities be created in one form or another: a *Poolco operator,* to buy power from competitive bidders and run the system fairly; a *Power Exchange* to let buyers and sellers of power transact business; an *Independent System Operator,* to operate the system in a way fair to all users.

11.2 THE WHOLESALE POWER MARKETPLACE

A Completely New Concept: The Power Marketplace

Under de-regulation, some system must be put in place where competitive sellers of electricity can offer their product (power) and transact sales. There are three basic ways this can be done, poolco, bilateral trading, or power exchange. Often they are combined in different ways into a composite system.

Poolco, or Agent Buyer

The Poolco is the only buyer in this truly "one buyer" approach to running the wholesale market. It is a governmental or quasi-governmental agency that takes bids from all sellers, buying enough power to meet the total need, then taking the lowest-cost bidders. Usually, the Poolco operator also has responsibility for running the power system, and it is thus a combined buyer/system operator.

Bilateral Exchange

In this type of multi-seller/multi-buyer system, individual buyers and sellers "make a deal" to exchange power at prices, and under conditions they agree to "privately." However, they may be required to disclose publicly some or all details of their transaction. For example, marketers from Big State Power Generating Company (a Genco) might meet with the managers of the City of Greenville Electric Department and agree to sell it up to 500,000 kW, every hour, for the next decade, at a price that varies from 2.0 to 4.3¢, depending on the season of the year and time of day. These two parties might agree on other, special terms, such as a purchase escalation clause that permits the city to raise the amount it wants to buy to as much as 600,000 kW, by giving one year's notice of its growing need.

Power Exchange (PX)

The government sets up, or causes the power industry to establish, a trading exchange for electric power, which operates much like a stock or commodities exchange. Buyers enter their needs into the power exchange *("I need up to 330,000 kW tomorrow. I'd like to pay 2.3¢/kWhr)* as do sellers *("I have 500,000 kW I'd like to sell at 4¢")*.

When they transact business with the power exchange, buyers and sellers are really talking to "the marketplace," not other individual sellers and buyers. As in a stock exchange, the power exchange constantly updates and posts a market clearing price (MCP), which is the current price at which transactions are being made. Thus, the buyer who'd like to buy at 2.4¢ kWh and the seller who'd like to get 4¢/kWH both know that the current MCP is, perhaps, 3.17¢, and that to make any transaction, they'll have to adjust their offered price closer to that. Thus, the power exchange becomes a power commodities market, with its price point fluctuating depending on demand and supply just like the markets for other commodities.

Note that when buyers and sellers communicate to the power exchange, they don't know whom they may be doing business with, just as when people buy through a stock exchange, they do not know for certain who will be selling the stock they buy. Nor is that matter particularly important to them.

These three market structures are not mutually exclusive. In fact, a bizarre

Table 11.1 Three Types of Marketplace Mechanisms Used to
Establish a Market for Sellers to Offer Power Competitively

Type of System	Number of Buyers	Buyer Knows Seller?	All Buyers Pay Same Price?
Poolco	One	Yes	Usually
Bilateral Trades	Many	Yes	No
Power Exchange	Many	No	Yes

combination of all three could be made to work, but seldom has been proposed.[2] But it *is* common for two of these three mechanisms to be in place simultaneously. For example, bilateral transactions are permitted in California, but the Western Power Exchange (WEPEX) was created to permit buyers and sellers to do business with "the marketplace" on a real-time, next-hour, or day-ahead basis. Similarly, some other de-regulated national or state systems permit bi-lateral trading with Gencos only among large buyers of power, e.g., over 100 MW, and designate a Poolco operator to do the buying, in aggregate, as an agent for the smaller customers.

The details of implementation of these three mechanisms can vary a great deal, too, from one political jurisdiction to another. For example, in systems where bilateral contracts are permitted, how "secret" can these contracts be? Some jurisdictions force the parties to any bilateral power sale agreement to disclose publicly the quantity, the place, the time, and the price of their deals. Others don't. This disclosure requirement affects the strategies buyers and sellers adopt in the marketplace.

Likewise, the time period of power sales traded through the PX varies from one de-regulated system to another. Many power exchanges permit trading of power for only day-ahead and hour-ahead trading. Anyone wanting longer-term purchases must find an entity with which to make a bilateral deal. Other power exchanges permit buyers and sellers to make deals of power for longer periods – even months or years.

---

[2] For example, a government could set up a Poolco that was responsible for running the entire national grid, but also for selling power as "the local utility" in certain parts of the nation (e.g., all rural areas) and which, therefore, takes bids for those needs through its own bidding system. There might be sense to it, since a competitive system typically has no mechanism to encourage investment in the building of a power system in a rural area (see Chapters 2 and 8), and it could thus shoulder that effort itself. Simultaneously, the government could permit bilateral transactions, and set up a PX, in order to gain the advantages of competition in the developed, populated areas of the country.

But regardless, every competitive power industry establishes a "power marketplace" with some form of one or more of the three structures discussed above. Table 11.1 summarizes their key features.

This power market is almost entirely a *financial* function. It has very little to do with electrical engineering or power system operation, but a lot to do with banking, accounting, and transactions tracking. In terms of function, daily tempo, personnel skills, and computer resources needed, power exchanges resemble very closely commodities and stock exchanges, as previously noted.

11.3 DO BUYERS SUBMIT BIDS?

One of the largest differences at the wholesale level, among the different de-regulated electric industry structures set up around the world, is whether buyers (people who want electricity) submit bids, or only demand forecasts to the wholesale market. For example, although California and New York State differ in how the power grid is run and managed, in both systems buyers submit bids for demand. When they contact the local system to buy electric power or reserve capacity, etc., they identify both *how much they want* and *what they are willing to pay.* This is just like a buyer contacting a stock or commodities market *("I need 500 MW of firm capacity from 10 AM to 11 AM and will pay up to 3.4¢/kWH for it").*

By contrast, buyers in the New England system (Massachusetts, Connecticut, Vermont, New Hampshire, etc.) of the United States system, or the England and Wales system, submit only their forecast of *demand ("I expect to need 500 MW of power").* Implicit in this submission is that the buyer will pay whatever the prevailing price will be, at that time.

Clearly, a system where buyers and sellers submit bids is more complicated from a marketplace standpoint, as well as a transaction accounting and "power exchange management" standpoint. Such complexity can be accommodated, however, without too much difficulty in most systems. (The only exceptions are in some power grids where transmission constraints are too tight, as will be described later.)

Fundamentally, though, having buyers submit bids, rather than just forecasts of demand, changes the nature of the entire power marketplace, as well as the concept of electric use. Buyers are part of the competitive process. Use of electricity has a price that the user can vary depending on need and willingness to take risk.[3]

---

[3] It is vital to keep in mind here that this is a discussion of the *wholesale* market. Most individual homeowners, small businesses, etc., do not understand how to buy competitively by submitting bids, and are not interested in "playing" in this arena. Hence, the retail level (see Chapter 12) is quite different. Those dealing in power at the wholesale level presumably have the systems and knowledge of what they are doing. If

Power Exchanges Require Buyers to Submit Price Bids, While Poolcos May or May Not

Almost by definition, all true Power Exchanges require both buyers and sellers to submit price bids, as well as volume (amount) for their requests. Thus, in California, buyers contacting WEPEX have to identify both how much power they want, and the price they are willing to pay.

Usually, a Poolco system is the opposite in this regard: Buyers do not submit price bids. Both ISO-New England and the England-Wales systems are essentially pure Poolco systems, whatever they may be called locally. The job of the system operator's (SO) is to run the system to assure the best reliability and lowest average cost for users. The SO will try to satisfy all buyers by purchasing power as cheaply as possible and averaging that cost over all buyers, so that everyone gets equal benefit from this optimized buying. But some Poolcos require buyers to submit bids. For example, in the New York State system, buyers announce the price they are willing to pay to the combined PX-ISO system operator when they request power. Essentially, New York is a Poolco system, wherein the SO is responsible for operating the power grid.

But New York also runs a type of power exchange, wherein the SO matches buyers and sellers who submit bids using a "non-discriminate auction," a "stock-market" type of bidding process where everyone transacting business at any one moment does so at the same price, and everyone knows that price because it is posted. However, the combined New York PX-ISO can also adjust seller and/or buyer volume as needed using "contracts for differences," which essentially gives it the flexibility necessary to operate the system as it deems necessary (see Chapter 10), so that it is still, overall, somewhat of a Poolco system, both buying power and operating the system from a coordinated, macro, perspective.

Advantages and Disadvantages of Each System

The question of whether buyers should submit price bids has both functional and philosophical/policy implications. Where required, the jurisdictional authority is often trying to encourage as much competition and market efficiency as possible, and wants users to respond to "efficient price signals" as often as possible. Generally, most observers believe that the California system was set up to maximize competition in every aspect of the marketplace.[4] Not surprisingly, it requires buyers to submit price bids when buying power.

Where buyer price bidding is not allowed, the jurisdictional authority may have taken a slightly more "socialistic" perspective on the power system

---

not, they will be displaced in the marketplace by those who do.

[4] Experts, particularly economic theorists, may quibble with this assessment, but the authors believe that except for a few decisions on details, made for pragmatic reasons of simplicity, the evidence shows this is quite true.

operation, believing that it should try to do the best for all users, and not leave everything up to the individual buyer. Sometimes this view is simply political: The government believes that it has the responsibility to assure that demand is met, and believes that individual buyers will not be able to do so as well as its coordinated approach will. It may also be concerned about small buyers being forced out of the system if it becomes too competitive.

Often the decision to set up a system in which buyers do not submit price bids is pragmatic, driven by grid considerations, even if the regulatory authority would probably prefer a more competitive system. The power system in the northeastern United States is such a tangle of interconnected transmission lines, and so constrained by the myriad inter-related operating limits of the many transmission lines involved, that just operating the power system to enable the lights to stay on is a major challenge. In this environment, ISO-New England and its participants decided that the complexities of a "buyer bid" system would result in an intolerably complicated, and probably unmanageable wholesale market system. Regardless, it certainly would take much longer to get such a system up and running. Thus, in New England, at least initially, buyers submit only demand, not price bids, to the local electric system operator.

11.4 BUYING ENERGY VS. BUYING CAPACITY

Both *electric energy* and *electric capacity* are being bought and sold in the competitive generation market. Both are measured on the basis of quantity (100 MW, 500 MW), and both are sold on the basis of time, by the hour or half-hour, although usually one can buy for longer periods, such as a year, or a decade.

In most de-regulated structures, electricity sales are also tied to a location, by identifying a *point of transaction* for each sale. This location might be the seller's generating plant, the buyer's site, or some intermediate point agreed to by both *("I want 500 MW delivered to the high-voltage transmission bus at the California-Oregon border")*. Still other regulatory structures define locations implicitly, so that they are not an issue. A Poolco "takes delivery" of the power at the generating plants where it is produced, with the Poolco operator worrying about getting it from there to where it is needed.

Buyers can buy only electric energy, only electric capacity, or both. This is done in a "sideways" manner, by specifying different levels of *firmness* and a power sale contract. Firmness is the *reliability* behind the power's availability.

Firm Power

Firm power contracts mean that the seller is assuring the buyer that he will maintain sufficient *capacity* to supply the buyer with the electric energy ordered, even if the buyer is not using all that power at the moment, and regardless of any opportunities or contingencies that occur. Such a contract might have financial penalties for non-delivery.

Interruptible Power or an Identified Degree of "Firmness"

Contracts for power sales between buyer and seller often include very explicit terms defining the conditions under which the seller will "be there" for the buyer. For example, a seller might be willing to sell power at a tremendous discount, perhaps as much as 80% below normal price, with the understanding that he can halt the flow of this power at any time, without prior notice, if he finds someone willing to pay more for the power.

While this might seem like a poor deal for the buyer, if he has a *variable need* for power, the very low price quoted might be attractive enough to put up with the inconvenience. For example, maybe the user is a metal recycling center that operates metal shredders, and needs only two hours – any two hours – of operation per day to meet its needs. Interruptible power makes sense in such a situation.

In purchasing firm power, the buyer is acquiring both energy and capacity, but in buying interruptible power, the buyer is procuring only energy, not capacity. If this distinction seems confusing, one final category of sale, reserve capacity, will make it clear.

Reserve Capacity

To illustrate reserve capacity or backup power availability, we suppose that a hospital – a type of buyer that typically demands an extraordinary level of reliability – contracts with Bald Eagle Generating Company for 10 MW of firm power. Bald Eagle has a track record of good service, but just in case of trouble with Bald Eagle's generators, or the transmission system leading out of its power plants, the hospital makes a capacity reserve contract with Old Reliable Generating Company of New England.

In buying reserve power, the hospital is ordering *capacity*, not *energy*, from Old Reliable. If things go well, it will not need the energy that Old Reliable's generators can produce, and thus it is not really ordering energy. Both parties expect that none will be delivered. But Old Reliable will keep 10 MW of generation available just in case, and in exchange, the hospital will pay it a reserve capacity fee of whatever they agree to in their contract.

In this situation, Old Reliable now has two options. It can simply keep 10 MW of generation available on standby for the hospital. Or, it can use that generator to produce 10 MW of energy, *and sell that energy to someone else on an interruptible basis.* The hospital bought only the capacity of that plant, and as long as it is available when needed, the energy is not of any concern to it.

Figure 11.2 shows the three basic combinations of energy vs. capacity purchase that can take place. A contract can specify whether the buyer is purchasing energy, capacity, or both, resulting in the three types of contracts shown. The fourth combination, no purchase of either, is simply "no sale." If a company is not buying energy or capacity at a minimum, it is not buying anything.

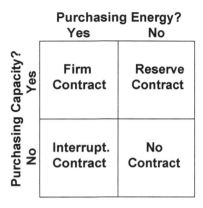

Figure 11.2 The basic types of power sales contracts and their relation to the purchase of capacity and/ or energy.

Grades of Interruptible or Firm Power

The discussion above and Figure 11.2 show only "firm" and "interruptible" types of power, but, in fact, a competitive marketplace encourages all manner of innovative grades of firm, semi-firm, limited-interruptible, and "very interruptible" power contracts.

For example, a Genco could offer various levels of commitment to a series of buyers. Suppose it owns ten 100 MW generators. It can offer one customer top priority among all its customers *("As long as I have one generator available, it's yours. We never expect to have all our generators fail at one time, so we'll commit to always having power for you, period.")* A second customer gets next priority, which is still outstanding reliability *("We figure the likelihood of having so many failures that we can't serve you will be only once in every 62 years")*.

At the opposite end of the scale, the Genco would make a contract at a very low price for a "last in line" customer *("Each of our generators will be off line for maintenance at least one week each year – that's ten weeks at least, plus we'll drop you anytime there are unexpected failures with any of our units that are up and running and needed for other customers. We figure you'll get energy under this contract only about 39 weeks out of the year")*.

There are innumerable other levels of firmness and interruptibility: Contracts can vary in commitment by time of year: firm in winter, interruptible in summer, or even day of week: firm on weekends, interruptible on weekdays, or vice versa. Some contracts might have a limit on the amount of interruption (number of times per month or year) or other conditions, as buyers and sellers try to do business in a way that meets each other's needs.

Definitions Vary: One System's "Semi-Firm" May Be Another System's "Interruptible"

Usually each wholesale marketplace system develops its own very explicit definitions of the various levels of capacity and energy commitment, and how they are interpreted. Thus, "firm" in one system may mean nothing in another, or have a very different meaning. Similarly, "capacity," "energy," and "interruptible" as well as "standby" and myriad other terms will be explicitly defined for use in that operational and regulatory jurisdiction.

11.5 HOW IS WHOLESALE POWER PRICED?

In a marketplace that allows bilateral contracts, there will be very few generalizations possible about how power is priced. Innovative pricing schemes, bartering or energy swaps, and complicated deals will be determined only by rules or limits set by the regulatory process, and those will vary from jurisdiction to jurisdiction. Some bilateral contracts might not specify a price in normal terms, but a trade for another commodity *("I'll supply your refineries all over the United States with interruptible electric power, and in return you'll provide me with 400,000 cubic feet of natural gas per hour delivered to my Lappet, Texas generating plant.")*

Table 11.2 Different Ways that Electric Power Is Offered and Sold on the Wholesale Market (in Generally Decreasing Order of Price per kW)

Commodity	What Can Be Offered and Sold
Control of output	The Genco turns over control, of a generator's energy and capacity, so that the buyer can track load .
Power	The energy produced by a generator is offered for sale, subject to other commitments of its capacity.
Capacity reserve	Availability of the generator "immediately" (i.e., it is already running). Capacity and energy together = firm energy.
Reactive power	Generators set to produce "leading VARS" instead of power, solving reactive power problems, but using up capacity.
Standby service	Availability on short not immediate notice (15 minutes). It is not running, but someone is at the site ready to start it.
Cold reserve	Availability on a longer basis (4 hr. Someone has to drive to the site, check it out, and start it).
No run	Payment not to run (lost opportunity payment).

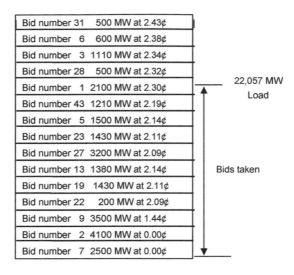

Bid number 31	500 MW at 2.43¢
Bid number 6	600 MW at 2.38¢
Bid number 3	1110 MW at 2.34¢
Bid number 28	500 MW at 2.32¢
Bid number 1	2100 MW at 2.30¢

22,057 MW
Load

Bid number 43	1210 MW at 2.19¢
Bid number 5	1500 MW at 2.14¢
Bid number 23	1430 MW at 2.11¢
Bid number 27	3200 MW at 2.09¢
Bid number 13	1380 MW at 2.14¢

Bids taken

Bid number 19	1430 MW at 2.11¢
Bid number 22	200 MW at 2.09¢
Bid number 9	3500 MW at 1.44¢
Bid number 2	4100 MW at 0.00¢
Bid number 7	2500 MW at 0.00¢

Figure 11.3   The Poolco takes bids in order of cost, but charges buyers and pays all bidders the market clearing price – the highest price bid by any bidder whose bid was accepted. Here, a set of bids has been sorted in lowest to highest cost, against a need for a 22,057 MW load. The market clearing price is 2.3¢/kWH. All successful bidders (numbers 1, 43, 5, 23, 27, 13, 19, 22, 9, 2, 7) are paid at 2.30¢/kWH. Bidders 31, 6, 3, and 28 make no sale because they bid too high.

Sales generally deal with energy, capacity, and other commitments, as shown in Table 11.2. These sales are measured in two ways: How much (kWhr of energy, kW of capacity) is sold, and how long for the duration of usage (by the hour, or quarter-hour). Usually, power is identified as to amount, and price, on a quarter hour, half-hour, or hour basis. However, in some electric marketplace systems, participants actually buy and sell "schedules." For example, while the amount of power is specified on an hourly basis, the "transaction" between a seller and a buyer involves 24 hourly demands for an entire day, or some longer period.

Competitive Bidding

In a competitive system, sellers (and perhaps buyers) submit bids: Basically within the rules of the local system they decide what price to bid. Thus, price is really set by "the market." Individual buyers and sellers determine how much risk to take by bidding high (if a seller) or low (if a buyer).

In Some Systems, Every Seller Gets the Same Price

In some systems, every seller gets the same price. For example, in the England-Wales system, competing Gencos submit bids for the amount of power available and the price they are willing to take. The Poolco operator then "stacks" the bids in order from lowest to highest cost, and takes them in order of increasing cost, up to an amount of generation needed to satisfy the load (see Figure 11.3). The cost/kWh of the final bid taken defines the *market clearing price,* which all bidders are paid.[5] Thus, operators of "must run" plants and any others that want to take "whatever I can get," simply bid *zero*. In this system, competitors still have a compelling reason to bid as low as possible: Bid too high and you are not included among the sellers. This system, although perhaps bizarre to the lay person, makes a good deal of sense. When all the rules are known to all participants, as is the case here, it makes for a very "fair" marketplace.

The Trading Floor: Electricity As a Commodity

In the de-regulated structure discussed in the previous sections, electricity is essentially a commodity: goods bought and sold by quantity, and (at least within its categories such as firm, non-firm, etc.) indistinguishable in quality or characteristics from one batch to another.

Electricity is bought and sold on *trading floors* in essentially the same manner that wheat, pork bellies, soy, and other commodities are transacted. All of the familiar aspects of commodities trading are involved, including forward contracts, a futures market, and price hedging. Many former electric utilities, such as Louisville Energy (the de-regulated energy trading arm of what was formerly Louisville Gas and Electric), expanded rapidly into de-regulated trading. In fact, Louisville Electric moved more power across its unregulated trading floor than through its (still regulated) T&D system.

Price Volatility

One of the advantages of a regulated power industry is price stability and predictability. Buyers and sellers know the price of power in advance and can depend on their price estimates to be fairly accurate: prices do not vary quickly or unpredictably. By contrast, in a competitive power market prices fluctuate constantly (from hour to hour in some regions) depending on market conditions, and often drop or spike suddenly for inexplicable reasons, just as prices do in a stock market. This price volatility makes planning more of a challenge for users and suppliers and increases the business risk (hence costs) they perceive.

---

[5] In a non-discriminate auction, all bidders are paid the market clearing price. In a discriminate auction, they are paid only what they bid. A non-discriminate auction method has been selected for many Pools and power exchanges because, overall, it provides lower and more stable costs. A discriminate auction provides bidders with more incentive to "game" the system by bidding higher.

Under many de-regulated marketplace structures, particularly those that maximize "competitiveness," the price of power can be extremely volatile, changing with market or system conditions in a very dynamic, unpredictable manner. In addition, transmission prices (for wholesale delivery, see Chapter 10), while regulated and based on cost, can be somewhat volatile in their own right, changing quickly from hour to hour as power flows interact with system constraints (congestion).

Figure 11.4 shows the peak cost of power by day for one de-regulated competitive market. Note that prices rise and fall sharply for many reasons, among them concern if not panic among uninformed buyers, e.g., a sharp rise in the beginning of November, when a major generation plant went out of service unexpectedly, causing a sharp reaction in the market, although there was still plenty of power. (Prices are at their peak during peak season, beginning in late November.) The dotted line shows annual average cost before de-regulation: Average electric cost has declined. Volatility in power prices is due to market conditions: demand, supply, and contingencies in supply or demand.

Figure 11.5 shows the cost of power delivery from outside the system to a particular bus inside an ISO control area, for a location in coastal Texas. As congestion occurs, the price for delivery skyrockets, ultimately reaching a cost far greater than the price of the power itself. Various economic and operating brakes can be placed on price to try to control its volatility, but these will likely work as well as they do in the stock market: Prices are less volatile than they would be without controls, but ultimately some price volatility has to be accepted in the marketplace.

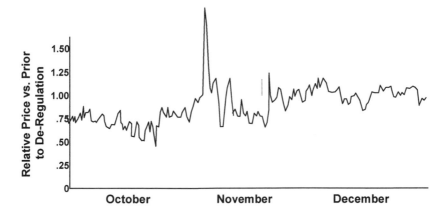

Figure 11.4 Relative daily peak price for one de-regulated market in its second year of operation. Costs average 7% less than before de-regulation but are quite volatile.

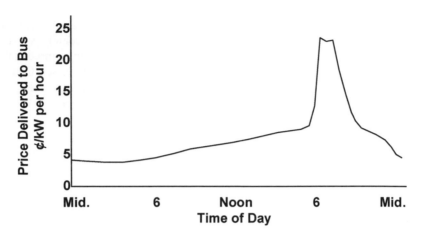

Figure 11.5 Price of power delivered to an industrial plant on the edge of a large grid. Late in the day, during system peak (6 – 8 PM), transmission constraints ("congestion") tax the system and, under locational based pricing rules, create high costs for the plant.

11.6 SUMMARY

De-regulation's goal is to create competition at the wholesale generation, or bulk power level. In *every* de-regulated, re-structured system, the wholesale generation marketplace is competitive. In some systems only sellers submit bids, with buyers merely submitting their needs, and a coordinating (Poolco) buyer purchasing power for all and allocating power to every buyer according to his need. In what are typically considered to be more competitive systems, buyers bid for power, identifying how much they are willing to pay. In both types of systems actual function of the marketplace is complicated, with firm, non-firm, and other forms and nuances of "interruptibility" defined in contracts. Competition in the wholesale sector has spurred considerable activity in the United States, as well as elsewhere, as borne out by the following statistics:

- Competitive power accounted for *half* of all capacity growth in the United States between 1990 and 2005.

- More than $220 billion has been invested in the competitive power industry as of March 2005.

- An estimated 50,000 MW of new competitive power capacity, capable of providing wholesale power at prices in the range of 2¢ to 2.5¢ per kilowatt hour, will come on line through the year 2005.

12

The Power Grid in the De-Regulated Industry

12.1 INTRODUCTION

In order for the de-regulated, competitive power market discussed in Chapter 11 to work, buyers and sellers must be able to move the power they have transacted from the point of production to the point of consumption. They need access to the transmission grid in order to do so. Thus, in every competitive power marketplace, some manner of assuring access to transmission usage is given to all users. In Poolco systems, this is not really an issue: The Poolco operator manages the interconnected power grid so that all users get power. But in other systems, including those with Power Exchanges and those that allow or encourage bi-lateral trading, the transmission grid must be available for use as needed by all users, on some equitable basis.

Beyond this, however, the power grid must be kept operating. As is covered in Chapters 9 and 16, keeping the disparate parts of a grid interconnected, and functioning as a whole, is not an easy task. It requires considerable skill to balance the multiplicity of operating considerations, to keep equipment in synchronization and balance, and to anticipate in advance every contingency.

Thus, in every de-regulated power system, a "System Operator" of some sort "drives" the system, keeping it on a sound operational basis, transporting the power that has been requested to be moved, and providing equitable access to all qualified users, according to the rules of its particular system.

This chapter looks at the interconnected system, basically the power grid, under de-regulation. Section 12.2 examines the evolution of the power pool operations center into the modern, de-regulated "system operator" and the various mechanisms used to implement that. Section 12.3 discusses the services actually being sold, or given, at the transmission level, including power transmission and ancillary services. Pricing methods for transmission, and several of the major issues involved in setting price, are discussed in Section 12.4. Section 12.5 discusses locational pricing, by far the most complex approach but one that appears destined to become best practice. Section 12.6 gives some final comments.

12.2 THE WHOLESALE TRANSMISSION LEVEL

An Old Concept: The Power Pool, or System Operator

Moving power is a *service,* provided at the wholesale level by the *transmission system* (power grid) run by a *System Operator (SO),* a control center staff of expert engineers and power system operators. The SO must run the power system, making sure that it works in a stable and economical manner, and that all buyers and sellers have "equal access" to it. In some de-regulated structures, these duties are performed by the Poolco operator, but they are an identifiably separate set of tasks from the Poolco function of buying bulk power, tasks dealing purely with *moving bulk power.* Unlike the power exchange, which is different from anything seen in the power industry prior to de-regulation, the *System Operator* is very recognizable, a form of the traditional power pool operations center.

During the middle half of the 20th century, the advantages of economy of scale led utilities to pool their resources by tying their individual transmission systems together with high-voltage interconnection lines. This resulted in large power pools that mingled the electric flows from all utilities in a state or region. *Power pools assured high reliability at low cost.* When something failed in one system, another utility's system could cover the shortfall in power until corrections were made. In order to assure that the power pool worked well, the member utilities created a jointly-owned "pool operations" center, from which power engineers and operators monitored and controlled the "grid" over their entire region, e.g., Texas, the Northeastern United States.

Power pools provide so much more reliability without detriment, and sometimes with an increase in economy of operation, that no one has ever proposed anything except continuing their existence in some form or another. In addition, a power pool that covers a very wide region permits more buyers to do business with more sellers, creating a larger "marketplace," which, according to many economists, ought to create more competition.

Thus, nearly all de-regulation re-structuring plans call for a single System Operator (control center) that monitors and operates the interconnected power system that extends over a wide region – England and Wales, California, Texas, the Northeastern portion of the US, the Southeast. Under de-regulation, this pool, or regional, control center metamorphoses into a slightly different entity than it was in the past, but one still quite recognizable as a classic "power system operators center."

One familiar aspect is that the System Operator *runs, but does not own,* the power system. Traditionally, power pool operations centers were jointly owned by the utilities involved, who kept title to their individual parts of the power system, but turned over some aspects of operational control to the power pool operator. Under de-regulation, the transmission grid might be owned by either the government (England and Wales), or by many different utilities (Transcos and Discos) just as it was in the past (California). Either way, the System Operator merely runs the system.

The actual name given to this entity varies. But whether called the Pool Operator, the System Controller, or Independent System Operator (ISO) in the United States, it has four major responsibilities regarding the power grid: system security, delivery of power as promised, cost recovery, and fair (equitable) access.

System security

The systems operator's top priority, overriding everything else, is to keep the power system operating as an interconnected system. This means keeping all the transmission lines, substations, and generation "on-line" operating in synchronization.

This is basically the same work that power pool operators have done for decades, complicated somewhat by the fact that now there are more players buying and selling and moving power through the system than in the past, making the system slightly less predictable, and adding a delay in gaining the co-operation of all involved during system contingencies, etc. In one sense, system security is the most important function of the System Operator: If it is not made secure, the power system does not run, there is no market for power, and none of the other issues discussed here matters at all.

Delivery of power

The open access power grid exists to deliver power for its users (the buyers and sellers moving power across the grid). And except during emergencies that force curtailment of service, the System Operator will try to deliver all the power transmission services requested by all users. A large power grid may have hundreds, if not thousands of individual power delivery transactions occurring simultaneously: User 34 moving 350 MW from point 371 to point 64; User 71 moving 84 MW from point 123 to point 512; User 6 moving 890 MW from point 2 to point 7, and 700 MW from point 45 to points 300 and 295; and so forth.

Cost recovery

Along with delivering power for the users, the system operator's responsibility is to bill these users for both cost recovery of his operating expenses, and for the use of the network – the owners of the lines deserve reasonable payment for the use of their lines. There are many ways transmission prices can be structured– zonal pricing, marginal locational pricing, "pancake rates," and many others that will be discussed later in this chapter. But regardless of what pricing structure is used, someone must track who uses what, and send them the bills, collect all fees and payments, and pass appropriate revenues along to the owners of the transmission lines.

Fair transmission market

The System Operator must ensure "fairness" in access to and use of the grid. If a pure Poolco marketplace is selected, this isn't an issue: There is only *one buyer*, and thus *one user* of the system for whom its availability and performance can be tailored.

But in any de-regulated industry structure, where there are many buyers, competition for usage of the best parts of the transmission system is probably inevitable. The System Operator must set up a mechanism whereby everyone with a need to use the grid has an equal chance to do so. Usually, this is by some form of "first come, first served" reservation – a system whereby people can reserve transmission use an hour, day, or perhaps weeks ahead, implemented in a "public disclosure" manner where every user knows pretty much all the facts about who, what, when, where, and how much other people are using and paying for the grid.

Table 12.1 summarizes these four requirements. The System Operator faces many daunting challenges in performing them. But the SO has many resources, including a large staff of experienced power system operators and engineers, a fully computerized control center, and a massive remote data collection system to monitor, analyze, and control the power system.

12.3 WHAT IS BEING SOLD AT THE TRANSMISSION LEVEL?

Primary and Ancillary Services

The System Operator, in conjunction with the transmission grid, is providing a *service*, i.e., moving power from one place to another. Without this service, a competitive market cannot exist and no one can get power. Buyers and sellers have no other way to move the commodity from production point to point of use, except to use the transmission grid. But while moving power is the primary

Table 12.1 Four Major Goals of the System Operator of an Open Access Grid

Responsibility	Type	Comments
System Security	Electrical	Operator must ensure that the power system continues to operate in a stable, economical manner, a traditional, if difficult, power engineering function.
Power delivery	Electrical, financial	Provide the power transportation services requested of it by buyers and sellers in the competitive wholesale marketplace.
Transmission pricing	Financial	System Operator must determine and post the prices for transmission usage, offer to reserve or sell usage, track, bill and settle with users, and pass on revenues to transmission owners.
Assure open access	Public	System Operator must run a "reservation" system for use of the grid, which does not discriminate for or against any competitive player in the market, but offers everyone equal access to the Grid.

Table 12.2 Primary and Ancillary Transmission Services and Their Mandatory vs. Voluntary Nature Under United States FERC Requirements

Transmission Service Provided	System Operator Responsibilities	Mandatory* for S.O.	Buy	Prov.
Power transportation	Moving power from one location to another	X		X
System operation	Monitoring and control of the power system	X	X	
Reactive power	Balancing reactive power needs of system locally	X	X	
Losses makeup	Providing power to move the power	X	X	X
Energy imbalance	Making up for supply-demand shortfalls	X	X	X
Load following	Compensating instantaneous load fluctuations	X	X	X
Operating reserves	System backup in case of generation failure	X	X	X
Supply reserves	Backup supply to load in case generator fails			
Dynamic scheduling	Monitoring and control signals for load following			
Black start	Providing help in starting "cold" Genco plant			

* An X under S.O. means the System Operator is required to provide this. An X under Buy means that the transmission user is required to buy this from the System Operator. An X under Prov. means the user has the option to provide this himself (perhaps buying it from a third party) or to buy it from the System Operator.

service, it also involves a number of *ancillary services,* additional ones to support the main one, moving power. Many are essential, without which the interconnected grid won't function, or the power won't flow. Ancillary does not imply these services are of secondary importance. Instead, it means their sale to transmission users is subsidiary to, or contingent upon, the sale of the main service, moving power.

Capability vs. Capacity

Every transmission line, transformer, and other equipment unit in a power system has a *capacity rating* – the maximum electric load, that it can sustain. Usually, the capacity rating of any particular transmission route is defined as the minimum capacity rating among all the equipment in the chain of line segments, transformers, etc., from one end to the other. But the *capability* of the power grid to move power from one end of that line to the other might be much less. Often, system conditions, like voltage, security, or stability concerns, or reserve commitments to customers limit the power flow on a line to much less than its rated capacity.

In other cases, the available capability on a line can be more than its rating, at least in one direction: If a line with a 360 MVA rating is moving 300 MVA in one direction, a user might schedule as much as 660 MVA of flow in the other direction. The first 300 would just reduce flow on the line to zero.

Customers of an ISO or other type of transmission system operator are buying *transfer capability* – the ability to move power between points. They are not buying *capacity*, even if they buy or reserve capacity, i.e., a firm commitment that the capability, as it is called by some utilities, will be there.

Unbundling of Transmission Services

Under the de-regulatory orders and re-structuring mandated in the United States, the Federal Energy Regulatory Commission (FERC) ordered providers of transmission services to *unbundle* ancillary services that are part of moving power. Unbundling means that the System Operator must identify each of several different services that are part of moving power, but are recognizably distinct functions or tasks. The provider must price each one individually, and offer it for sale as a separate item. [1]

---

[1] The auto industry provides a good analogy here. Air conditioning and power windows are items ancillary to the basic product, a car. They can be bought, but only in an ancillary manner, i.e., one can't buy the power windows without the car. Some manufacturers, such as Cadillac, bundle these items with the car, i.e., they are "standard" equipment and the car's price includes them, but the amount they contribute to the overall price is not identified. Other manufacturers *unbundle* these items, listing their prices separately, and usually, but not always, offering them as *selectable options* ("AC

Mandatory vs. voluntary nature of the various ancillary services

Separate identification and listing of services in an unbundled manner does not necessarily mean that transmission users have to buy each unbundled item, or that they can buy that item alone. In the United States, FERC distinguishes ancillary services by three categories of mandatory or voluntary nature, as shown in Table 12.2. First, there are some services that are so vital that, without them, the power system will not be able to function. The System Operator is required to offer these services (X under S.O. column in Table 12.2). Requiring the System Operator to provide them assures that ancillary services will not be a barrier to open access for any user ("Gee guys, we're sorry, but since you don't have load tracking capability, we can't allow you to use the system."). The System Operator is *required* to offer for sale all ancillary services that are *necessary* to operate the system: They are a mandatory offered service.

Since these services are needed for good operation of the system, the buyer is required to make arrangements for them, but not necessarily required to buy them from the System Operator. He can provide them himself, purchase them from the System Operator, or from someone else, perhaps a Genco.

Several services are considered unique, in that no one else but the System Operator can provide them. These are mandatory purchases on the part of the buyer, who has no choice but to buy them from the System Operator. The major services are described below.

Transportation of power

The primary service provided by the System Operator/Transmission System combination is the transportation of electric power from one location to another. This is accomplished by routing the electric power onto high voltage transmission lines and controlling that flow from the point of origin to the point of delivery. The transmission system is basically a railroad for electric power. In both industries there are many "lines," interconnected at key points along paths where the routing can be changed. In both industries, they are called "switchyards" or "switching stations," and there are usually many transactions, i.e., cargo trains, or quantities of power, using the system at the same time by dint of the system operator's careful scheduling.

---

costs $895, we sell the car with or without it") or sometimes requiring them, even though the price is separately identified ("Sorry, the $210 emissions control package is required.")

Similarly, power delivery can be priced and sold with or without ancillary services, such as reactive compensation, losses' makeup, reserve backup, etc. The US government ordered that prices be listed for all items, and that in some cases transmission users are not to be forced to buy the options along with the primary service.

System operation (scheduling, system control, and dispatch)

As stressed earlier, it takes skilled personnel, large amounts of monitoring and data communications equipment, and a massive computer setup to operate the power system. Providing this service requires money, and any user of the power grid is billed a share of the cost. In the US, and everywhere else, as far as the writers can determine, the service is always mandatory: The buyer of transmission services has no choice but to purchase and pay for it.

Reactive power supply and voltage control

Reactive power is nearly impossible to explain to non-electrical engineers. The best simple analogy is that given in Chapter 2: As power flows through an electric system, turbulence created as its passes through wires and transformers causes a type of "electrical foam" that takes up space in the wires, creates higher losses, and lowers the voltage. This electrical foam is called VAR (volt-amps reactive) power, and is more than worthless. At best it is a nuisance, and often a hazard to safe system operation. At worst, it leads to poor voltage, instability, and, in severe cases, voltage collapse and blackouts.

Certain electrical equipment, along with some active control measures, can diminish the creation of VARS on the system, or remove them. This apparatus costs money, and its operation requires personnel, data communication, and control systems, all of which the System Operator must bear. But without these measures, things don't work very well. Thus, VAR-makeup, or reactive power loss compensation, is another mandatory service in the United States: Buyers have to pay the system operator to provide it.

Real power losses

Moving a train through a railroad system requires fuel to power the locomotive as it drags its cargo cars from one place to another. Likewise, moving power through a grid requires energy, but with power, electrical energy is needed, not gasoline or diesel fuel, i.e., power delivery lines are run by electricity.

For example, under certain conditions, and using a particular electrical route, moving 500 MW from Moscow to St. Petersburg might require 15 MW of power. Thus, one would put 515 MW into one end of the transmission line, in order to get 500 MW flowing out the other. This three percent difference, 15 MW, powers the transmission line, but is referred to as *electrical losses.* The energy isn't really lost. Electrical engineers can explain exactly where it went, and why, but one can understand the term, because that 15 MW isn't there when the power comes out the other end.

A buyer-seller combination that has transacted a 500 MW sale, and wants to move power from Moscow to St. Petersburg, would need 15 MW of additional power to ship the power. If it is not provided, the power system either will not deliver the power, or will automatically take it out of some other flow on the line, the missing 15 MW no doubt distressing some other seller-buyer combination.

The system operator makes certain this doesn't happen by offering a losses' monitoring and make-up service. By monitoring the system the operator informs all parties in advance of what their losses are likely to be. During transmission the operator actually measures the losses, making sure they are accounted for. These duties are part of the standard (and mandatory) service of running the system, described above.

In cases where a shortfall occurs, e.g., the buyer-seller combination moving 500 MW from Moscow to St. Petersburg does not provide an extra 15 MW, the operator provides it at a nominal charge. In many de-regulated industries, this is a voluntary ancillary service. The system operator offers to furnish the 15 MW of losses' make-up, but the buyer-seller combination does not have to buy it. Rather, they can contract just for delivery of 500 MW, along with the mandatory operations reactive power services required. Perhaps they intend just to inject 515 MW at the point of sale.

Energy imbalance

In the case cited above, a buyer in St. Petersburg fell short of need because of a failure on the part of the seller's generator. If he expects to have power in those situations, he has to arrange for backup reserve.

But another, more typical way that he can fall short is if he simply underestimated his demand. Perhaps the day was colder than expected, electric heating was used more, and instead of the 500 MW St. Petersburg Light and Power expected to need, their customers actually use 513 MW. There is not very much that St. Petersburg Light and Power can do about this at the time. They have thousands of customers connected to the grid, using power. They don't have much direct control of their heaters, other than to "pull the plug" by opening breakers, causing a local blackout and dropping all 513 MW. Nobody wants to do that, particularly since the sudden change in flow, 513 MW being a lot of power to suddenly trip off the system, might create instability problems that would prove difficult for the System Operator to handle.

The buyer can contract with someone (the system operator, the seller, or another Genco) for *energy imbalance*. This will provide a small increment of generation, the amount depending on what the buyer wants to purchase, that stands by ready to fill in any shortfall of actual demand as compared to the Genco's generation, making up the 13 MW. Since the power system will not work well if such imbalance is allowed to occur, the System Operator is required to offer it. But since the buyer can obtain this power from several sources, this is usually a voluntary purchase, the buyer being free to procure this wherever he wants, from the System Operator or some place else.

Load following (regulation)

A power system must be kept very tightly controlled in two ways. *First*, power coming into the system must be exactly balanced against power flowing out, at every moment. Even a minute's imbalance can lead to instability and a system shut down. *Second*, the system frequency (see Chapter 3) must be held precisely constant. It, too, can wander as power flow changes, and as a result the system can become unstable.

In any buyer-seller power transactions, if the buyer's demand or the seller's production output fluctuates from moment to moment, it will affect the energy balance of the system (incoming from seller versus outgoing to buyer) and the frequency. Now, demand and supply always do fluctuate slightly. Appliances in homes turn on and off, industrial machinery starts and stops. Compensation must be made through control of various kinds of equipment like phase shifters, or "swing-bus" generation, i.e., a generator kept under tight control, accelerated and decelerated from second to second to change its output to provide the small increments of power makeup needed to keep the system in perfect balance. This is often called load following. It is similar to energy imbalance (see above), except it is happening on a much shorter time span, seconds rather than minutes. It is one thing to note that an extra 13 MW is needed and ramp up a generator to produce it. That is *energy imbalance*. It requires another order of magnitude in effort to sense the fluctuations of load on a moment-to-moment basis, as equipment like pumps and heaters turn on and off, and to react instantly by adjusting generation to track these fluctuations. That is *load following*.

A certain amount of load following control is part of normal operation and is included in the mandatory system operation. But providing a controllable generation capability to track variations in load is a service. The System Operator can provide this, or a buyer-seller combination can. Therefore, in some de-regulated industry structures, load following regulation is a voluntary service. The System Operator monitors a buyer-seller's flow, and offers to provide them with control services and resources for a price, but they are not obligated to buy this.

Dynamic scheduling

In order for a power seller-buyer combination, i.e., an owner of a generator and the operator of some load(s), to provide their own energy imbalance and load tracking services, themselves, they have to perform several measurement and control functions on a rapid and on-going basis. These are lumped together under the title, *dynamic scheduling,* and can be offered by a System Operator.

To determine the amount of mismatch between seller's input and buyer's output from the system, so that they can correct the imbalance and track load themselves, the buyer-seller combination must measure simultaneously the power flowing out of the seller's generator and into the buyer's load, even though these two sites may be hundreds of miles apart. They must bring that information to a common point, where a computer or other control system can compare them, and if they are not in agreement, determine what changes in generation to make, and send appropriate signals to ramp-up or ramp down the seller's generator. Worse, determining mismatch is not a trivial task, because the analysis must account for electrical losses between buyer and seller.

Suppose at a particular instant, 515 MW is leaving the seller's generator, 15 MW is being lost in transmission, and the user is actually drawing 503 MW from the system. Then, despite the fact that generation is > load, the seller's generation should be ramped up by 3 MW. Similarly, if his load were 498 MW, the unit should be ramped down by 2 MW, because having too much generation on the system is just as damaging to stability as not having enough. Such analysis must be updated constantly, almost on a second by second basis, because load and losses can change rapidly depending on system conditions. Dynamic scheduling consists of measuring flows at the generator and load points, determining any mismatch, and sending appropriate adjustment signals to the generators. It does *not* include the actual running of the generators, which is energy imbalance and load tracking.

The System Operator is in a position to offer this service with very little additional effort. His control system usually has the required measurement and data communications equipment in place, and his control computers are monitoring the losses, and determining actual mismatch. After all, he has to be able to offer energy imbalance and load tracking services to his users, which means he must be able to perform this dynamic scheduling and collect the required data himself. The only additional duty required of the System Operator to provide dynamic scheduling as a service is to transmit the information he collects and computes to whomever and wherever the buyer-seller requests so that the energy imbalance/load service has the information it requires. Providing *just* the information to enable energy imbalance to be done is basically what dynamic scheduling service is all about. Under United States FERC mandates, this is a voluntary ancillary service for both System Operator and system user. The System Operator is not required to provide it, and the buyer is not required to buy it if it is offered.

Reserve margin

Suppose that a Genco and a municipal buyer have made an agreement for 500 MW, the entire output of a large generating unit. They have ordered transmission services from the System Operator, and are moving this 500 MW from the Genco's site to the municipal utility's location, when the impeller in a main feed pump on the Genco's generator suddenly shatters, forcing the generator to shut down within seconds.

The buyer, however, is still connected to the power system, still using electricity, and thus still drawing power from the grid. As a result, the sudden loss of the large generator creates a serious mismatch between demand for power, and power being produced. This can cause what is called a *cascading blackout:* the remaining generators on the system sense the mismatch, and try to produce more power. But if there are not enough of them to cover the mismatch within a matter of seconds, they overload, their automatic equipment takes over and shuts them down, making matters worse. In a matter of a minute or less, the entire grid can shut down. The System Operator avoids this catastrophe by keeping enough reserve generation available so that a sudden failure will not overload the remaining generators. There are two categories of reserve.

Operating reserve. To prevent a system collapse (blackout) when a failure occurs, the System Operator will require every transmission user to provide a bit of operating reserve, so that if his generation fails, there is enough extra generation on the system to avoid such stability problems.

Based on power system calculations done by his computer control system, the System Operator will inform each transmission user in advance, at the time the user informs him that he wishes to use the system, of how much additional generation will be required. This will be updated from time to time as conditions change.

This additional margin of operating reserve is usually far less than the amount of power the transmission user's transaction is producing.[2] The user can provide this himself, purchase it from someone else who can supply it, or buy it from the System Operator. Since it is essential to the operation of the system, the System Operator is required under US law to offer it, and every user must either provide it, or purchase it somewhere.

Supply reserve. This occurs when an equipment failure affects only parties to the particular transaction that broke down as a result of this equipment failure. The buyer might be in real need of power, e.g., a municipal utility that wants to kept the lights on in the city, but is now 500 MW short. This municipal utility might want the System Operator to provide a backup supply until other arrangements can be made.[3]

---

[2] In a system where 10 users are all moving 500 MW, the likelihood of two or more generators failing at one time is very remote. Therefore, the System Operator might decide that one 500 MW backup unit is enough to assure adequate system security. That works out to only 50 MW of reserve per user, or a 10% backup. Each user must therefore run, or pay someone else to run, or contract with the System Operator to run, 50 MW of available generation that is not putting out power, but is instantly available. Large power pools cover such wide areas that they distribute risk over many more than ten customers, and therefore the required reserve drops to about 3% per user.

[3] When such contingencies occur, and they often do, the buyer (or the seller if he signed a contract guaranteeing delivery) would go to the "spot market" – the power exchange –

Supply reserve is really not a transmission level issue. It is a generation (power market) problem, and many de-regulated industry structures prohibit the System Operator, who is restricted to providing only transmission services, from providing this. But in other de-regulated industry structures, making supply reserve available is a service provided by the System Operator. The buyer is not required to buy or provide it. Maybe it's not that important, or he is willing to take the risk, or he has bought capacity from another supplier, as described in Section 10.2).

Regardless, both services, operating reserve and supply reserve, are assured through having sufficient generation reserves, i.e., spare generation on line. "On-line" means that the generators are running, just not producing maximum power, like a motor idling. The System Operator, acting through the power marketplace, just like any other buyer, contracts for certain amounts of capacity or operating reserve and standby generation *("I'll pay you 1.2¢ per kW, per hour, to keep your generator turning and ready to produce 500 MW on 5 seconds notice.")*.

The amount and type of reserve the SO buys will depend on the system operating condition in any hour, and whether he is interested only in keeping the system operating, or whether he has sold backup supply reserve services to the system users.

Black-start reserve

Many large power plants cannot be started unless they have a small supply of electric power from the grid to begin or "bootstrap" the process. It is both the energy itself that is required, as well as a "signal" or sense of the voltage from the grid, that is needed to awaken a larger generator and start it into action. For example, a 500 MW power plant may require 15 MW of power just to begin operation, i.e., to power its many control systems, to run its boiler feed pumps, fuel conveyers, pressurizers, cooling systems, etc.

A System Operator can offer "black start" capability to any Genco. For a fee, he will assure that power and alternating voltage supply, for synchronization, is available at a site, so that whether as scheduled, or due to an emergency, the Genco can start its unit. Usually, this is a voluntary service, since the Genco might have other methods of obtaining the required power and signals.

What Is Actually Bought and Sold, and How Is It Sold?

A Poolco operates its transmission system by buying from competitive

---

and immediately buy 500 MW of additional power, paying whatever the going price is, and then contact the System Operator to have that moved to St. Petersburg. But this could take up to an hour, and meanwhile, the buyer might want continuity of service.

generators, and delivering power to the contracted wholesale delivery points. In a "Poolco system," the operator alone cares about the transmission system, running to best accomplish the task of buying from the least expensive generators possible, and moving the power to the required wholesale delivery points to satisfy all demand. No generator or wholesale customer is involved in the operation of the grid or in asking for power delivery: they only ask for power at their site and let the operator worry about "where it comes from and how it gets there."

In some systems, the Poolco operator will charge wholesale customers for "transmission," assessing a fee to cover the shipping of the power to each site, but the wholesale customers do not have to manage or plan their transmission usage, whereas in other de-regulated structures (e.g., California) they do.

In some de-regulated systems – a California is a good example -- users who want power moved must arrange from transmission delivery services from the Independent System Operator (ISO) by whatever means of ordering transmission the operator has in place. Usually, this is an electronic bulletin board/bidding system in which the ISO posts the supply of transmission services available (available capacity) and buyers can inquire about prices, and reserve transmission capacity for their use. Often this is called an (Open Access Same Time Information System). The ISO also posts on the OASIS the available transfer capability (amount of power that the system can move), based on analysis of future (next hour, next week, next year) system conditions.[4] Capability, as previously noted, is the ability of the system to move power. The OASIS postings list by location both "network service" and "point to point service:"

> *Point-to-point service* involves moving a specific amount of power from one location to another, usually over a long distance. The ISO will identify this between major points (control areas) in its pool. "It is possible to move up to 750 MW from Moscow to St. Petersburg." Point-to-point service can be thought of, not entirely correctly, but generally, as "long-distance" transmission service.

> *Network service* involves moving power *from* many different points, *to* many different points, and covers all combinations among them. For example, a Genco may have contracted with a

---

[4] In California, and other states that adopt "California-like" approaches, the ISO does not use an OASIS *per se*, nor does it post and sell transmission capability in the purest sense. Instead, it accepts what might be termed a "balanced generation and load schedules" submittal and reservation system, which it then accepts and "dispatches." The specific California model will be discussed later.

municipal utility to delivery 2,700 MW of power. The Genco has five generating sites, the municipal utility six major transmission-level buses. Network service means getting 2,700 MW produced in some combination at those five generating plants to some combination of 2,700 MW of demand at those six municipal buses. In network service, the "some combination" is not specified precisely and in fact may not be known beforehand.

The OASIS lists all of this information on availability of network and point-to-point transfer capability, and ancillary services, and price by hour, for the next 168 hours, and by week, month, and year farther out, typically going up to a decade into the future. "Available" means what remains to be sold, i.e., maybe the system can move 1,150 MW between Moscow and St. Petersburg, but if 400 has already been reserved, then only 750 remains available. Point-to-point capability has to be listed in both directions. If the system can move 1,150 MW in one direction, that does not imply it can move 1,150 MW in the other – it might be able to move more or less depending on conditions.

The data listed on an OASIS daunting in quantity, both for the ISO to produce and manage, and for potential users to wade through to determine what they want to buy. However, by preparing in advance, everyone can take their time and then reserve either long-term or short-term use of the network. Finally, similar to the manner in which power was sold on a firm or interruptible basis, transmission capability can be bought in various levels of "firmness" or priority of service.

Reserving or Buying Transmission Capability

Potential buyers of transmission services log onto the OASIS, scan the available capacity, and then reserve what they need in advance of use, between an hour to several years ahead. Users can reserve on the basis of the amount of power to be moved, ancillary services, the from and to points, and the time involved. Sales are usually made on an hour-by-hour basis, and users can buy multiple hours *("I'll take 500 MW of transfer capability from Moscow to St. Petersburg, beginning at 6:00 AM on April 15th, to 5:00 AM on Monday, the 22nd. I want to buy firm capability, along with losses, operating reserve, and load tracking. In addition, I want to buy 500 MW of network capability in St. Petersburg, as I'm delivering that power to 17 local distribution substations, with varying loads, and which in aggregate never exceed 500 MW").*

When a user reserves the use of capacity, he essentially buys its use at the time reserved, and commits to pay for it.

Re-Selling Reserved Capability

Some regulatory systems permit a user to re-sell transmission capacity he has reserved, but others do not put limits on the sale price, or otherwise try to restrict speculation and profiteering for transmission access reservation. As an example, suppose St. Petersburg Light and Power reserves 500 MW of wholesale capacity from Moscow to St. Petersburg, for the week of April 14-21. Subsequently, a large nuclear power plant belonging to a key Genco fails unexpectedly, and additional generation from Moscow is needed by other utilities in the region. They are willing to pay more than SPL&P agreed when it reserved that capacity.

If permitted, SPL&P might decide to sell that 500 MW of capacity at a higher price than it will have to pay. Remember, since it bought it when the price was quoted low, it can make up its shortfall by running its own local generation, which it normally doesn't do because power from Moscow is less expensive, even after paying normal transmission charges.

Should SPL&P be allowed to make a profit in this case? Some people argue no and have proposed regulations in some jurisdictions that prohibit "profitable re-sale." But in that case, its best course of action is to say, "I'm sorry that you other guys are in deep trouble, but I reserved it and to give it up would cost me money, so forget it – it's mine."

Some regulatory proposals would cap its profit in this case at the level of its lost opportunity cost, i.e., it can make enough on the sale to cover its higher costs for running its own generators.

12.4 HOW IS TRANSMISSION SERVICE PRICED?

Purposes of Pricing

All industry re-structuring plans, regardless of the jurisdiction, envision only one transmission system run "for the public benefit" as a type of monopoly franchise.[5]   Therefore, the tariffs (prices) will be set by government regulation, which means that transmission tariff structure, and the way it is determined and applied, are elements of policy. In general, policy makers and those giving them advice, i.e., economists, engineers, utility executives, consumer advocates, see pricing/cost allocation as having three goals:

(1) Recover costs–Fees for transmission use must produce revenue to cover all the expenses of investment, operation, and maintenance, as well as provide a small (regulated) level of profit for the owners.

---

[5] In any regional pool, there are many transmission owners, each a monopoly franchise holder of local transmission facilities inter-tied into the regional grid. These various Transcos do not compete with one another, since they service non-overlapping territories, but instead, acting in unison, provide a single regional grid.

Table 12.3 Five Important Goals of Transmission Pricing

Recover costs of the transmission system and its operation
Encourage efficiency of use and sound investment in transmission
Equitable treatment and opportunity for all users
Understandable price structure: users must understand it
Implementable (workable) in the real world

(2) Encourage efficient use–The price structure (relative cost as a function of the service sought, e.g., amount of power transported, distance, etc.) should give incentives for using the transmission system efficiently. Whatever "efficiency" means is the subject of much disagreement, but everyone agrees pricing should encourage it.[6]

(3) Encourage efficient investment–The price structure and the way money is paid to owners should provide an incentive for investment in new facilities where they are needed.[7] Otherwise, how can one expect the transmission system to be expanded as need grows? This is really a payment to owners issue, but some pricing issues impact it

In addition, some requirements can be defined for a pricing system, based on practical stipulations of the ISO and its customers. A pricing system must be:

Fair. Exactly what is fair and what isn't might be debatable, but most people will agree that any pricing system must be fair or equitable to all users. Most people think this means it must not unduly penalize or favor certain classes of customers or certain types of buying behavior or usage.

Understandable. Any pricing system must be understandable by its customers. If a pricing mechanism is so complex that its users do not trust their ability to make good buying decisions, they will not use the product, and will seek other ways to satisfy their needs.

---

[6] One can talk of *economic efficiency*, meaning that users are encouraged to use transmission to maximize value and societal benefit, or *engineering efficiency* (low losses, high reliability), or efficiency as viewed from any of several other perspectives.
[7] In the United States, the Energy Policy Act of 1992, title 211, has provisions mandating that new transmission be built under certain conditions where need is demonstrated by users. However, that law is untested, and a law, alone, will not provide a workable system if that system does not provide sound financial incentives for investment in it.

Workable. A pricing system so complicated that it cannot be implemented economically is impractical. Many apparently good ideas in pricing require measurement of power flow at so many points, and system analysis in so much detail, that it is questionable if they can be implemented successfully, and it is doubtful if their cost is justifiable.

Table 12.3 summarizes the requirements for a good pricing method.

There Are Many Different Opinions on How Transmission Pricing Should Be Done

No other subject in de-regulation – in the entire power industry for that matter – provokes even half the controversy and argument as this particular subject. Opinions differ on the grounds of basic philosophy: What is really being sold? Who is the grid supposed to benefit? There is also no consensus on very technical details: Just how does one compute the cost of load tracking?

Questions on very pragmatic issues also arise: Do the practical advantages of a simple tariff structure – understandability and ease of billing – outweigh any disadvantages created because it is approximate, not exact? This debate is very likely to continue long after de-regulation is implemented and the industry is re-structured, for the government can and will make changes in transmission pricing from time to time.[8]

About the only thing generally agreed to is that the charges for transmission system use should cover all the costs and provide a small, i.e., regulated, level of profit for the owners of the facilities. The transmission system, elements of which are the property of many monopoly-franchise transmission owners involved in the regional power pool, is a regulated utility service provided by its owners, and:

$$\sum \text{fees charged} = \text{cost of all parties} + \text{regulated amount of profit} \qquad (12.1)$$

Thus, the real debate over pricing boils down to how the cost of the system is *allocated* among the users – who pays what portion of the overall cost? Some of this debate is due to genuine differences of opinion among politicians, policy makers, economists, and engineers. But a majority of the arguments derive from baser motives. Early in the de-regulation process, every utility, Genco, and future Resco studied the various pricing structures, and became a proponent of the method giving it the best advantages. Each marshaled all manner of

---

[8] Often, changes are made on a "running basis:" FERC's announcement of its order 888 de-regulating wholesale power in the United States contained a set of transmission tariff guidelines. Simultaneously, FERC announced its intention to change the way transmission tariffs were calculated.

technical and theoretical arguments in favor of "its" approach. Still other utilities took a purely defensive approach, favoring tariff structures that would create high costs for potential competitors to access their service areas.

Three Serious Pricing Debates

Actually there are four serious, debated issues regarding transmission operation. However, the fourth, *transmission congestion* (See Chapter 2), is a complicated *business*-related issue, although its satisfactory resolution requires a technical solution. It will be dealt with later. The remaining three issues are among the most frustrating types of issues: Debates where both sides have some, and in some cases each side can cite a lot, of validity to their arguments.

Displacement

Many people argue that power actually doesn't move long distances through a power grid, even when it appears to do so. For example, suppose a Genco inputs 800 MW into the grid in Chicago, and its customer takes out 800 MW in New Orleans. Has 800 MW actually moved across the United States? Some pricing schemes, and their proponents, say no. They liken the transmission grid to a large pool of water, or lake (Figure 12.1), in which the goal is to keep the water level, i.e., voltage level, constant: Pour 10,000 gallons into one end of the lake (Chicago), and it will raise the water level enough that about 10,000 gallons will flow out the other end (New Orleans).

But that 10,000 gallons didn't actually flow across the lake from one end to the other: The addition of water merely caused a *displacement* of water in the pool, making it possible for water to flow out where needed. And, it could have been drained out anywhere in the lake. Similarly, adding power to a power pool grid, full of millions of MW rather than millions of gallons, just raises the level of the system enough that one can take the power out anywhere. Thus, it is argued, transmission users should pay a flat fee regardless of distance between points, because their power really isn't moving very far.

There is some truth in the view that, like water, power "flowing" in a network doesn't actually travel far, but opponents point out a flaw in this argument. Adding the 10,000 gallons to a lake makes an incremental disturbance in every gallon in the lake: 10,000 gallons may not move from one side of the lake to the other, but the millions of gallons in the lake are each moved by a small amount.

Adding 800 MW in Chicago and taking it out in New Orleans displaces power through the network, in a chain of re-distributions, or displacements of flow, as shown in Figure 12.2. The 800 MW of power pushed into the network in Chicago might actually never leave Chicago. Very likely it will be consumed by users taking power from the grid in the Chicago area. But that will release power generated elsewhere, which was heading to Chicago, to flow elsewhere,

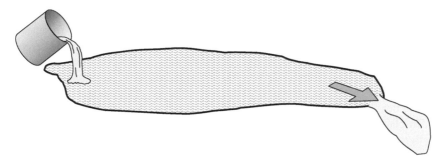

Figure 12.1 The concept of displacement. Pouring water into one end of a pool or lake makes an equivalent amount flow out the other end, but the poured water did not actually travel from one end to the other. Does electricity pushed into a power pool at one end, and withdrawn hundreds of miles away, behave similarly? If so, transmission tariffs should not really charge users more for transactions that involve longer distances.

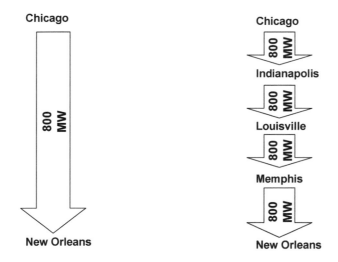

Figure 12.2 Displacement in a power system. Perhaps 800 MW of power doesn't travel all the way from Chicago to New Orleans if it is injected at Chicago and removed at New Orleans (left), but by the process of displacement (right, see description in text) the transaction creates a large displacement in the network equivalent, perhaps, to 800 MW traveling that far.

perhaps slightly south, to Indianapolis, where it is consumed, thereby pushing 800 MW that were previously feeding that city toward Louisville, and so forth, all the way to New Orleans. While 800 MW doesn't move all the way from Chicago to New Orleans, *an equivalent total change in system flow has been effected.* The user should pay *as if* his power had been moved over this distance.

Both perspectives given above have some validity, and a proponent of either view can find proven examples in the real world to bolster his arguments. But in order to move power a greater distance, one has to build longer and larger power lines, and everyone acknowledges that costs money. Thus, while the burden placed on a transmission grid by a wholesale wheeling transaction is not strictly a function of the distance that power is moved, distance is usually an important, and often a critical determining factor of power transmission cost.

Distance

Very much related to the question of displacement is the issue of *whether and how price should vary as a function of distance.* The displacement issue revolves around whether power actually travels the distance between two points. Irrespective of that, there is a great deal of debate over the issue of whether and how much transmission delivery price should depend on distance. This is a classic "economist vs. engineer" conflict of perspectives.

First of all, it is worth noting that traditional regulated utility cost-recovery rate methods did not base T&D cost recovery on distance. In fact, it was important in traditional rate-making that utility prices did not reflect differences in distance from generation to customer. Within the monopoly franchise framework, to achieve minimum total revenue requirement, the local utility is supposed to locate generation plants and build lines in the manner that most benefits its customer base as a whole. Electric customers who just happen to be close to a generating plant have no right to benefit from this coincidence. Similarly, those who end up far away from power production points, due to a utility siting decision into which they had no input, shouldn't be penalized.

But in a de-regulated electric industry framework, wholesale buyers have a choice of where their power is produced. They can buy from local or distant producers. Therefore, many people argue that wholesale transmission customers should pay for transmission usage on the basis of distance. Buy from far away and it will cost you more to ship your power to your site than if you buy locally.

Disputing this argument are many politicians who point out that their cities are far from any generation sources, because that's just the way it worked out. Many large commercial and industrial power users also note that their plants are far from any competitive generation. These people maintain that setting up a transmission pricing system that charges a customer for long-distance power transmission will penalize them unfairly. Thus, one aspect of "the distance issue" is fairness, at least from the standpoint of some customers. The other relates to price signals and efficiency: market vs. electric efficiency.

From the economists' standpoint of encouraging competition among power producers, the largest possible "marketplace" is desirable. Many competing power producers selling to lots of customers means an efficient, truly competitive marketplace. If transmission prices vary only a little bit as a function of distance, then producers in Chicago can compete with those in New Orleans, and vice versa, both offering their product to customers in both areas. Low price as a function of distance creates a wide market. This encourages competition, which, after all, was the whole purpose of de-regulation.

On the other hand, high long-distance transmission prices would encourage the construction of local generation where there is none. It also tends to discourage long-distance transmission of power, with its higher losses. High prices as a function of distance are conducive to efficiency from an engineering sense, short, i.e., reliable, power delivery paths, and lower electrical losses.

Again, as with the issue of displacement, and as is the case with so many other real controversies, both sides of the "don't charge a lot for distance" argument have valid points in their favor.

Parallel flows

One thing every expert agrees on is that the 800 MW flowing from Chicago to New Orleans would not travel over any one transmission line. Instead, it would distribute itself over as many as two dozen different routes that lead through the network, some rather direct and others more circuitous, as illustrated in Figure 12.3. But while everyone acknowledges this, especially the experts, they disagree about whether or why this matters, and how it should affect pricing.

The real problem is the utter complexity of tracking power flow through a network. For a moment, suppose that the 800 MW actually does flow from Chicago to New Orleans. Along each of the two dozen routes it takes, it mingles with the flows from hundreds, perhaps thousands, of other transmission users. In some cases it might "travel alongside" another user's flow for only a few miles. In others, it might flow parallel with it for hundreds of miles.

Proponents of exact pricing schemes want to track everyone's power on a segment by segment basis throughout the network, determine which users are loading what portions of each, and thereby compute very precise allocations of network cost to everyone. Opponents of this approach argue that the cost of tracking, computing, and billing to this degree of detail would be high, and claim that the computations, while detailed, would still be inexact, because many of the computations involved would be based upon assumptions about unmeasureable system conditions.

A fundamental problem though, is whether a transmission user who is wheeling power through only that portion of the grid near him is using just that portion of the grid, or more of its resources? In answering this, one must look at

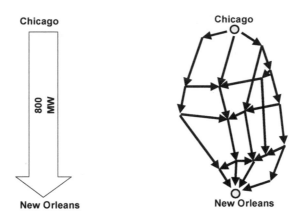

Figure 12.3 Power flowing through nearly any transmission system does not take just one route, but instead distributes itself over a variety of *parallel paths,* as illustrated here. Its flow mingles with thousands of other power transactions, making determination of *who* is responsible for the loading on any one line a difficult undertaking.

the system and any particular transmission line within that transmission system that exists only for reserve reasons. Under normal circumstances (no other lines out) it is carrying no load. Yet this line could be a vital reserve backup to the system as a whole. Without it, in the event of any of a number of generation or large transformer failures, the system would cascade into a blackout.

How should the cost of this "backup" line be allocated? Since no one is using it, and yet it benefits everyone, perhaps they should all pay a portion of its cost. Based on this observation, one could argue, as many do, that the transmission system as a whole supports the user base as a whole. Why not just divide the total cost by the total number of users, perhaps weighting each user's bill according to the amount of power shipped, the length of time for delivery, and/or the distance moved, and leave it at that?

Pricing Methods

For transmission services, price is a function of cost

Since transmission pricing is based on the concept of regulated cost recovery, the price is considered a function of cost. In a de-regulated power industry, however, generation prices do *not* depend on cost. Rather, they are determined in a competitive demand-supply, buyer-seller transaction manner. By contrast, transmission services' pricing use cost recovery allocation methods. Basically, they split up the transmission cost among the various users.

Therefore, each of the pricing methods in use, or proposed for application in a nation's or state's re-structuring plan, assumes that the cost of transmission is known. The level of detail needed in the cost information upon which the pricing method is based varies greatly, however, depending on the particular method. For example, a pricing method that charges transmission users based on their time of use (i.e., peak usage costs more than off-peak usage) must have information on how costs and usage vary as a function of time. A method that does not vary price as a function of time does not need this feature.

Thus, one of the important issues to examine in any pricing method is not only how it distributes cost, but the degree of difficulty the System Operator will have in determining cost to the required level of detail.

Listed below are several pricing methods, in order of increasing complexity of information required and cost variation.

Flat fee

Flat fees are the simplest approach to cost distribution over a large number of customers. If the cost is a million dollars, and there are ten thousand users, then everyone is charged one hundred dollars. Flat fees are generally acknowledged to be "unfair," but are used in cases where the difference isn't worth the trouble of computing and applying a variable tariff, which is usually true when the total cost per customer is small and inequities would not be burdensome. No one has seriously proposed a flat fee for transmission pricing (at least recently), both because the amount of revenue needed per customer is significant, and the information needed to allocate total cost on the basis of at least one significant user characteristic, e.g., how much was used, read from the customer's power metering equipment, is always available.

Postage stamp method

Postage stamps, used for mailing letters and other light correspondence, are an example of a pricing schedule where the "unfairness" created by a simple and admittedly inexact approximation is accepted in return for ease of use. In this case, the approximation is to *ignore distance* in allocating cost among users, which is effectively how postage is priced, i.e., users pay the same amount to mail a letter across the street or across the continent. A postage stamp rate applies a flat fee for use, irrespective of the distance involved.

Postage-stamp transmission tariffs set a price on use of the grid that depends only on the amount of power moved, the duration of use, e.g., two hours costs twice what one does, and perhaps the time of day or season of use, prices during peak being higher than off-peak. Given the cost-recovery goal of pricing (see above), this means that postage stamp rates effectively divide cost among the users on a *pro rata* usage (kW) share basis, regardless of location or distance of transmission usage. Proponents of postage stamp pricing favor it because of its simplicity, and/or because they want to move power over above-average distances (anyone who does so probably benefits from postage stamp rates, as compared to paying a rate that allocates usage based on distance).

License plate rates

Fee schedules for vehicle registration (license plates) are typically more complicated than for postage, but are still a simple schedule with prices not intended to bear a close relationship to usage of costs caused. "License plate access" fees call for different costs in different areas and for different size users, just as fees vary from state to state and by type and size of vehicle.

Pro forma transmission tariffs

Whenever and wherever pricing is unresolved in the United States, FERC requires that it be done using traditional regulated power industry methods. Each user must pay a capacity fee, based on the installed cost of the transmission system as a whole, allocated on a per kW basis, as well as other fees for use associated with the variable operating costs incurred at the time of their use. Often, this approach very nearly results in "postage stamp" rates for the capital and fixed cost portions of the charges.

The main advantage of this approach is that it is based on a proven method of assuring that costs are recovered. During transition periods, this is the chief priority of all parties: Pricing signals and other issues can wait. The chief disadvantage of the *pro forma* approach is that it is unwieldy and not really suited to the "commodities" market in a competitive industry.

MW-mile

Pricing that simply sets a wholesale wheeling price proportional to both amount and distance is called MW-mile pricing. *("It's 900 miles from Chicago to New Orleans, and our rate is 1¢ per MW-mile/hour. Moving your 800 MW will therefore cost you $7,200 per hour.")* A fixed price throughout the network can be used, e.g., 1¢ per MW-mile/hour, anywhere, anytime, or different rates can be charged for different routes or times. *("The charge is 1¢ per MW-mile/hour on peak, from Chicago to New Orleans, but only 45¢ per MW-mile/hour in the other direction. Rates off-peak are less.")* Or, the rates can be computed and updated hourly. *("Due to load increases, in the next hour, Chicago to New Orleans will have from/to rates of $1.05 and .47¢ per MW per mile.")*

MW-mile methods can be based either on the straight-line distance between the two points, or the transmission route distance between them. The former is easier to determine, as it avoids the problem of resolving the different distances of the many multiple routes (see Figure 12.3) that usually exist between any two points in a network.

MW-mile is simple to understand, easy to implement, and fairly simple to apply to real power systems. It is only approximate, however, and shares some of the problems of contract path pricing.

Contract path

Contract path pricing that calls for the price of transmission from point A to point B is to be based on the cost of a single identified path. The parties moving 800 MW from Chicago to New Orleans might choose a particular route between the two cities as "the contract path." Even though they know the actual power flow will split itself among many parallel paths (Figure 12.3), they compute the price to be paid on the basis of this one line. *("This segment of the route has 800 MW capacity, and you're moving 800 MW, so we'll charge you for the whole cost of the line. This other segment in the route has 1,000 MW capacity, so we'll charge you 80% of its costs.")* The price usually includes a capacity charge to cover the capital cost of the equipment, and energy charges based on losses and other operating costs.

Contract path pricing was widely used before both the widespread sale of transmission services and full de-regulation. When wheeling contracts were rare, the contract path method was often used in the written contract between the transmission owner, i.e., utility, and the customer buying the service (often another utility). Hence, the word "contract" in the method's name. As usually applied in bi-lateral contracts for wholesale wheeling, contract path pricing implies that the payments for use of line go to the owner of the contract path. This brings up the method's major failing. Suppose two parallel paths are owned by different Transcos. A Genco makes a contract for one, but the power actually flows over both. The contracted Transco gets compensated for more than the actual usage on its line, while the owner of the parallel path, on which a portion of the Genco's power is flowing, gets nothing.[9] Proponents of contract path pricing point out that the approach is simple, that it means contracts for power delivery are based on *an identified physical asset*, a specific line, and that it allocates cost somewhat proportional to use. Opponents argue that contract path pricing is selling capacity, not capability, which could get a power system operator into trouble by overloading the grid.

Rated system path

This method bases the cost on a computed set of parallel paths for a particular path. Here, a rated system path from Chicago to New Orleans would be identified by load flow and other engineering studies of the grid. Looking

---

[9] Contract path pricing does not have to work this way. A system whereby routes were reserved for later usage, and priced on a contract path basis, but with all revenues distributed to the transmission owners on the basis of actual loading (use), would avoid most of this problem.

something like the left side of Figure 12.3, this would consist of a weighted average of the various parallel lines involved. Wheeling from Chicago to New Orleans is then based on the cost of the equipment used. *("Your 800 MW transaction from Chicago to New Orleans will use 15% of the total capacity of the rated path, so the price will be 15% of its total cost.")* This arrangement, used in some cases in the western United States, addresses many of the objections to the contract path method.

12.5 THE EVENTUAL WINNER: LOCATIONALLY-BASED PRICING?

Complicated, But Good to Have

This section, which is greatly expanded from the first edition, reviews what many people now agree is both the most complicated realistic pricing method for grid usage, and also the one most likely to become widely used. Locationally-based pricing (LBP), particularly locationally-based marginal pricing (LBMP), is more complicated than other pricing systems. That complexity has led many people to seek other, simpler systems, particularly for the first implementation of de-regulation. Beyond that, in some cases where LBMP was adopted, rule-makers and system operators had a good deal of trouble fine-tuning it before they "got it right."

But running a de-regulated regional electricity grid seems to be a case where the illness's symptoms justify such an unpalatable medicine: it appears that without something like LBMP, the industry will run into problems, maybe not in the short-term, but definitely in the long run. This section explains LBMP, transmission congestion, and the advantages LBMP seems to have, from several perspectives and with a lengthy example.

Locationally-Based Pricing of Transmission Service

Locationally-based pricing varies the price users pay for use of the power grid as a function of both location in the grid and time. While zonal pricing varies transmission costs a bit from location (zone) to location (another zone), LBP does this with much more detail, setting prices on the use of individual transmission lines in some cases.

Thus, despite the fact that two transmission lines in a regional power grid are identical in design and construction, and cost the same to build and maintain, etc., the price for moving power along one might be quite different than for the other, particularly at certain times. The more expensive of the two would be the one more heavily used and in an area of heavy use of nearby lines. It would be assigned a relatively high price, particularly in times of very heavy demand. The other might be lightly loaded at that same time and in an area of light grid usage. The system pricing authority would quote a much lower cost for its use. Locationally-based pricing is very much a "demand and supply" pricing system.

This variation in pricing from one line to the other is difficult to implement because it requires the system operator to deal with a lot of data on current (this minute!) system operating conditions in order to compute and post the line by line costs for buyers and sellers using the grid., And doing that requires extensive computations that must be updated frequently. LBMP can be equally complicated for buyers of transmission service to understand and use to make sound business decision: for example the price to move power one way on a line might be much more than to move power the other way, and both the prices and this difference could shift a great deal during the course of a day, etc. But all that complexity and bother seems justified to many, because LBMP is a "fairer," or at least more consistent, pricing system that encourages reasonable and workable buying, selling, and investment decisions by all involved parties.

To make locational pricing work, the system operator must determine transmission prices, based on *current cost of usage,* for power at each bus (location where there is a generator, or user, of electricity) in the system, and revise these figures as conditions (demand, supply, equipment in and out of service) change – i.e., constantly. Usually these costs are computed on a quarter-hour or hourly basis – for every time increment in the day – and posted on an electronic bulletin board or internet site where grid users can review the prices and use in their business decisions. In this regard LBMP shares a characteristic with many other pricing mechanisms: it would be impossible to implement without considerable use of computers, database systems, and electronic data communications.

The system operator's need to constantly re-compute and post transmission prices based on cost brings up a critical point: What *are* the costs of transmission use? Certainly they include all costs associated with recovering the investment to build the lines, for their operation and maintenance, and to cover any electrical losses on them, etc. But a key factor in LBMP is transmission congestion cost, something that used to be called "out of merit" generation cost. An advantage of LBMP is that it can address congestion (which could and sometimes does occur in any grid with any pricing mechanism) in a meaningful and workable manner.

Out-of-Merit Generation Dispatch

Congestion costs are perhaps easiest to grasp quickly by looking at how the larger vertical utilities operated before de-regulation of the power industry. Traditionally, what were called vertically-integrated utilities (see Section 1.4) owned both generation and transmission-distribution systems. As regulated utilities, they were expected to operate their systems on a lowest-cost basis at all times. For the sake of example, assume that one such utility, Big State Electric, owned 80 generators of various designs, fuel types, and capacities, with an aggregate capability of 10,000 MW. However, on a certain day and at a certain hour, demand for power was only 8,000 MW. Big State was obligated to operate the lowest-cost set of 8,000 MW out of its 10,000 MW.[10] Obviously, that set included the generator, among the 80 that it owned, that was least-expensive to operate (most efficient with whatever fuel had, that day, the lowest price, etc.), and the next least expensive, and the next, and so forth, up to a total that met the 8,000 MW need. Specific generators among the 80 it owned were dispatched (selected for operation) in order of "merit" – operating cost – in order to meet the 8,000 MW demand. And as demand changed from hour to hour, Big State would cycle up or down those generators so, at any time, what was running was the lowest cost set of units that could produce the required power.

But suppose that on a particular day, Big State's system operator, while planning the day's operation, ran into a snag. The Operations Center's EMS/SCADA computer told the operator that if one of the generators in this least-expensive set were to actually be put on line, it would overload a particular transmission line near its location. Perhaps there had been two lines leading out of that generating plant, but one of them is out of service for maintenance today. Regardless, the operator cannot permit an overload to occur, so the operator would select another generator in that unit's stead. Big State would be dispatching, or running, its system "out of merit" – in something other than the lowest cost manner. During this period, costs go up, but overloads are avoided. Most important to this example, note that *generation* costs go up because of a *transmission* limitation.

Traditional vertical utilities were expected to make certain that out of merit situations almost never happened, by designing their transmission grids with enough lines or enough capacity in enough locations, that except in rare instances of multiple simultaneous equipment outages, any combination of least-cost generation would never create a transmission overload. As a result, "out of merit dispatch" situations were extremely rare in the traditional regulated power industry. Big utilities could make sure this was the case for two reasons. First, a regulated vertical utility owned and controlled both the generation and the transmission system: It could assure itself and others that the two were designed and operated in a compatible manner. Second, it had a way to pay for transmission upgrades required to achieve this requirement that transmission would never limit generation merit dispatch: Any line that was needed in order to assure Big State could operate at lowest cost was a justifiable expense – that line's cost could be put into Big State's rate base.[11]

---

[10] In actuality, Big State would keep more than 8,000 MW operating, so it would have backup in case one generator should suddenly fail, and so it could handle changes in demand as the day progresses. But the concept is the same: it would be obligated to operate the least expensive set of generators that satisfied all these requirements, too.

[11] Of course, this was only the case when the transmission line's cost less than the money

Transmission Congestion: "Out of Market" Dispatch

Under de-regulation, much of the foregoing structure has changed. First, the concept of "out of merit" no longer really applies, although there is something analogous. Under de-regulation, there are many competing generator owners, rather than just one. These competitors bid or offer power to the electric marketplace at prices not precisely related to cost. There is no single operator who selects generators to run. Now, different buyers decide whom to buy from and what they will agree to pay; the entity that determines what is "next" is not the system operator, but the market in general. "Out of merit" changes to "out of market," a situation where some buyers cannot buy power from the bidders they want, at a price they are willing to pay, because transmission limitations prevent that from happening.

There is another difference under de-regulation that makes this issue quite important to the industry. "Out of market" situations are much more frequent. In some systems they occur every day. This happens because generation planning and operation is no longer coordinated with that for transmission. As discussed earlier, the traditional vertical utility controlled both, and could plan, build, and operate both so they always worked in a coordinated manner. Now, the independent, de-regulated, competitive generator companies do not coordinate their plans with the regional transmission grid owner/operator or one another; they are reluctant to share information about their intentions with any outsider (they are unregulated companies whose plans often constitute competitive advantages). Further exacerbating this situation, no one, generators, buyers, or grid operator, can know what prices and operating plans some generators and buyers will make even a few days ahead. (Keep in mind that under de-regulation, prices are volatile at times – the hour-to-hour and day-to-day pattern of generation "dispatch" changes a lot more than it did over similar time periods when the industry was regulated.)

What happens frequently – in fact every day in some grids – is that the set of generator owners who want to run that day because they successfully bid low enough to obtain buyers for their power would create a transmission overload, *congestion*, somewhere on the grid: demand for power flow there exceeds the grid's capability there. These situations are usually caught in time. The grid operator's EMS/SCADA system reviews, ahead of time, the day's requests for power flow, just as the traditional utility operator did under regulation. The grid operator identifies where and how bad any overloads would be, and determines how to "solve" the problem by changing the operation of the grid, or by simply

---

it saved in generation dispatch costs. In cases where the line would not save as much as it cost it would not be built. But generation out-of-merit costs are so high that all but extraordinary transmission lines were nearly always justified.

denying or ordering changes in the requested schedule of generation (i.e., who will be allowed to produce power). The system operator may have to step in to prevent an overload by saying in effect, "No, generators A and X can't run even though they were low bidders and many consumers are depending on them." To make up for the capacity shortfall this causes, so that no one suffers a lack of power, the system operator would also order other generation – perhaps companies G and M, which will not cause an overload to run, even though both bid too high or for some other reason were not able to find a buyer and had not expected to run.

With this "intervention," or *congestion management* by the system operator, the grid can function without an overload. Buyers who need power get it. But now there are a host of new issues to resolve. First, the cost of power has gone up compared to the "merit" case of taking the lowest-bid generation – who pays for that cost increase? Shouldn't the specific users of the transmission grid who caused the overload pay? But who did cause it? There might have been a dozen companies moving power over that would-be-overloaded line. How is a fair allocation of that cost to be determined? Furthermore, if this is going to happen every day, shouldn't someone build a new line or upgrade the congested line so this limitation is removed and the congestion doesn't happen any more? Who would do that? How would it be paid for?

Before moving on, the reader should make note of the fact that congestion can occur in any grid, regardless of what mechanism (zonal, postage stamp) is used to price transmission. Similarly, it is important to keep in mind that there *are* ways to resolve congestion regardless of the pricing system being used, zonal, postage stamp, etc. But none of these methods are simple – some require massive cost tracking combined with sets of rules that see nearly Byzantine. Regardless, it is impossible to resolve all those questions above in a way all concerned consider "fair." Someone will always complain they were inequitably treated.

Locationally based pricing makes congestion part of the overall demand/supply paradigm that de-regulation applies at wholesale level. In other words, it lets the market take care of it. And while not 100% successful in this regard, it both reduces and "makes fair" most of the congestion issues in a grid. Many of the issues and problems the industry has had during its transition to de-regulated framework revolved around congestion and how to manage it. Thus, over time, in spite of its complexity, LBMP has come to be regarded as a preferred pricing mechanism from several standpoints, including that of congestion management.

Marginal Pricing

One way to address a lot of the issues raised above is by pricing the use of transmission by its location and loading in the grid, and then charging users by the location of their power flow in the grid – what is called locational pricing. That way, high prices can be attached to a congested line: Those who use it are going to pay more than those who use other, non-congested lines. This both discourages use of the line (if the price is high enough) and provides revenue to "fix" whatever problems are caused by the congestion and its out of permit solution (more on this later).

Making this concept work seems to require that the prices posted for transmission service as a function of location be based on *marginal* cost: The price posted for use of a particular line at any moment is for the *next small increment* of power put through the line. Why this is the case, along with a number of other important aspects of congestion and LBMP, will be explained in the course of the example below.

Conceptual Example of Congestion and LBMP

Figure 12.4 shows a small regional power grid (only some of the locations and electrical pathways in the region are shown). All the demand centers shown are using power. The generators shown as circles are producing power at this time. Those shown as dotted lines are not. The power being pushed into the grid at various generation points is flowing through the grid's transmission circuits, shown as lines in the drawing, to the demand centers. None of the grid's lines or electrical facilities are overloaded at this time, and all the power that is wanted at every point is being supplied.

This example will focus on one particular line in this grid, the route shown from location B to E, and follow its operation and usage through part of a typical operating day. In early morning, when loading is light and power generation far less than it will be at peak later in the day, this particular line is not heavily loaded. The generator at B is producing power, which has been ordered through the regional power market the local retail distribution utility serving the city at E. The generator at B bid and the utility serving retail customers at E took the bid – either because it was the lowest cost or because other terms (payment schedule, etc.) were to its advantage. Regardless, B is selling power to E.

Similarly, the generator at A is producing power which has been bought by the local retail distribution utility serving the city at F. The power from both of these transactions, B to E and A to F, is flowing through the line from B to E, but at this time in early morning, the total power flow requested is not near the line's maximum rating. Similarly, everywhere else this regional grid is operating well. As is typically the case in most grids in early morning, demand and line loadings are not near peak conditions.

An important point about this example is that at this time, early morning, the pattern of generation is defined *entirely* by the de-regulated power marketplace: the generators running are those that were able to bid low enough

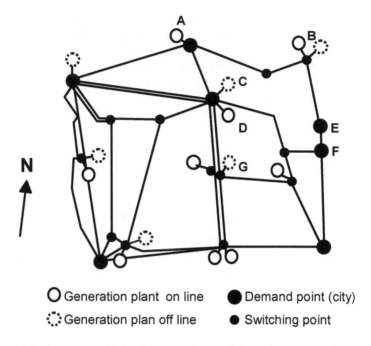

O Generation plant on line   ● Demand point (city)

⦂ Generation plan off line   ● Switching point

Figure 12.4 The power grid for this example, consisting of generators that are running (circles), demand centers (filled circles), switching or control points in the grid (small black dots), and generators that are currently not producing and selling power (dotted line circles). Letters refer to points discussed in the text.

or otherwise make arrangements with buyers for their electricity. Those not running either could not or would not make deals to sell power at this hour. Two of the generators who did make deals, and are running, are the aforementioned generator at A, which contracted with demand center F, and the generator running at B, which is selling to demand center E. Two not running are those shown at C and G.

In all cases, once a buyer-seller combination made its deal, they informed the System Operator prior to this current hour. The System Operation's computer system studied and evaluated each and all requests for "power transactions" for this morning, and was prepared in advance to operate the grid to accommodate their needs. Thus the grid operates now, and everyone gets pretty much what they wanted and expected.

All those buyers and sellers had one complication they had to deal with as they were negotiating their deals. They had to consider the price for delivery of the power they were buying/selling. They did this by consulting the grid operator's computerized bulletin board, where, since this example involves LBMP, the system operator posted prices for usage of each transmission line in the grid (how and why this is done will be discussed in more detail later). Each buyer-seller combination considered that they would have to pay that cost, too, and that cost often made a difference: buyer E had received a lower price *for power alone* from the generator at location C, but when the cost of transmission – transportation from that farther away location – was considered, that made the power more expensive than that bought from B over the shorter route from there.

Regardless, the prices quoted for the use of various transmission lines in the grid, computed and posted by the operator, always reflect costs of using each line as determined under whatever rules (fair or not) that the regional grid operator applies. They are updated constantly, on a 15-minute basis, throughout the day, as conditions change. Buyers and sellers have to take those prices into account, and eventually pay for the usage they do request and obtain from the grid.

So that is the situation in early morning, which is business as usual for the grid operator and the buyers and sellers moving power over it. As demand increases during the day and power usage approaches peak conditions, more buyer-seller combinations execute more and bigger contracts and more and more power is moved across the grid. In company with most of the other points on the grid, demand at E and F increases, and thus E buys ever more power from B, and F more from A, increasing the loading on the B-E line.

All this time, the system operator's computer system is constantly (perhaps on a quarter hour basis) re-computing and posting the cost for the B-E line and all other lines in the system, telling all users of the grid what their next increment of usage on each will cost in the next fifteen-minute period. Price for the B-E line escalates throughout the day as demand and loading on it increase, but not by a tremendous amount: price increases on an un-congested line as it nears its peak capability are relatively modest.

At some point, but in this example somewhat before peak demand is reached, projected load on the B-E line, based on requests for the next time period, exceeds the maximum allowable. That limit might be defined by the capacity of the line (any more and it would be damaged) or by any of several other inviolate criteria: perhaps the additional power flow would lead to unacceptably low voltage in this portion of the grid. This is shown in Figure 12.5.

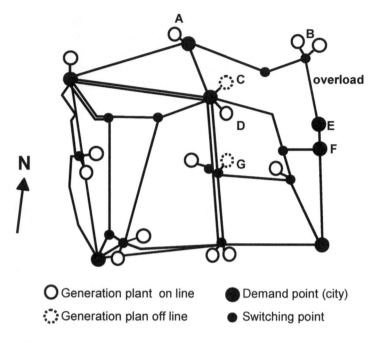

Figure 12.5 The situation requested by buyers and sellers near the peak time of day that would lead to congestion. Demand at most load centers has increased, and several additional generators are running (compare to Figure 12.4). The local utility at E wants to buy power from the second generator at location B (note it is shown as on line here). But this situation cannot be allowed. It would overload the line from B to E.

Regardless, at this point, if demand is to increase at B in the next time increment, it must be supplied through some other means than running the second generator at B, which is the deal that the utility serving demand center E has made. The "congestion management" sub-system in the grid Operations Center's control system looks at this situation. It tries to determine how to avoid overloading that particular transmission line, as well as any others, while still serving the demand, by re-adjusting the generation pattern to something the grid can tolerate.

In rare but really difficult situations, this can't be done – the grid has run out of capacity and users will have to be told "No!" The operator is prepared for this and there are procedures in place to assure such orders are followed. But usually, a way can be found: the power will simply have to come from somewhere other than at the north end of the B-E line. Generation will have to be re-dispatched, from the pattern of units defined by the business arrangements the buyers and sellers request (Figure 12.5), to one that serves all the demand, but that the system operator finds tolerable from the grid's perspective. The solution found by the system operator is shown in Figure 12.6. A generator at G is substituted for the requested, second, unit at B. The additional power needed for E would come from another direction over the set of lines leading from G to E. Neither those nor B-E overloads with this generation pattern.

In essence, the grid operator has determined a type of "out of merit" or "out of market" arrangement for generation in Figure 12.6, one that satisfies all the demand and stays within limitations of the grid. If this solution is implemented, a generator that bid low and thus made contracts to sell its power

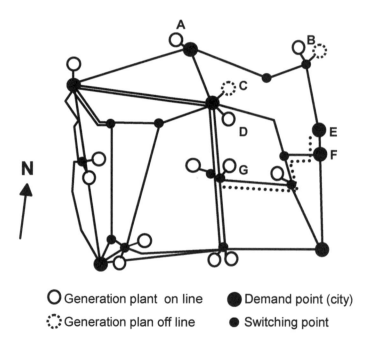

○ Generation plant on line   ● Demand point (city)

⋯ Generation plan off line   ● Switching point

Figure 12.6 The solution determined by the Grid Operator to avoid the overload of line A-B in Figure 12.5. The generator at G is ordered to run and the second unit at B is not allowed to run. This plan will not necessarily be implemented. It is used at this time only to compute "out of merit" costs for the congestion prices. It will only be implemented if the market does not respond by changing requested power flows.

(e.g., the second unit at B) will nonetheless be ordered by the system operator to back off on fulfilling the contract it made. Similarly, to make up for the shortfall in power generation that creates, another generator – one that bid higher and was not successful in finding buyers, in this case at G – will be ordered to run anyway.

Up to this point, this story could apply to a grid being run under just about *any type of pricing method* – zonal, postage stamp, contract path, etc. Regardless of the pricing system being used, congestion can occur, and the system operator must always be looking ahead for congestion. What differs with locational pricing is how the operator will try to resolve the congestion. The reader should note that at the moment being discussed here, the overload and the ordered re-adjustment have not yet occurred: *if* demand increases as expected in the next time increment, based on the information submitted by buyers and sellers, then the overload will occur.

If zonal, or postage stamp, or many other, non-locationally-based pricing mechanisms were being used, at this point in the day the system operator would probably have to step in and order this change in generation for the next time increment, or take other, equally draconian measures. That would create a mess – a manageable one, but one that few stakeholders like. The fundamental reason that no stakeholder will really like this solution is that the "out-of-merit" pattern of generation will be more expensive, but the objections are not just or even mainly pertaining to cost alone.

The problem that mandatory congestion management creates

Substituting unit G for the second unit at B avoids the congestion of line B-E, but raises costs (G is a generator that bid a higher cost than the second unit at B). Someone has to pay that higher cost: but who? The utility serving E made the deal with the second generator at B, the one ordered not to run, so someone could argue that it should pay the full price for the unit now running at G. But E has a very good argument against that. It is not the sole user of the line B-E. The utility serving F is buying from the generator at A, and that power is also flowing through the line. Shouldn't that utility share some of the added cost?

The situation is even more complicated. The second generator at B, which had contracted with E to run, had made plans to run and has costs associated with making those plans: maybe it has already started that unit and run it up to speed, calling in workers who it has to pay for the day, burning fuel to warm up the unit and bring it on line so it is ready to produce power, etc. It believes it should be compensated for those costs, which means more money has to be paid by someone. But even being compensated for those costs doesn't entirely satisfy G, because G's management believes that if it had been informed in advance, they could have found another buyer, somewhere else that didn't cause congestion, so they could have sold power and made money this day.

Thus, intervention by the Grid Operator will avoid the overload. But it will create an additional cost that must be allocated among users in a way that many will consider "unfair." And regardless, frequent intervention will create a huge burden of accounting and shifting costs among users to resolve the "broken" contracts. The reader should realize that the situation outlined here is relatively simple compared to situations that develop in large regional grids, where there can be dozens of buyers and sellers involved, and several lines that are congested and which interfere with one another (solving one line's congestion plan might make that on another more difficult to solve, etc.). Again, computers can deal with all this. It is people who can't, or would prefer not to. Ultimately, someone has to pay that higher price, and no matter how the rules are made, someone will feel they were unfairly treated.

The reader should appreciate that in some grids, situations like this happen just about *all the time,* producing hundreds, thousands, or even tens of thousands of congestion adjustments to track and resolve each month. Furthermore, in an atmosphere where plans and contracts will frequently be altered by the Grid Operator, buyers and seller can't make dependable business plans. They become frustrated with the unpredictability of their projected costs and revenues. It's difficult to do business in such an environment.

Thus, nearly everyone would prefer a way around this situation, even if it involved a good deal of complexity in dealing with prices. Locational pricing, specifically locationally-based marginal pricing, is a way of letting "natural market forces" – by which one means the buyers and sellers themselves – resolve the vast majority of congestion situations. Cost still increases in order to avoid congestion (that is inevitable) but buyers and sellers make their own decisions about how they will resolve and live with the situation, and if the rules are set up and applied correctly, everyone has a "fair" opportunity to compete on equal terms.

Posting Congestion Prices

Under locational pricing, as with every other pricing rule set, the system operator does everything outlined above throughout the morning: always looking ahead to the next time increment, determining if and how overloads will occur and what out-of-merit generation adjustments would be required, and always ready to step in if it does become necessary. But with locational marginal pricing, instead of actually stepping in and ordering the changes in generation, the system operator adjusts the prices for transmission by:

Determining how costs would increase if the computed changes were ordered, i.e., the second unit at B was told to shut down and the unit at G were substituted for it: generation cost would go up by X.

Revising the posted cost for use of the almost-congested line B-E:

essentially adding the cost X to the line's posted usage price. (Price adjustment is a bit more complicated than this, but for the moment those details will be passed over.)

Buyers and sellers of power, looking at these newly posted prices for use of the grid in the next increment of time, will now scramble to adjust their contract patterns in a way that seems best to them. Knowing they were working within an LBMP price framework, they knew this could happen, and the smarter among them were both ready and willing to make changes as needed.

Taking the higher posted price for use of the B-E line into account, neither E nor F will be quite as willing to contract for power that can be supplied only over the B-E line – one way or the other, whether the grid's rules require them to pay the higher price of the line, or whether the seller just bundles it into the price they pay, they will have to pay it. Thus, they will seek "higher priced" bids elsewhere, which, when the cost of transmission from those sites is considered, will actually prove to be lower in total cost. Similarly, generators located near the upstream end of this line (those at A and B) will seek buyers elsewhere in the grid to avoid having to pay or share in the high cost of that transmission. *The market takes care of adjusting for the congestion.*

Thus, aware of the higher cost the use of the line (buyers and sellers, in this example mostly the generators at A and B and the utility buyers at E and F) will their contracted agreements so that the pattern of generation requested for the grid avoids any congestion on line B-E. Their modified plans are submitted, the system operator determines those requests will not congest the line, and that pattern is run. The congestion doesn't occur. The grid continues to operate smoothly. In this example, what happens is that E decides to buy from the generator at C, as shown in Figure 12.7, and not G as the system operator would have ordered (Figure 12.6). It shopped around and got a better deal from C.

What happens if the market (buyers and sellers) does not adjust? The System Operator steps in, just as it would in the situation where other pricing mechanisms were being applied, and orders generation changes, and tracks cost increases and so forth, just as would be done under those other systems.

Marginal Pricing

The alert reader may have noticed a flaw in the foregoing simplified example. Once the market has adjusted, i.e., once E has shifted contracts so it buys from G and not generator number two at B, the congestion isn't going to happen. Thus, when the Grid Operator is informed of this change, the cost it computes and posts for used of the B-E line in the next time increment would drop. Thereafter, buyers and sellers would again see an advantage to using that line, and contract requests that overload that line would again be made.

Marginal pricing largely resolves this problem – at least if implemented in company with a number of small procedural details about bidding and operating that will not be discussed here. The system operator posts the *marginal* cost of the line – the cost for the next small increment (say 5 MW of power) that anyone wants to move along that line.

Marginal pricing makes the whole system relatively stable, because users of the line see the impact of increasing usage further. If usage on line B-E is far below its capacity, the marginal cost quoted is quite low, and buyers and sellers see a "price signal" that encourages them to use it. But as B-E nears its limit the price signal for it says "back off." Buyers and sellers know that this price does not mean "stop doing what you are doing" but that it does mean, "don't use any more than you are." In essence, marginal pricing, along with knowledge on the part of transmission buyers that the posted prices are marginal, means that the market knows that it can use the full capacity of the line, but also knows when it has reached that limit.

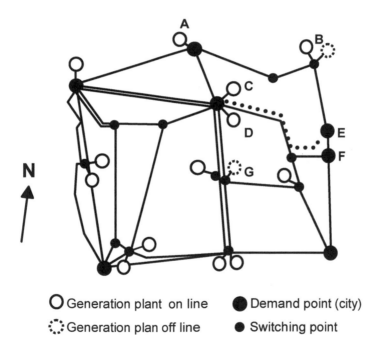

O Generation plant on line   ● Demand point (city)

⦂ Generation plan off line   ● Switching point

Figure 12.7 The solution found by the marketplace. The utility buyer at E decides to buy from the generator at C (compare to Figure 12.6). It was offered a better deal by generator C than any other when power and transmission costs were taken into account. Power flows along the path shown by the dotted line. B-E is not overloaded.

Why LBMP Works

Congestion situations where the operator must make major adjustments seem to occur a lot less under LBMP than under other pricing systems. The market does resolve many if not all congestion situations. Furthermore, the pricing system is "fair" in a very demonstrable way. Those paying higher prices (ultimately buyers E and F in the example) are those whose unaltered use would overload the line. But they had freedom to seek any acceptable way they wanted to handle their dilemma. No system operator stepped in and told them "This is how it will be." They could decide to buy higher generation from here or there, or they could decide to conserve energy and not buy power during high-cost periods.

This system works for a number of reasons, but two stand out as contrasts to other pricing systems that don't work as well. First, locational pricing permits the system operator to set prices in a way that single out lines or areas that will be overloaded, so as to discourage usage of them. This permits the market to avoid much of the congestion that would otherwise occur.

Second, marginal pricing – revised constantly, always anticipating the consequences of the next small increase in usage – sets prices in response to expected conditions, to discourage *further* use of system resources that are near their limits. It provides some stability to the pricing, avoids overloading lines, but promotes efficient use of (up to but not exceeding) system capabilities.

Is the Complexity Worth Its Cost?

The summary given above ignores many messy details as well as a host of small but important ifs, ands, or buts, some of which will be discussed later. But it does cover the major issues associated with congestion, its mitigation, and the reasons why locationally-based marginal pricing are seen as a good approach to power grid pricing. Simple as it might seem as described here, operating the grid with LBMP involves a tremendous amount of monitoring, computation, and back-and-forth communication among buyers, sellers, and system operator, as well as careful consideration of costs and some up-to-the-last-minute re-re-negotiation amongst buyers and sellers, and a good deal of intricate cost tracking and accounting. That is the case no matter how locationally-based marginal pricing is fine-tuned or implemented.

But artful, well-designed use of computers and data communications systems can streamline all of that work and make it straightforward and even routine. If one assumes that that has been done, and when all other considerations are taken into account, LBMP seems to work better for all concerned. The system operator has to step in and order changes much less frequently. Buyers and sellers can make business plans that are more predictable and over which they have more control. And there are additional advantages, to be discussed later, as seen by regulators and policy makers.

Some Important Details

There is no common, generally accepted way of implementing some of the big aspects as well as any one of the details of LBMP: it is too new for any uniformly accepted "best practice" to have evolved within the industry. Furthermore, needs vary from region to region and grid to grid, so that there may never be just one way of doing LBMP. Finally, in most regions where LBMP is or will be used, rules and procedures will be fine-tuned over time, as the Grid Operator and the buyers and sellers in the region, through an RTO (Regional Transmission Organization), refine and improve the rules governing operation of their grid. That said, there are some common characteristics that apply in nearly all cases.

Cost-based pricing. The prices computed and posted for transmission service are based on the total cost of transmission including the cost to own and maintain the grid lines and equipment,[12] the cost of operating the grid (including the cost for the Operations Center, its computers, and the system operation staff), and certain other operating costs.

Constant re-computation and posting. While the cost of ownership and operations is relatively constant (it can be computed once and allocated across all lines and hours of the year) those other operating costs vary as a function of demand, generation pattern, equipment outages and other conditions, and therefore have to be re-computed and re-posted periodically, usually on a 15-minute or 60-minute period basis. These other operating costs include the cost of electrical losses, the cost for the grid operator to contract for a sufficient amount of standby generation in case there is an emergency (e.g., pending blackout), the cost to obtain reactive power (voltage support), and other costs of maintaining a smooth, reliable grid operation. And of course, congestion needs to be anticipated, too, and its mitigation costs computed so that appropriate "price signals" can be posted.

Look ahead price posting. While conditions are evaluated and prices re-computed and posted on a 15- or 60-minute period basis, the system operator might perform this analysis and post hours, days, or even weeks ahead, estimating future prices so that buyers and sellers can make plans, then update those price estimates as the timing becomes closer to actual operation. This means the System Operator needs to request or forecast

---

[12] The Grid Operator does not own the lines, but computes and collects and passes on a fee to the owners of the lines. This fee is usually regulated by local utility commissions (or FERC) and is fair compensation for the cost of building and maintaining the lines in good condition.

information from buyers and sellers of their expected contracted amounts, locations, and times far in advance, and that if buyers and sellers and the system operator make plans far in advance, they might have to revise them from time to time. But most stakeholders in any grid prefer this: it permits longer-term planning and cost consideration.

Lines are not priced – point-to-point service is priced. The example covered earlier had one very significant difference from reality. Costs are not really computed and prices set for the use of specific *lines*, as was outlined there. Instead, they are computed for movement of power between specific *locations*. The system operator computes and posts, and grid users see prices for, transportation of power from one point to another, not for the use of specific transmission lines. This is shown in Figure 12.8. The reason is that often there will be several lines or paths (a set of lines in series) that connect two places, for example a large

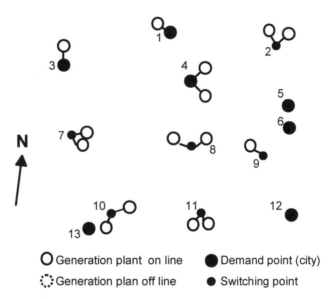

Figure 12.8 Prices are not actually posted for transmission *lines*, but only for transportation of power between transmission *points*. Here, the points in the example regional grid (earlier figures) which have either demand or generation have been numbered 1 to 13. Two 13 by 13 tables (one giving prices for transmission from one location to the other, the second giving prices for power flowing the other direction between each set of points) posted by the system operator would give all the price information buyers and sellers needed. How the power actually flowed would be the system operator's business only: buyers and sellers really would not care.

generating plant and major demand center (e.g., a large city), as shown. The grid users don't really care about which line the power goes over as long as it gets from the generating plant to the demand center. In fact, as discussed in other chapters, the laws of physics dictate that in reality a bit of it travels along all the various paths between the two locations. Thus, the determination and posting of price is done on a between points, not a line, basis.

Long-Term Incentives for Growth of Supply

LBMP has one additional advantage that is appealing to some rule and policy makers, and to people like the authors, who prefer to see rational consistency and balance in any system. The "price signals" LBMP sends to all concerned provide a long-term incentive for someone to invest in eliminating the root cause of the congestion. Consider the case in the foregoing example, in which line B-E overloaded under the preferred buyer-seller arrangements.

Readers unfamiliar with electric grid operation and industry practices might conclude that the root cause of the congestion in that example was an insufficient capacity in the line B-E, and that the solution to the problem is to increase its capacity, or build another line so that it can handle whatever loading buyer and seller combinations will request. Upgrading the line or adding other lines will eliminate the congestion. However, the real cause of line B-E overloading is that demand at E exceeds power available at E, i.e., there is no generation at E, or at least not enough, to satisfy location demand. And there is a similar situation at F which creates its problems and adds to E's congestion.

If the congestion in this example occurs frequently, for example every day of the year for some amount of time each day, the higher costs seen at E and F for power create an incentive for someone to build generation at E or F to satisfy that demand locally. They will have a market hungry for power and a cost advantage because competitors (generators at A, B, and C and G) face higher transmission costs serving that demand. LBMP is not alone in sending price signals that provide an incentive to build generation where there is demand. Other pricing systems also provide incentives that point in this direction. But many economists (people who deal with the details of price signals and their implications) see LBMP as providing price signals that are more consistent with other prices and price signals, that balance short- and long-term needs, so that they lead to better use of resources, including both existing grid capability and new capital investment.

12.6 SUMMARY OF KEY POINTS

Transmission service is a necessary element of every competitive wholesale power marketplace, whether it is provided *en masse* by a poolco structure, or through open-access availability of a common-use transmission grid. The interconnected grid is operated by a system operator, often called the independent system operator (ISO) to identify its impartiality to all users.

The transmission level can be quite complex, and operating it reliably, while making it both responsive to the market and fairly accessible, is perhaps the greatest challenge facing power industry de-regulation. In some de-regulated systems, bulk power transmission services are unbundled, and users are free to pick and choose those they want. At the transmission level, prices are usually determined based on cost, and allocated to users based on any of several pricing methodologies.

Some readers might think that, although this complexity can be handled, and that, even though most of the complexity can be handled smoothly and dependably by computers, there is no reason why the smooth, stable regulated world should ever have been disturbed. That not-uncommon opinion explains much of the debate that swirled around de-regulation when it was first implemented, and that continues to this day in somewhat reduced degree. But in fact, by just about any and all reasonable measures, the complexity and "mess" really are justifiable. Power grids *are* more efficient and power prices are a bit less, as measured by some very important metrics, than they were prior to de-regulation. De-regulation at the wholesale level works, at least if the arrangements for operation and pricing of the transmission grid are done well.

FOR FURTHER READING

J. Finney, H. Othman, and W. Rutz, "Evaluating Transmission Constraints in System Planning," 1996 IEEE Summer Power Meeting, Institute of Electrical and Electronics Engineers, New York.

S. Hunt and G. Shuttleworth, "Unlocking the Grid," *IEEE Spectrum*, Institute of Electrical and Electronics Engineers, New York, July 1997, p. 20.

C. Pleatsikas and B. Turner, "Electric Competition in New Zealand: Putting Last Things First," *Electric Utilities Fortnightly*, June 1996, p. 26.

R. D. Masiello, "It's Put Up or Shut Up for Grid Controls," *IEEE Spectrum*, Institute of Electrical and Electronics Engineers, New York, August 1997, p. 50.

M. Shahidehpour and M. Alomoush, *Restructured Electrical Power Systems – Operation, Trading, and Volitility*, Marcel Dekker, New York, 2001.

R. D. Tabors, "Lessons from the UK and Norway," *IEEE Spectrum*, Institute of Electrical and Electronics Engineers, New York, August 1997, p. 45.

H. L. Willis and G. B. Rackliffe, *Introduction to Integrated Resource T&D Planning*, ABB Power T&D Company, 1994, Raleigh, NC.

H. Willis, J. Finney, and G. Ramon, "Computation of Unbundled Transmission Costs," *IEEE Computer Applications in Power*, Institute of Electrical and Electronics Engineers, New York, October 1996.

13

Power Distribution in a De-Regulated Industry

13.1 INTRODUCTION

Electric industry de-regulation and re-structuring is aimed at creating competition among power producers at the wholesale generation level, and in many cases at the retail level, as well. In the latter, competition is done through an "open access" distribution system that permits *direct customer access*, or customer *choice of electric energy suppliers*. Often called *retail wheeling,* open distribution access provides much the same effect at the retail level that open transmission access created at the generation level. Here, as at the transmission level, there is only one "set of wires," run by a regulated, monopoly franchise distribution company. The same wires, transformers, breakers, voltage regulators, and other equipment that comprised the distribution portion of the traditional "integrated utility " under the previous monopoly franchise paradigm, or that constitute the distribution system of an LDC when only the wholesale level is de-regulated, form a local delivery "electric roadway" system, shared by all competitors at the retail level. The "local distribution utility" owns this system as a regulated monopoly franchise, and operates it on an open access basis. As is the case at the transmission level, the operator of the open access system is not allowed to be a player in the competition that uses it. Thus, the company who would have been distribution operator and retail seller (an LDC) if only the wholesale level is de-regulated is now not a retail seller of power.

Many aspects of distribution operations remain the same as they were under the traditional utility structure – after all, it is the same equipment, connected in the same manner, and serving the same electric consumers. But many other aspects change, particularly with respect to metering, operation, investment, and customer service.

This chapter looks at the basic concepts behind power distribution systems in a de-regulated environment. Sections 13.2 and 13.3, respectively, consider "open" and "closed" distribution systems, examining them from both the *electrical systems* perspective – how it will function and be controlled, and the *business and managerial* viewpoint – how transmission assets are paid for and how their use is allocated. Section 13.4 speculates on the changes in performance, investment incentives, and managerial perspectives that are likely to result from this approach. Finally, Section 13.5 provides a brief description of how the combined Disco-Resco infrastructure might work from the *customer's* perspective.

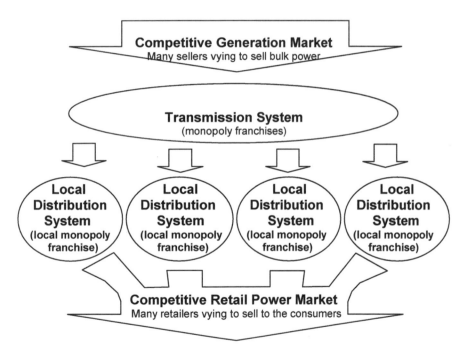

Figure 13.1 If the competitive retail market is permitted by local regulatory jurisdiction, it will be reached through local distribution systems, each a monopoly franchise "wires company" that provides retail energy service companies with access to customers throughout its service territory.

Distribution and the Retail Market

Just as transmission was necessary in order for wholesale buyers and sellers to transact business, so is the distribution system the key to the retail market. Competing retailers cannot sell electric consumers their product, unless they can deliver it to them, the distribution system being the roadway to do so. Figure 13.1 is a redrawn version of Chapter 11's Figure 11.1 showing more detail in the distribution level.

Just One Set of Regulated Wires

As mentioned elsewhere in this book, no one has ever seriously proposed competition in the task of delivering power, and while local distribution systems overlap slightly in many locations, for example, near Carrolton, GA and White River, MO, in general there is only one distribution system in any locality. In most nations, including the United States, there are many more local distribution systems than there are ISOs, or "transmission systems." As Figure 13.1 illustrates, each local distribution company is the monopoly franchise supplier in one particular area – usually a city or town, or a small portion of an agrarian region.

This sole local distribution system will be managed, one way or another, as a monopoly franchise for public use, either *directly* under government supervision, through governmental or municipal ownership, or *indirectly* through government regulation of investor-owed Discos, to assure it is operated fairly and within reasonable financial rules. Many of these distribution systems belong to utilities that also own a portion of the transmission grid. For example, Pleasantville Power and Light might own several dozen miles of transmission linked into, and essentially controlled by, the Independent System Operator in its power pool area. It also owns and operates the distribution system in and around Pleasantville. It has a single monopoly franchise to own and operate power delivery facilities in and around Pleasantville.

What Is Transmission and What Is Distribution?

The U.S. Federal Energy Regulatory Commission (FERC) has identified seven distinctions that test whether a part of the power system is under its jurisdiction, and thus available to open access under FERC rules, or whether it is "local distribution," and thus under the jurisdiction of the states (Order 888). These seven characteristics of local distribution (Table 13.1) are a fairly good test of whether equipment and facilities are transmission or distribution in any power system.

Table 13.1 Differences Between Distribution and Transmission According to FERC

1. Local distribution system is normally in close proximity to retail customers.
2. Local distribution system is primarily radial in character.
3. Power flows into local distribution; it rarely, if ever, flows out.
4. Power on the local distribution system is not reconsigned or transported to another market.
5. Power on the local distribution system is consumed in a comparatively restricted geographic area.
6. Meters at the interface from transmission to distribution measure the flow onto the distribution system.
7. Local distribution is of reduced voltage compared to the transmission grid's.

"Closed" and "Open" Distribution Systems

The distribution system is the pathway to the retail market – without access to it a company that wants to sell power cannot get its product to the consumer. Therefore, only if local distribution is "opened" to access by companies competing for retail electric sales will there be customer choice among various retail sellers.

Strictly speaking, open distribution and competition at the retail level do not have to happen just because competition and open access have occurred at the wholesale transmission level. In the United States, the manner of de-regulation, if any, allowed in a particular area of the country will be determined by *state regulators*, not by the federal ones of FERC. In many other nations, local power delivery is similarly turned over to local government authorities, thus giving local politicians the power to make the decisions about de-regulating their local distribution systems and permitting competition in the sale of power to their constituencies.

Thus, there will be great variation in whether and when local distribution systems throughout the world are de-regulated, and in how local distribution companies will be re-structured to permit competition.

Closed distribution: Business as usual

Under regulation, the local monopoly-franchise utility company, usually vertically integrated, owned the distribution system and was the sole power retailer. Any other entity wanting to move its power to a customer over the system – or even a customer wanting to generate his own power at one of his sites and move it to another within that system – was prevented from doing so.

Open distribution: retail wheeling = direct customer access

In some retail markets, local distribution may remain "closed." The only seller permitted will continue to be the local monopoly franchise utility that owns and operates the distribution system, and sells consumers power. Other systems might be "open." The local monopoly franchise distribution company owns and operates the wires, but other retailers are permitted to use that system to deliver power to consumers who prefer to buy from them, rather than from their local distributor.

The service of delivering electric power to end-users, through an "open access" distribution system, is called *retail wheeling*. It is in every sense analogous to wholesale wheeling at the transmission level. But a better term might be *direct customer access,* because that phrase gets to the heart of the matter: Distribution is both the key to competitive access by electric customers, and the key to *consumer choice.*

Direct Customer Access, or Retail Wheeling Seems Almost Inevitable

"Closed" distribution systems will undoubtedly continue in limited areas, but direct customer access, or something similar, seems inevitable in a majority of jurisdictions where de-regulation of the transmission system has already occurred. This is likely for two reasons:

1. No technical barriers stand in the way of retail wheeling, because it can be done on just about any distribution system.

2. Politics: Voters will demand it.

De-regulation of the utility industry is largely a political process. Open transmission systems, in conjunction with closed distribution systems, provide *choice among energy suppliers* only to very large users of electricity, those industrial and commercial buyers whose demand is great enough for them to buy at the transmission level. Only these very large businesses directly receive whatever real and perceived benefits accrue from de-regulation. Homeowners and small businesses are supposed to benefit, because their local distribution utility is now buying on the competitive market and getting better prices, etc.

However much these savings are, and however practical the arguments in favor of keeping local distribution "closed" in a particular area might be, they will very likely not impress local voters, residential customers, and owners of small businesses who will want *choice*. De-regulation and unbundling of other industries has led to considerable improvement in quality of service.[1] Voters will expect it in electricity, too. Eventually, political pressure will force a move toward freedom of choice for *all* consumers, which means retail wheeling, or "open access" distribution. Whether retail wheeling actually provides lower costs and better service is immaterial: The ability to "shop around" is a fundamental part of the freedom consumers want.

Closed Distribution: Business Pretty Much As Normal

In many regulatory jurisdictions, re-structuring will adopt a "two-tier" approach in which the transmission grid is open to use as the conveyance mechanism of a competitive wholesale (bulk) power market, and the distribution systems on a local level will be closed. In such a framework, LDCs (local distribution companies) "shop around" for power, buying from the lowest cost producers on behalf of their customers and shipping the power over the open transmission grid to their substations (see Chapter 2).

LDCs sell the power they bought at the wholesale level to their retail customers over their distribution lines, with electrical consumers having no choice in their retail company. From all standpoints, inlcuding the ultimate energy consumer, the distribution business managers in the LDC, and most of the electric distribution system engineers and operators, their environment is "business as usual," very similar to how distribution and retail operated under the traditional regulated structure.

Each local retail distribution company has a monopoly franchise in its exclusive service territory, and the attendant obligation to serve. As at present, the line between delivery and power sales is blurred for both utility and customer, and traditional ways of operating and planning distribution systems and billing for power will generally prove sufficient.

In this type of de-regulated environment, large vertically-integrated utilities will very likely disaggregate – functionally or actually – their *combined* distribution and retail assets and functions into a Local Distribution Company (LDC) that buys power at the transmission level, distributes, and resells it to consumers, covering the cost of both delivery and the power, itself, from what it charges for the power. These LDC businesses would be regulated, much as the entire vertically integrated utility was under the traditional structure. Even so,

---

[1] For example, the phone industry. It is often argued that de-regulation of long-distance service in the United States resulted in no cost reduction. It certainly resulted in far less cost escalation, and major improvements in service, as well as lower costs, for numerous commercial users. It is also very doubtful if a fully regulated phone industry would have been able to keep pace with the explosion in technology and volume created by the data communications (Internet, fax, modems, etc.) needed by today's society.

there will be changes that affect distribution. The following five impacts are expected in an industry where distribution remains "closed."

Increased demand: Competition at the generation level will lower the cost of power, and these savings will be passed on to the consumer. Since the retail distributor companies are regulated utilities, they will have no choice but to pass on those savings. Thus, demand will rise in response to lower prices, perhaps by as much as 5%.

Worsened load factor: With lower prices overall, there may be less incentive to flatten load curves and schedule some uses off-peak. Load factor will probably worsen: There is certainly little reason to expect it to improve. Thus, peak demand will probably grow slightly faster than energy usage.

Higher volatility and risk in power pricing: Today's non-generation-owning municipal and REA utilities buy power on an annual or even ten-year contract basis. Future retail distribution companies will be able to buy in an *hourly* market within a competitive power exchange. Prices will probably be lower overall, but more volatile and less predictable over the short run, which means slightly greater opportunity and risk on the power purchase side of the business.

Expectation of improvement: Local Distribution Companies (LDCs) will be part of an electric industry that holds all elements to a higher standard of cost effectiveness than in the past, and all the talk of de-regulation and the expected lower prices will reach electric consumers, who will expect extremely high levels of value, i.e., lower prices.

Competition for customers: LDCs will see limited competition in the electric sector, and lose some customers to other suppliers. Open transmission access, alone, creates opportunities for niche suppliers and some types of "load aggregators" at the retail level.

Thus, in a "closed" distribution environment, distribution utilities will do business in much the way they always have, but load levels will probably rise very slightly and cost-reduction pressures will be greater than ever.

13.2 OPEN ACCESS DISTRIBUTION

While it is too early to predict the details, it is clear that open access distribution (retail wheeling) will cause sweeping changes in distribution systems and their operation. For distribution, the largest change from the traditional "closed" industry structure will be *the separation of the distribution of power from the retailing of the power delivered.* LDCs, as described above, both distributed and sold power at the retail level. In a framework where distribution is open, a local distribution company (Disco) owns and manages the wires and equipment that distribute power, while separate companies, Rescos, sell power at the retail level. Figure 13.2 shows the overall structure expected of a "retail wheeling" industry. The players in this industry are:

Genco: Owns generation and "manufactures power." Sells competitively into the wholesale (bulk) market or may be affiliated with a Resco to market at the retail level.

Transco: Owns and maintains transmission facilities that move power in bulk. May be the local area operator of the grid, or may be run by an ISO. Not affiliated with any Genco.

Disco: Owns and operates the local distribution system. Owns no generation. Sells no power. May be allied with, or owned by, a Transco. Such a joint T&D company is often called a "wires company."

Resco: Sells power directly to the end-users in a competitive retail environment. A Resco (sometimes abbreviated ESCo) may be a subsidiary of a Genco or may be only a "buy wholesale, sell retail" company.

LDC: A monopoly franchise combination of Disco and Resco. Essentially, it operates the way a non-generation-owning municipal utility did under the traditional regulated structure: Buys bulk power on the (now competitive) wholesale market, and resells it through its distribution system. It may or may not permit other suppliers to use its system.

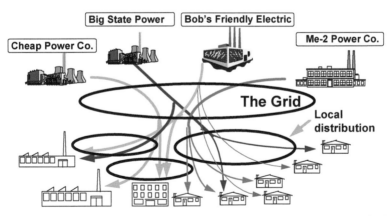

Figure 13.2 The ultimate "de-regulated" industry has open access at both the transmission and local distribution levels. Producers, sellers, "middlemen," retailers, and customers alike would contract with distribution companies (Discos) to move power at the local level. Some customers would buy from two or even more suppliers, and every Disco would be providing wheeling services to several retailers simultaneously.

Except when an LDC permits no competition, the distribution system, like the transmission system, is available for use by any qualified party. Users would reserve capacity on it and pay fees based on their reservations and use.

Many Different "Solutions" to Open Distribution Access

There will be more variation from region to region in how de-regulation works at the local level. One reason is that retail distribution access will be regulated at a state level, in the US, and a more local level in general than transmission and generation. Also, there are thousands of distribution companies instead of only a few dozen power pools, as at the transmission level, there will be more diversity in how open distribution access is implemented and accommodated by utilities.

Impacts on the Distribution System

As mentioned above, the single biggest change in distribution will be that the distribution system and its operation will split off from the other parts of the utility into a separate company, which, unlike the combined distributor-retailer, does nothing but own and operate the distribution system. It does not sell power.[2]

Electric consumers will buy their power from a Resco, a retail energy services company, and they will likely have a choice of many suppliers, just as they now have a choice of many telephone service companies. Many Rescos will be allied with or will be part of a large Genco, for example, the retail division of a company that owns and operates 20,000 MW of generation. However, some will be pure retailers, companies that buy power in bulk, arrange for its shipment across the transmission and distribution systems, and then sell it directly to consumers.

Dis-aggregation and divestiture

To begin with, many vertically integrated utilities will dis-aggregate, either legally or functionally, putting their distribution assets and functions into separate Discos, far apart from their retail activities, which will be put into a *non-regulated* Resco. Very likely, these dis-aggregated Discos will own only those assets essential to power distribution and handle only those functions fundamental to distributing power. All other equipment and services will most likely be held as part of an unregulated "service company," or Resco. This approach is taken in order to move as great a portion of the business as possible into the unregulated, and potentially more profitable, arena.

---

[2] In fact, it is a *buyer* of power, for it must buy the power to make up for the electrical losses in its system. As transmission operators do in the de-regulated environment, the Disco will buy power to cover the electrical losses that occur on its system. It will do so as other companies and users buy power, by bidding for it on the competitive market.

Two examples of such services are metering and line construction/repair. A utility might spin off its meters, meter maintenance, and meter reading into an unregulated service company, or as part of its Resco, which could then sell metering services competitively, perhaps back to the Disco. Similarly, line trucks, crews, and equipment might be owned by this same Resco, or another service company, and be leased or contracted as a service to the Disco.

From a business standpoint this step makes a good deal of sense in many cases. The utility has technical and managerial expertise in both metering and construction, from which it can probably earn more than a regulated rate of return in a competitive environment. It is not inconceivable that the entire distribution operations process, i.e., inspection and repair management, trouble call answering and processing, as well as the outage restoration process, could be removed from the Disco and offered as a competitively priced service by a utility services company. The Disco would contract for these services to be performed, as many small REAs do at present.

The large investor-owned utility, having broken off its distribution operations in this manner, might bid not only for the contract to manage its former distribution assets, but for management of other distribution systems, too. Likewise, planning and engineering could be contracted by the Disco, which may well be a firm with a very small staff, basically an asset management company similar to large real-estate holding companies, which own billions of dollars in assets, but have only a dozen on staff.

Distribution becomes a business in its own right

One way or another, distribution will have to justify its existence and cover its costs on the basis of providing the basic service of power delivery, as well as whatever optional services it still has, e.g., meter reading, if still a part of the Disco. The costs of providing distribution services can no longer be covered by an allocation from the revenues recovered in the sale of the power itself, since that is now a separate business.

An entire pricing structure for power distribution must be created, including access fees, demand and energy prices, and other service tarrifs that relate to distribution of power. This is a significant change, and will be discussed later. But much more important, it means a change in culture and focus, because distribution has always been viewed by utilities as ancillary to the primary goal of selling power. For the Disco, this is no longer the case. Its identity is purely that of a mover, not a seller, of power.

What happens to the "obligation to serve?"

Currently, as part of its franchise in its service territory, an electric utility has an "obligation to serve." It is not clear what happens to this requirement under open distribution access, and with many different regulatory jurisdictions involved, there will undoubtedly be many different solutions. Under fully open access T&D, the local distribution company clearly has no obligation to serve, for it does not sell power; it only provides connection. It would probably have some regulated "obligation to connect."

Regardless of how this obligation would be implemented, it creates some interesting and very basic questions about why the Disco exists and what the purpose is of the distribution system it manages. For example, does the Disco have only *an obligation to connect*, or does it have *an obligation to connect with sufficient capacity?* If the latter, how is sufficient capacity defined? Who covers the costs, if it turns out that building to this target overspecifies the capacity need and the Disco spent more than necessary?

On the other hand, if the obligation does not carry some capacity target, then is running #24 wire to each customer sufficient to meet the minimal obligation to connect? Perhaps some basic minimum, e.g., 5 kW, will be established, and customers would be billed for the actual cost of any additional capacity running to their property. This is similar to how local telephone service is regulated in many states, where one incoming line (or in the case of some newer systems, two) is run to each address, and anything more is only at the request of consumers, if they are willing to pay the additional cost.

Another possibility is that the Disco will extend service to new locations or upgrade it to existing locations, only for a Resco, not a consumer. Thus, to obtain new service extension, a consumer would first have to be under contract to a Resco, which would tell the Disco how much capacity to build, and contract to pay for it.

One important issue is how to assure that electricity is available to all consumers if there is no obligation to serve. There *will* be customers whose load patterns, location, payment record, or other characteristics make them unattractive to most Rescos. They could be handled as drivers with poor driving records are treated by the automobile insurance industry, i.e., the assigned risk category, through government-regulated prices at which any Resco must provide minimal electric service to a customer/site.

The concept of the distribution system's "customer" changes

The dis-aggregated distribution company's (Disco's) customers will not be electric consumers, as in the past, but the Rescos, a relative handful of retail energy suppliers that negotiate and buy access and services from the Disco. Electric power consumers will be their customers, not the Disco's. These retail energy sales companies are the entities who will pay the Disco's bills. It will not take long for the Disco to realize that its incentive is to keep *these* customers happy, and not to respond directly to the interests (real or perceived) of the electric consumers. This is, in many respects, a reversal of recent trends, which sought to make all employees in vertically integrated utilities more "customer aware," i.e., consumer aware. In the future, the Disco will become "customer aware," but this will mean running it as a business serving the Rescos.

Multiple Users of the Distribution System

Several vendors of power will make simultaneous use of the power distribution system to sell power at the retail level. These Rescos will buy access to the distribution system and sell their power to individual consumers, channeling it through the distribution system, which they must share. This will have many impacts on the Disco, the first being to force more explicit record keeping and documentation of all distribution functions.

No doubt one of the major retailers in any area will be remnants of the vertically integrated utility that once owned the distribution and sold all the power through it. However, the Disco will be separate, at least functionally and perhaps completely, from its former sister division. It will be prevented from "unfair" communication and cooperation with that or any other retail company, and it will probably have to be able to prove that it has followed the rules with regard to that. It will face new challenges in handling these multiple users of its system and will need to have:

Improved Metering and "Flow Analysis": Multiple vendors on the system will bring about a need for much improved and detailed metering of the flows through its system, and for a whole new concept of pricing, or more correctly cost-based allocation methods.

Auditability:. The Disco must be impartial with respect to all Rescos. All procedures and information transfers, both internal and external, will be audited to assure propriety. It will have to maintain rigid policies on communication and data handling and put formal documentation procedures in place to prove it has followed "the letter of the law."

Limits on Sharing and Use of Data: Finally, it is worth noting that the Disco, as well as the Rescos, will have limitations, legal and proprietary, on the use and sharing of data. Businesses are always reluctant to share data on their customers with their competitors, and most Rescos will hesitate to give information on their customers to anyone, particularly other Rescos. However, the Disco may be required by law, or in order to prove impartiality, to share all information it receives with all its customers. This will lead to Rescos not sharing any of the data on their customers, the consumers, with the Disco. As a result, planners may not have access to, or be allowed to use, some data they traditionally have depended upon.[3]   Alternatively, the Disco may have access to such

---

[3] As a result, a Disco could find itself in a bizarre situation. As an optional service, it

information, but with very stringent conditions of use, under which it must document its activities carefully and take prudent measures to keep the information confidential. Either way, paperwork and documentation effort will increase greatly.

Some Consumers Will Buy from More Than One Vendor

At least some electric consumers, and perhaps many, will buy power from two or more vendors, either simultaneously, or by schedule, one on-peak supplier, one off-peak supplier, etc. Non-generating municipals and large industrial customers do so now. In many cases, other consumers will, over time, find this very useful. Beyond the complexities that multi-vendor power buying presents to the consumer, although not to the distributor, it complicates further the issue of metering, cost allocation, and billing for distribution services.

Overall, Electrical Usage Will Probably Increase

As mentioned earlier, de-regulation and competition normally lead to reductions in the cost of power, despite those who point out anomalies. The weighted average for power prices has either dropped or failed to track expected "regulated industry" escalation trends in every case where the electric industry has been de-regulated. As electric power becomes less costly, it looks more attractive as an alternative to gas, fuel oil, and other energy sources. Some forecasts predict a long-term increase of up to 12% in annual electric energy usage in the United States. Thus, loads on the distribution system will increase.

Distribution Reliability May Worsen – By Design

As previously noted, present power distribution systems in the United States provide a service availability of about 99.98%, a nationwide average of slightly less than two hours without service per customer per year, excluding major storms. This is an incredible feat, one distribution engineers should be proud of, but in an open access distribution industry, it may be an unwanted one for the following reasons:

- While about 40% of respondents in most surveys of electric consumers say they want improved "power quality and reliability," an equal or higher number state that "cost is a major consideration."

---

might read meters under contract for its Resco customers, and even do billing services, for that matter. That information on customer usage will be considered proprietary data by each Resco. This Disco could thus find itself reading meters and yet be unable to use that data in planning and analysis of its operations, or if it does use it, being responsible for making sure the information is guarded and kept secret from other Rescos.

- Slight reductions in reliability of service often result in noticeable reductions in cost. In one distribution system, 99.93% reliability, or 5 hours of interruption per year average, could be had for only 95% of the cost of 99.98% reliability. To many users, a 5% savings on their annual electric bill is worth more than three additional hours of interruption each year.

- Where needed by a consumer, high availability service can be provided using UPS (uninterruptable power supplies) and power conditioners, installed at the home or business, at a lower cost than similar reliability can be provided by reinforcing the utility power system.

In a retail wheeling environment, the Disco may find its customers, the Rescos, clamoring for the lowest possible prices, even if that means lowering distribution system reliability. First, many Rescos will want to compete on the basis of price: They will negotiate the lowest costs possible and will accept marginal service in some cases and in some markets to get it. One option is to offer a basic *economy electric service* that does not accept trouble calls outside of normal business hours.[4]

Second, many Rescos will have a vested interest in seeing average distribution level reliability slip to marginal or even poor levels. They will want to sell premium services such as UPS, power monitoring, and on-site generation, rather than pay the Disco to provide equivalent reliability. Poor distribution reliability produces a market for such services.

Since the Rescos are the Disco's customers, ultimately it will listen to them. Beyond this, letting distribution level reliability fall will be the path of least resistance. As mentioned above, demand will be growing and construction will be necessary at times to keep pace, but the Disco's customers will apply tremendous cost-containment pressures. The Disco may find the best course is to build the leanest system possible, to be described in "budget constrained planning" later.

It is important to realize that as discussed here, reliability at the distribution

---

[4] The authors are aware that even if many customers subscribe to this offering, it would do little to change the trouble-management/restoration process, since even if just one "non-economy customer" were on an outaged line or transformer, it would have to be restored on weekends. However, Rescos, and some economists and policy-makers, will argue that those customers who want only such a service should have it available and that the cost of taking and responding to trouble calls outside of normal business hours should be borne by only those who want it. A small insurance broker's office, open only 8AM to 5PM Monday through Friday, would probably prefer such service, seeing little value in "weekend and night-time" restoration.

level can drop, while that at the customer level, at least for critical loads, can improve due to installation by the Rescos of UPS and other customer-location reliability augmentation devices. The bottom line, however, is that *customer value* will rise. The price for electricity should drop, and reliability, where needed and paid for, will be higher than at present.

The reader should realize that this really must happen with retail de-regulation. A key goal of retail de-regulation is to give customers "choice." To be meaningful, this has to mean more than just choice among companies who offer precisely the same service and quality and price. There has to be a something they can really choose from, and that will no doubt be price versus service quality and options.

Although price will be a competitive battlefront among retailers, they will be buying from the same wholesalers and paying the same (regulated) rates for use of the distribution system. There will not be room for a lot of cost control and the price competition that could go along with it. They will all be delivering over the identical distribution system: reliability of power as delivered from the distribution system will not be a product distinction area, either. What will develop is competition in packaged offerings: of equipment like UPS (premium power) and at the other end "economy power" using demand limiters and load control. This will be discussed more in Chapters 14 and 15.

Pricing and Billing: *Much* More Complicated

Pricing and billing of distribution services will be far different than at present. Distribution-level pricing may or may not follow the precedent(s) set by transmission access pricing, but fundamentally new ways of evaluating the cost of services and of allocating these to customers will be necessary to bill distribution on its own merits. Since prices in a regulatory framework are nearly always "cost-based," it is anticipated that these would be also. Therefore, accurate knowledge of the actual costs of providing delivery services, particularly costs that vary as a function of time, location, or load pattern, is vital to a system operator for a number of reasons:

> *Pricing*: Accurate knowledge of cost is thus required to determine prices that are defensible to both regulatory scrutiny and customer challenges.
>
> *Unbundling of services:* The Disco may prefer, or may be required, to unbundle (price and offer separately) services such as delivery, losses, voltage regulation, and power factor correction.
>
> *Billing for "inadvertent services":* This applies to losses on a utility's system caused by unscheduled retail wheeling.

Operating decisions regarding economy of operation and commitments to customers require knowledge of how costs relate to alternatives available to the transmission system. *Business decisions* regarding operations, investment, and commitments to customers all require accurate knowledge of costs and how they vary, if for no other reason than for the operator to make certain that the sum of all revenues exceeds the sum of all costs.

The pricing formula used at the distribution system will depend on what is worked out between the Disco, its customers, and the local regulatory authorities. It might be quite simple, such as a flat fee for use of the system, regardless of the exact amount of usage, and based only on the class of consumer (e.g., all homeowners pay the same amount, all small businesses pay the same amount, etc). Distribution-level pricing formulas could also be quite complicated, if the utility, customers, and regulators believe that complexity is warranted because it allocates cost more equitably among users. A research and development project completed recently by ABB and four electric utilities in the United States formulated a method of pricing for open access transmission based on very exact determination of "who is using what." The methodology has been successfully applied in tests to distribution systems, and appears to work well at that level, too.

Distribution's Radial Nature Simplifies Much of the Pricing Controversy Seen at the Transmission Level

Chapter 10 summarized several of the most debated issues in determining how pricing responds to transmission cost, and what happens during wholesale wheeling at the transmission level. In particular, displacement, distance traveled, and parallel paths are all issues that have considerably different interpretations, and whose effect on delivery costs varies greatly from one system to another.

The radial nature of power distribution (see Figure 13.3) frees it from most of these controversies. Power in a distribution system flows "electrically downhill" along a single radial path. This is a much simpler flow pattern than in transmission systems. With respect to each of the three serious pricing debates at the transmission level (see Chapter 10), this simplicity makes them all moot at the distribution level:

Displacement is simply not an issue. Power on a distribution feeder flows out from the substation to customers. Adding any additional load on the system means transporting that power the full distance from the substation to the load's location. There is no displacement possible.

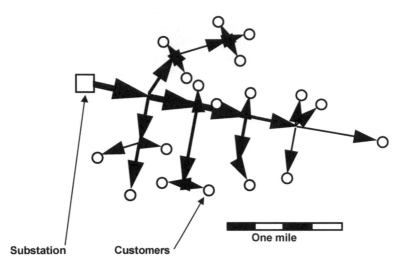

Substation **Customers** One mile

Figure 13.3 Distribution systems consist of many radial feeders. This is a simplified map of one feeder, which would normally serve 500–1,000 homes. A feeder would normally have hundreds of segments (arrows) and serve hundreds of customers (circles), not the few shown here. Arrows show major line segments and direction of flow, with line width illustrating the amount of power flowing. In a radial system, there is only one source (substation) and power travels the full distance from source to customer, along only one, not multiple, pathways. Unlike in a network, there is no displacement, no question of the distance the power has traveled, and no parallel paths to create confusion about cost allocation.

Distance covered is the electrical distance from substation to load. Again, except for cases of distributed generation, counterflow cannot occur. Like displacement, any complexities of distance at the transmission level do not occur in the distribution system.

Unbundling of Distribution Prices and Services

One aspect of pricing at the transmission level that is sure to follow to the distribution level is "unbundling" of services. Unbundling means that various aspects of power delivery are separately identified, priced, and available for purchase. Thus, at the transmission level, a buyer of power delivery can purchase the basic transportation of the power, but not buy the energy for losses in moving it, or voltage regulation: Legally, he must provide these if he does not buy them, but he is free to buy from another source or provider other than himself. Unbundling of pricing and availability of services creates noticeable complexities in price schedules and required business procedures. It also increases the need for detail and accuracy in the analytical tools used for cost-based pricing.

At the distribution level, similar unbundling is technically and practically possible, but it may not be worth the trouble. Distribution could be billed by taking into account the following services and costs:

Primary service, or delivery capacity

This includes the provision for and maintenance of delivery capacity to a particular customer. It is a capacity charge designed to recover the Disco's cost for the equipment required to meet the customer's peak demand. This might also include the cost of no-load losses for the transformers required.

Ancillary charges

Ancillary charges include all other costs of operation:

Operating the system: The Disco can charge enough to cover its costs as the "System Operation" charge, on either or both of a per kW or per kWhr basis, if it can develop tariff formulas (rate schedules) satisfactory to its regulators. This corresponds to the charges levied at the transmission level by the "System Operator" as discussed in Chapter 10.

Losses: In the same manner as at the transmission level, power at the distribution level requires electrical losses in order to move. Energy required for losses in the power system can be provided by the Disco, Resco, or customers. One complication is that, strictly speaking, losses include no-load losses of service transformers, which is a significant factor. These are perhaps best handled as a fixed cost covered in the primary service fee.

Reactive compensation: Again, the Disco has the ability to provide this (shunt capacitors). However, there is a natural way to offer both an incentive and billing mechanism for customers to provide reactive correction: Billing by power factor has been widely used in the past and could certainly be applied in a de-regulated environment, too.

Voltage regulation: The Disco provides voltage within accepted industry standards. Any tighter regulation, or protection from momentary voltage sags, surges, etc., is a "power quality service," provided at extra cost by a Resco.

Reserves: Reserves in the normal sense are not required on the distribution system. Backup (primary feeder select switching) could be contracted as additional capacity.

13.3 CHANGES IN DISTRIBUTION OPERATIONS

Inspection, maintenance, and construction aspects of distribution operations will probably change only slightly under de-regulation, i.e., there will be far tighter budgets and a need for more documentation, but the overall emphasis and priorities will be the same. However, in the day-to-day operation of the distribution system, customer service, trouble call analysis, and outage restoration will change dramatically:

Customer interaction: Distribution operation activities are consumer-related and can be viewed as either Disco or energy retailer related. Whom does the consumer call when service is interrupted? There will probably be all manner of solutions here, from situations where the whole spectrum of present services is performed, as at present, by the Disco and priced as part of the basic "distribution" service, to unbundled pricing of meter reading, trouble call receipt, analysis, restoration, etc., as options to the basic service. Part of this diversity will be different Discos and Rescos trying different business practices, and part will result from the variety of different regulatory climates that will prevail among the different states.

New opportunities: These may emerge for performing services for the Disco's customers, the retailers. Meter reading and various "premium" types of trouble call and service restoration activities may be among these, but smart load control and energy management and other services will result.

Tiered-reliability customer service: As mentioned earlier in the discussion on reliability, the Disco may be asked to implement various levels of trouble call and outage response. This could run the gamut from economy customers, whose trouble calls are accepted only during normal business hours, to premium customers, with on-site "smart" power monitoring units (PMUs), computerized systems that circumvent human interaction, and communicate service trouble at the customer site directly into the utility's computerized outage management system.

Rescos "overbooking": Currently it is accepted practice in the air transportation industry for airlines to overbook flights, for example, selling 160 seats on a 150-seat aircraft. Airlines do so because this proves most profitable, and Rescos may act similarly for all the same reasons, basically because customers overestimate/over-reserve future usage. Airlines plan carefully just where and how much to overbook flights, thinking only of their bottom line profitability, not their customers' convenience or the impact this may have on other business, such as airlines with connecting flights. When they discover a net overage for a flight, they pay customers to delay usage by taking a later one.

Rescos will do likewise, planning their "overbooking" in a purely profit-driven manner, and very likely not considering the Disco's concerns at all. Thus, they may take contracts for delivery of 5.5 MW of peak demand, when they contract with the Disco for only 5 MW of delivery.

Their first tactic to "handle the situation" when demand exceeds contract will be to ignore the situation and see if the Disco tolerates it. If necessary, they will "delay passengers" by using load shedding and automated load control, which they will install where and when and in the manner that suits them. From the Disco's perspective, it must make certain that it is protected from overloads or adverse operational problems due to such practices.

Possible Competition for All but the Basic Delivery Service

If the Disco's customers believe they can do any of its activities, such as meter reading, customer service phone answering, trouble call analysis, dispatching, even repairing the wires, they'll want to send their crews out and bill the Disco for the work, and more, they will push to see that they are permitted to do so. In essence, the Disco should view itself as being in competition with the retailers in all of these areas, and:

1. Make itself "competitive."

2. Seek out its customers, the retailers, and find out how to satisfy their needs, so that they do not "do it themselves."

Documentation and "auditability": This will become important for the entire operations process. This means having a verifiable paper trail (or digital data trail) to prove what, how, and why all decisions were made and implemented. For example, after any storm, the Disco can expect:

1. To be unduly blamed by the Rescos for all outages.

2. To be challenged on the order and priority attached to restoration, often to the point of being accused by each Resco of restoring service to another retailer's customers first.

Beyond the benefits in response time, resources management optimization, and operating efficiency that computerized distribution management systems bring, they will probably be needed simply to provide the necessary time-tagging and archiving of distribution operations data and decisions, in order for the Disco to meet this requirement to document and defend itself against criticism.

Changes in Planning and Engineering Priorities

In a competitive retail power industry, distribution planning will develop into two distinctly different activities. First, the Rescos will perform comprehensive "distribution" marketing, investment, and cost-versus profit studies in order to evaluate their opportunities, determine how to take advantage of them, and plan both their marketing strategies and their commitments for the purchase of distribution capacity contracts from the Disco. This activity has no counterpart in the present electric industry: It goes beyond the "marketing" that regulated utilities did and mirrors the very detailed focus that auto manufacturers and other consumer product companies put into their retail planning.

This activity will involve targeted load forecasting and delivery price analysis, what one might term retail distribution market planning. It is quite interesting and will be an exciting new area of utility planning, but it is not germane to this discussion, as it will not be a function of the Disco, but of the Resco.

Second, Rescos will perform a type of distribution planning that is more traditional. The operator of the T&D system will need to project the total usage (load) on the T&D system and plan the system's expansion, for the very same reasons as in the past. However, in a multi-retailer environment T&D planning grows in dimensions, and presents several new features:

More work with fewer resources. This goes without saying and is not a new trend, rather a constant trend in the power industry for the last fifity years: improve productivity. However, cost-cutting and innovative efficiency increases on the part of competitive elements of the industry (Rescos, Gencos) will put additional pressure on Discos to follow suit as will their own competitiveness and market-share goals.

What does the Disco plan? As asked earlier, what obligation does the Disco have, and how are capacity targets for its planning to be set and used in planning? At a basic level, one can note that all the customers will be there regardless of which retailer serves them. However, the various programs and plans of multiple energy vendors will change the load patterns.

There is a very basic change here that could occur, a shift in what the Disco's planning goal is as compared to today's. Very likely, the situation will vary from state to state and utility to utility depending on local regulatory laws and consumer requirements, but every Disco will fall somewhere between these two extremes:

1. *Business as usual:* The Disco essentially plans as it always has, perhaps under "performance based" regulation by the state utility commission, which set the standards to be met.

2. *Build only on contract:* The disco builds only if a retailer says, *"We have a new customer and would like service extended to it and we will pay for it."* Only then does planning start, and only for the contracted amount of capacity.

Expansion will be needed. In the long run, Discos can expect a noticeable, if not significant, increase in electric demand wheeled through their systems: as electric costs drop, demand will increase.

Pricing studies will play a part in planning. Discos will evaluate new facilities and rank alternatives in the planning process on the basis of how their capabilities can be priced or their cost recovery allocated under price-based regulation. Studying the pricing of the services provided by, and the revenue gained from, all new additions will be important for planning.

Planning will be held against tougher and more intensely applied standards. The Disco will probably find itself "eating" the cost of any identified planning mistakes, including higher than expected losses as well as the capital cost of "planning errors," inevitable though they may be. This might be considered "unfair," but this trend can be expected anyway. An open access environment will mean the Disco's customers will be only a few large companies (Rescos), larger, and with more resources and skills than the Discos. In some cases they will simply "out-gun" the Disco in arguing their case, and win major, if "unfair," cost concessions.

Zero-based planning, in which *all* expenses must be re-justified on an annual basis, will be a likely outcome of the scrutiny and tight budget focus forced on Discos. Traditional regulated basically used "past practice based" planning. If they spent X on a particular activity in year N, then planners and management really only had to justify any increase above X in year N+1. This was part of the institutionalized "non-innovation" structure of regulation, discussed in Chapter 10 (p. 282). Zero-based planning assumes the "normal" level of all expenses categories is zero.

Less information, more partnering. Once de-regulation comes, the Disco may find itself cut off from some traditional sources of load, customer, and growth data. As mentioned earlier, the retail energy suppliers may not share all their plans, customer survey data, etc., with the Disco. Some of this lack can be made up from the extensive demographic and other types of data available in the public domain. Some adjustment can be made, because the consequences of any problems caused by not having all the information can be passed on to the retailers. However, the distribution planner of the future will find a far different information environment than in the past. Planning may become similar to the cooperative ventures

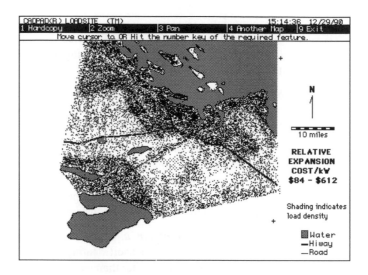

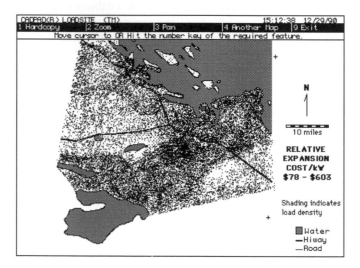

Figure 13.4 Maps showing the total T&D investment needed to deliver power to customers for two different ways the power system could be built in and around a coastal city, computed and allocated to customers on a locational basis. The two plans differ by less than 0.2% in total lifetime PW cost, but greatly in where their areas of lowest and highest delivery costs are. Regardless of the allocation mechanism the Disco uses to recover its costs, the Disco or one or more Rescos could benefit if decisions between alternatives like these are biased in their favor. Disco planners will need documentation showing that such decisions were not "gamed."

between auto manufacturers and tire companies. The latter work under proprietary and secret arrangements with auto manufacturers to plan new tire designs and production capacity for cars as yet unannounced. Similarly, the Disco might work with one retailer in planning the capacity or other expansion necessary to allow that retailer to increase its operations/sales, or to offer enlarged or new services. How extensive this practice becomes will depend on regulatory rules about information sharing and disclosure, and the success of early trial programs for such partnering.

Auditability and documentation. As discussed in the section on distribution operations impacts, the Disco will also find itself in need of copious documentation to "prove" that it acted fairly to all retailers and that it "did the right thing" in every case. For planning this will include far more than just substantial documentation to show that the utility diligently followed industry best practices, and that it conscientiously pursued every avenue to reduce costs. Inequalities in cost and capacity inevitably develop and, in fact, are impossible to remove from any least-cost power system (Figure 13.4). The Disco will need to be able to demonstrate that it played no favorites in its system design – that it did not select plans for the benefit of one retailer, but for the best overall. This will require much more documentation of data, planning procedures, and decision-making than is done at present.

13.4 WILL DISTRIBUTION PERFORMANCE IMPROVE DUE TO "COMPETITION"?

Designs and Procedures Will Improve Most

The design and manufacture of the equipment from which distribution systems are assembled – transformers, capacitors, switches, and the like – has been honed through nearly a century of intense competition among electric equipment suppliers. Barring unexpected breakthroughs in materials (i.e., development of a *perfectly* magnetic material, as opposed to the nearly perfect material used currently) or completely new processes (e.g., a "solid state" loss-less transformer/capacitor), future improvement will be slow, and mainly in the form of slightly lower costs for basically the same devices.

But system design – the ways in which many pieces of that equipment are assembled to form a distribution system – has never seen similar levels of competition. In fact, distribution system layout and design was quite parochial at most utilities in the regulated era. Every vertically integrated utility developed its own distribution design standard some time in the first half of the century, largely independent of much outside influence. These designs were used and re-used until institutionalized as "the way we do things here."

As a result, as the 21st century gets underway, otherwise similar electric utilities still assemble very different types of systems from essentially the same standard transformers and line equipment. Most of these different design approaches do an adequate job of distributed power to electric consumers, but the fact is that most are far from "best" and many fall short of "world class" performance in economy, reliability, or electrical performance by up to 30%.

Such marginal performance was seldom noticed, because distribution was blended into a vertically integrated system where it was nearly impossible to isolate and analyze separately. And, as described in Chapter 8, regulated utilities had no incentive to "make waves" by trying to innovate or improve systems that were stable and "doing their job."

In a de-regulated industry, this situation will change. Poor performance will stand out sharply. Many utilities will operate under performance-based rate regulation that specifically targets distribution performance. Others will simply feel tremendous pressure because they can be objectively compared to the best industry performers. As a result, distribution designs and operating procedures will improve: Those designs that are the best will see widespread use. The remainder will become obsolete. Distribution utilities will need to respond with a keener focus on both their customers' needs and their own finances. Lean financial structure, new types of planning methodologies, and maximized use of existing assets, be they equipment, human resources, or public image and good will, will be the keys to prevailing within the new industry environment.

Is an Intelligent Distribution System the Way to Go?

A number of separate but aligned initiatives within the power industry are examining the role of the distribution system in future power systems and assessing the extent to which automation, "smart" equipment, new technologies, and extensive control and data systems can improve distribution performance. A good deal of these efforts, including particularly that of the Electric Power Research Institute (EPRI), are aimed at achieving very high reliability of service to consumers, on the order of 99.9999% availability of continuous electric power (less than one minute without power per year). Modern needs for continuous power supply, created by the "digital world" with computers, robotic equipment, and on line systems that require near-perfect availability of power, are cited as the reason behind such research.

In the authors' opinion this initiative and others are missing a key point. While such high levels of reliability are clearly desirable, the cost to achieve them has to be considered. An intelligent distribution system with 99.9999% reliability would be expensive, even if the advanced equipment it requires were to reach full-scale production and thereby achieve significant economies of scale. The fact is that there are other, and perhaps better and lower cost, ways to achieve high reliability of supply to digital equipment

The answer to "how smart" or how reliable the distribution grid of the future depends greatly on the cost of competing ways to attain the required reliability of the end use. *One does not need reliable power, just reliable computers and appliances.* And one proven method to this end is has a cost quite far below that projected for highly advanced distribution systems: distributed resources (DR). DR includes uninterruptible power supplies (UPS) and distributed generation (DG), equipment known to work well and proven in installations and millions of homes and businesses in the US already.

For example, the UPS system shown in Figure 13.5 costs $49.95 at a retail business supply store. It provides outage protection to up to 500 watts of demand in the authors' home (a large home computer system with all its associated equipment along with two digital memory telephones). During five years of operation the electric load served by this UPS has never had a power outage, although there have been roughly 2 events per year during that period, one lasting over three hours. Assuming that, in the next hour, something would go wrong and there *would* be an outage of this UPS, the reliability would still be 99.99999% (one hour in ten years). Using the unit's cost and load capability, one obtains a marginal capital cost for this reliability of than $100 per kilowatt of protected load.

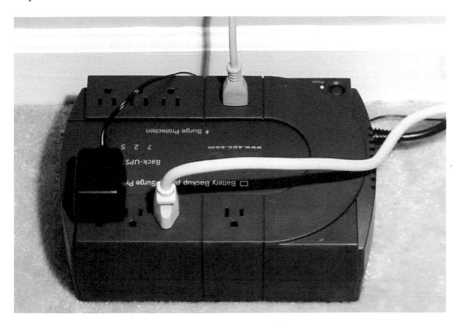

Figure 13.5 An uninterruptible power supply (UPS) system like that in the authors' home can be purchased for about $50. It provides better than "six nines" reliability (better than 99.9999% availability of power).

It is against figures like this that any proposal to spend billions on upgrading the nation's distribution systems to "six nines" reliability must be weighed. Such costs typically run on the order of $1,000 or more per household (about 4 – 10 kW demand each) for a computed marginal cost for protected KW of demand of between $250 and $100 dollars. Furthermore, a UPS solution is superior to that provided by the distribution system in two significant ways. First, it protects against failures inside the house (wiring or breakers fail, or lights go out due to a short circuit in the house). Second, it is selective. Homeowners or business people who want high reliability can buy it in this manner; those who do not can defer and select instead the lower cost that an "old and dumb" distribution system will provide. Among homeowners who would select the second option are the authors, who do not want to have to pay the additional cost (which they estimate would be about $12/month) for the additional reliability that an intelligent distribution system would provide (the UPS protects to only critical loads in the house).

What Does the Distribution System of the Future Look Like?

Overall, it is too early to reach any firm conclusions about how distribution will be affected by de-regulation and open access, and if and how new technology and advanced computation will permit big improvements in design and performance. However, this chapter has covered some of the basics, and tried to speculate on the most likely results and the responses that will occur in the power industry. One can anticipate growing demand, disaggregation of "wires" from merchant functions, unprecedented pressures to reduce cost, and a need to document every decision and aspect of operations and planning.

That said, the authors believe that the following is most likely to happen. Distribution systems of the future will be *somewhat* better in reliability than current systems, but only somewhat. The improvement will come about mostly through rebuilding distribution systems to largely traditional types of design (i.e., "dumb"), because of aging infrastructure reasons – many existing systems are close to being worn out. When rebuilding, utilities will make circuits a little bit more robust and a little smarter. But the result won't be "intelligent" systems with omnipresent automation and smart equipment throughout. Instead, distribution circuits will have very limited smart equipment installed at only a few important sites. Nationwide, SAIDI and SAIFI will probably drop to about 75 minutes and 1.1 events per year, better than the national average today (about 120 minutes and 1.5 events per year) but certainly not a major leap forward. Equally important, the total cost of power distribution will be slightly less than today's: most of the money spent on distribution improvement will go into reducing operating cost, not improving reliability.

Very high reliability of service – less than one outage every ten years – will be provided by the competitive retail service providers that distribution de-regulation will create (see RESCo in the Index) at prices far below what it could be provided through distribution improvements. They will use "smart" UPS systems and small distributed generator sets to provide this, to customers willing to pay for it. Those unwilling to pay will be able to select normal reliability at lower cost. De-regulated industry will thus be able to deliver one of the key components it is supposed to offer – *a wide range of choice* to consumers in what they buy and how much they pay.

FOR FURTHER READING

James J. Burke, *Power Distribution Engineering – Fundamentals and Applications*, Marcel Dekker, New York, 1994.

H. L. Willis, *Power Distribution Planning Reference Book – Second Edition*, Marcel Dekker, New York, 2004.

14

Retail Sales in a Fully De-Regulated Industry

14.1 INTRODUCTION

The entity that most electric consumers will think of as "their electric company" in a fully de-regulated industry, and that could very well be the strongest among all the players involved, is the retail energy services company, or *Resco*. Whether independent, allied with, or integrated with other functions, such as generating or local distribution, Rescos will be the direct end sellers of electric power. All but the very biggest consumers, such as factories and large institutions that buy on the wholesale market, will purchase their electricity from Rescos. They will exist only in a fully de-regulated industry, one with competition at both the wholesale – energy production – and retail levels. Thus, they will co-exist with retail wheeling and direct customer access.

Rescos will compete for their customers' business, probably in a very intense manner, similar to the way long-distance and cellular phone companies and auto manufacturers do so, through advertising, tailored service plans, improved technology, and attempts to give the customers exactly what they want at the lowest possible price.

This chapter looks at both Rescos and the retail level of the re-structured power industry. It begins with a discussion of the Resco and its functions and priorities, in Section 14.1, its place in the industry, in Section 14.2, and the way a typical customer will get power, in Section 14.3.

14.2 LOAD AGGREGATION AND SERVICES

Load Aggregators

The term "load aggregator" was coined during the early phases of discussions on industry re-structuring, to describe the Resco function as seen from the wholesale level: A Resco *aggregates* many small loads into a "wholesale size" purchase, as depicted in Figure 14.1 Seen from the ISO or wholesale power level, load aggregators are *buyers of power* – entities that satisfy their many small customers, or members, by buying power from Gencos, arranging for bulk delivery through the ISO or Pool, and arranging for its local delivery by contracting with the local distribution, or "wires," companies.

Regardless of the nation or political jurisdiction involved, power industry de-regulation has been almost entirely a top-down process, with most of the initial thinking and initiative directed at the wholesale level. This was certainly the case in the United States. During the very turbulent and sometimes contentious discussions about open access and competition at the generation level, the term "load aggregator" was widely used in many different guises as a convenient way to explain the function of de-regulation at the retail level, without much detailed thought given to its form or function.

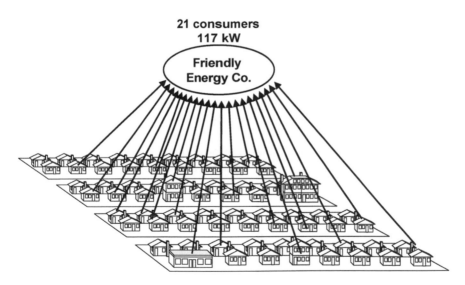

Figure 14.1 Here, a load aggregator, Friendly Energy Company, contracts with 21 homeowners and businesses in a neighborhood. When these are added to its other customers, Friendly Energy ends up with several dozen MW to take to the local wholesale marketplace.

For example, during the period when the U.S. power industry was grappling with the FERC "mega-NOPR," the notice of pending rule changes sent out prior to order 888, many people believed that community churches would form "load aggregator" organizations, pooling the loads of their congregation's home and business owners, and jointly negotiating with Gencos for power at the wholesale level and with the local Disco for local delivery. Thus, people said, this small group of consumers would enjoy such benefits as lower cost of electric industry de-regulation.

While such "church-cos" are possible within most de-regulated retail industry structures, they are unlikely to develop, or last long if they do, because most congregations would have little incentive to go through the effort. The church's members will be able to obtain similar savings from any number of competitive retail companies (Rescos) performing the same service. Very likely, many of these will offer prices below anything the church can obtain. Regardless, their prices will be low enough that the considerable continuing attention and resource cost required to run a "load aggregator function" well will not be worth the effort to the church congregation.

There Is Likely to Be Little Profit in Retail Sales of Electricity

The paragraph above brings up a significant factor at the competitive retail level. One supplier's electricity looks *just like* any other's: There is no difference whatsoever. Quality, or lack of it, as delivered to any specific customer or site is entirely due to the wires and equipment running to the customer site, which are the province of the Disco.[1] Electricity is a commodity. All of it is the same, with *price* being the only possible distinguishing characteristic in its sale.

The cost of electric power will be as low as possible, simply because any company that doesn't offer low prices will be bounced out of the market rather quickly by those who do. Profit margins on electric power will very likely be less than 3% of cost, the average level of profit in the retail grocery industry. In some cases, sellers may take a loss on electric power, in order to sell *services* along with it.

Emphasis on Services

Competitive power sellers will use the sale of electricity as the foundation upon which to offer services that will provide them with greater profit than just selling power. These services will run the gamut from basic "economy" electric service packages to premium levels of service superior to anything offered under regulation. For the moment, the important point is that offering additional

---

[1] Quality can be improved at a customer site with "custom power" equipment, but that is an additional service provided by the Resco, as will be discussed later in this section.

energy services, along with electric power, will permit retail sellers of electricity to accomplish two goals essential to success in a competitive market:

> *Product distinction:* Their electricity might be the same as everyone else's, but the services they provide with it will be quite different.

> *Profit:* Premium or special services may be able to garner a profit margin far above 3%, and in some cases, close to 100%.

As a result of this need for and focus on services, the successful, competitive "load aggregator" will be a fairly complicated entity selling much more than just power. To begin with, it will be much larger than the "Church-co" discussed earlier, aggregating not just one or two congregation's worth of consumers, but truly large numbers spread over many regions. The largest retail energy service companies will very likely have ten million or more customers. This will give them not only great economies of scale, but also tremendous "buying clout" in the wholesale market.

However, these companies will be most interested in offering many other support and energy related services, including, but certainly not limited to, custom power and premium reliability. These services will involve them in much more than buying and reselling electricity, spanning all phases of energy usage and all types of technology. Many will be immensely profitable, even at modest prices, because the marginal cost of supplying these services will be nil, given the infrastructure the retailer has built for power delivery and other services.

Selling Other Forms of Energy

There is no reason that an energy seller would restrict itself to electricity alone. It can sell gas, propane, fuel oil, and passive solar water-heating and space heating. *Consumers can get all their energy needs from one supplier, in one package, and receive one bill each month.*[2]

Energy and Energy Use Services

There are numerous special, and potentially very useful, energy services that a retailer can provide, all of which offer great value to some customers. Many people have trouble picturing these, because the regulated utility industry has created little if any incentive for electric suppliers to offer any of the following:

> *Improved Customer Service:* As any marketing textbook will explain, "customer service" includes all aspects of the customer's dealings with the

---

[2] Or, perhaps each week, or at some other interval. One option successful retailers are almost sure to offer is flexibility in how, and how often, customers are billed.

company, including the initial contact expressing interest, all dialogue between them, billing, and payment. Anyone who has sampled customer service quality of various traditional utilities, as the authors have, will have noted how much it varies from one to another.[3] Competition will bring forth service beyond the best traditional levels, including customer access by computer, and other options.

Flexible Customer Service: Electric utilities traditionally provided 24-hour "trouble call" service to all users, i.e., a phone number for reporting trouble, or "lights out" any time, any day. Some consumers do not need 24-hour, 7-day service, e.g., an insurance broker operating a one-man office in a single-room office building may be quite content with "normal business hours" service, if it brings him a small price discount. Other customers may need much more comprehensive trouble monitoring, e.g., the owner of an exotic fish store might worry that if the power goes out, and hence the heaters, while he is not there, many of his fish could die. Retail service companies will tailor service and trouble monitoring to meet these and many other needs.

Billing is an area of great flexibility. A business owner with 12 outlets could arrange for composite billing – only one bill not 12, to ease the accountant's work load, with statistics broken out by site for his study of site performance.

Direct End-Use Sales: Some retail companies may offer direct end-uses, for example, home heating and cooling. The supplier will install the type – electric, gas – and the appropriately sized unit, assuming responsibility for maintenance and repair, and billing the customer for the service of having the end-use.

Power and Quality Monitoring: Traditionally, electric utilities relied on their customers to telephone them if the power flow to their sites had failed. Instead, quite inexpensive equipment, power monitoring units,

---

[3] Of 61 utilities, with 100,000 or more customers, called from local pay phones during standard work hours on "normal weather" weekdays, i.e., not during storms or emergencies in 1996, roughly 15% answered their trouble line in three rings of the telephone or less. Two didn't answer within 15 rings of the phone. Utilities that answered quickly always had courteous and accurate responses to questions about services available, company policies, rate options, marketing department contacts, etc. Utilities that were slower to answer were not always as responsive. A survey of 23 smaller utilities (less than 50,000 customers) showed a similar wide variation in speed of answer and quality of customer interaction. In general, large investor-owned utilities had the best performance in answering, average four rings, while municipal utility departments had the worst, average eight rings.

costing about $25 each, can be installed at a site to tell the retailer instantly of a failure, before the customer could even place a call, or even if he or she is not at the site. Other smarter and more expensive monitors will report voltage surges, spikes, sags, or harmonics problems automatically, often allowing the retailer to fix an impending problem, even before the customer is aware of it, though caused by the customer's equipment, since most harmonics are generated by the *customer's*, not the electric utility's, equipment problems.

Equipment or Site Monitoring: Rather than monitor just the power, as described above, a retailer can offer to monitor both the *condition and performance of a consumer's electrical equipment.* For example, regardless of a power failure or another problem, is a water pump at a remote ranch site filling the water trough for cattle? Is the AC unit cooling properly? Is a vending machine empty? The retailer can deal not only with these matters, but also provide non-power-related monitoring, such as burglar and fire alarms.

Uninterruptible/Backup Power: A wide range of equipment is available to provide anything from emergency backup power supply that comes on for an entire building within one minute of a power failure, to uninterruptible power to just a single computer, so that it continues to function even if the power repeatedly flickers and goes on and off. Retailers can offer a number of options with regard to how long a period the backup will cover, e.g., UPS can be installed to cover outages of up to four weeks – for all of the consumer's equipment, or just critical devices.

Custom Power Quality: A wide variety of specialized electronic equipment can be installed, either at the consumer's meter, or on individual appliances and equipment units, to maintain one or more aspects of power – voltage or harmonics – within very stringent limits. For example, it is possible to control voltage to within 0.001% in cases where that is critical, as in certain types of integrated circuit manufacturing. Such service necessitates a unique type of controller, which is adjusted and maintained by using rather specialized skills, but retailers will make it their business, literally, to have access to everything required.

Automation and Control: Residential automation includes "intelligent" control of major loads, such as water heaters and air conditioners – control schemes considerably smarter than the "load cycling" used by electric utilities in their demand-side management programs of the 1980s. For example, an "adaptive" water heater system would not only reduce usage during afternoon peak conditions in order to lower the consumer's cost, but over time learn that peak usage was on weekday mornings, for which it would prepare by making certain there was plenty of very warm water available at that time, before the first faucet was turned on. Similar control and coordination of other major appliances can lower cost, increase their lifetime, and blend them into the consumer's schedule with minimum fuss and maximum value.

Weatherization and Efficiency: Perhaps offered as part of a "lower overall bill" or "environmentally friendly" package, it could include caulking and weather-stripping, sealing of ductwork, installation of awnings and "heat mirror" window film and similar building envelope sealing. Simply adjusting the duct vanes – the amount of opening or closing of individual ducts – throughout a home, using a coordinated analytical approach can make a surprising difference in comfort level and energy usage. A retailer could also offer efficiency replacement programs.

Demand-Side Management: Many of those services mentioned above overlap. However, Rescos can offer services, including the design of effective energy usage and efficiency programs, to homeowners and especially to small and medium-sized businesses. A Resco could provide analytical and decision-making services to determine just what types of appliances, level of insulation and weatherization, and load scheduling would result in maximum value to the customer. It could sell efficiency appliances and equipment, too.

Innovation is a continuing trait of competitive systems. Many, many more services than those given above are possible, and will be available.

14.3 RESCO IDENTITIES AND INDUSTRY POSITION

The Power in the Power Industry

In many ways the retail energy services company (Resco) is the most powerful player in the fully de-regulated electric industry, for two reasons. First, as depicted in Figure 14.2, it controls the money flow from the customer. Although it has business obligations that it must meet, which means it needs the Gencos, Transcos, and Discos, ISOs and power exchanges, it has considerable discretion in how it meets its commitments. It will have a significant say in the way business is done, and considerable clout in negotiations with other players.

Second, the Resco will be the most "vertically integrated" of the various players in the re-structured industry, as shown in Table 14.1. It, alone, among the Gencos, Transcos, Discos, ISOs, PXs, and regulators, has some reason to be involved in every level of the de-regulated power industry. Thus, it is in a very powerful position – involved in everything and in control of the money flow.

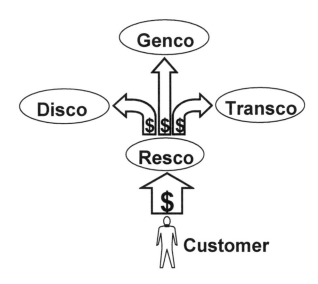

Figure 14.2   The Resco will have quite a bit of clout in the de-regulated industry, because the customer's money passes through it on its way to the other major players.

Table 14.1 Players and Their Involvement in Various Levels of the De-Regulated Power Industry*

Player	Generation	Transmission	Distribution	Retail
Genco	X	x		
Transco	x	X		
Disco		x	X	x
Resco	X	X	X	X
ISO	x	X	x	
PX	X	x		
FERC	X	X		
State PUC		x	X	X
Large Customer	X	X		X
Small Customer				X

* Size of X indicates relative interest or involvement.

Rescos in Combination with Gencos

A combined Genco-Resco is a permitted entity in most de-regulated industry structures: It is not barred by FERC, the state PUC, or similar regulatory authorities in most other nations. Such a company owns generation and sells power directly to the end user. As many large, vertically integrated utilities dis-aggregate their various functional divisions, they may combine their generation and marketing groups in such a "doubled ended" entity – a company that works in both ends of the de-regulated, competitive sandwich (see Figure 11.1).

A Genco-Resco does not have to restrict itself to selling only the power it generates, or vice versa: It can sell its generation on the wholesale market and buy power elsewhere, from other generators, in order to satisfy its retail obligations. Genco-Rescos must still use the "public" open access transmission and distribution systems to move their power. But they control resources at both ends of the de-regulated industry, and for that reason they ought to be able to achieve high levels of efficiency, if they monitor their customer loads precisely and control and carefully plan their generation accordingly.

Resco Marketing and Product Distinction

Like all companies in intensely competitive markets, Rescos will advertise, promote, and otherwise make every possible effort to put themselves and their capabilities before as many potential customers as possible. In this regard, they will closely resemble long-distance phone companies and auto companies in using print and media advertising, and in promotions that occasionally reach the level of silliness.

Product distinction and market niches

Each Resco will try to establish an image or name that differentiates it from other retail energy suppliers. Each will aim for name and brand recognition in the public's mind, usually in terms of a specific, identifiable characteristic, such as lowest price, or highest quality, or whatever else gives it the advantage it seeks in the marketplace. A distinction may or may not be real, but that is only incidental to the task of getting the public to accept and remember it.

However, some distinguishing qualities may be genuine. For example, some Rescos could specialize in meeting the needs of customers who must have very high power quality. Others might offer low rates, with admittedly limited service – bargain basement electricity. Table 14.2 shows several, but by no means all forms that these promotional efforts could take.

Table 14.2 Possible Forms of Market Distinction by Retail Power Companies

Market Approach	Method of Market Appeal
"Green Power"	"We buy power for consumers who want only power from wind, solar, and similar forms of renewable generation. It costs more, but we're sure you don't mind that in order to help save the planet."
Big and Experienced	"We're Big State Power. For 80 years we've brought over a million homeowners and businessmen reliable, low-cost power. We own the generators that make our power. We're the biggest, and the best."
Basic Economy	"We offer good power at a low price, and specialize in weatherization, load control, and other energy efficiency measures that can help you save on energy and lower your total energy bill. Nobody can beat the total energy price we can give you!"
Reliable Service	"We guarantee you exceptional reliability, and we pay you if we fail. We put power monitors on your meter. We put a UPS on your critical loads. No more than 5 min./yr. without power, we promise."
Premium Upscale/Conspicuous Consumer	"Hey, gadgets are neat! Get the electronic equivalent of a Mercedes from us – smart meter, a home automation system, UPS for your PC and TV/video player, appliance monitoring and control, security alarms, power monitoring, and an on-line analysis of your energy usage, any time, right from your home PC. It comes with a nifty looking control module to replace your thermostat, which will really impress visitors to your home. You get a premium service line, and your own designated service rep, who gets you priority service if you ever call us. We bill you electronically, of course."
Why Bother?	"Tired of the hype? We're the power company that sold you power, and your parents power. Good power, at a good price, from people who live right down the street."

Table 14.3 Residential Energy Usage Plans for a Hypothetical Resco

Market Approach	Method
Family Plan 1	Designed for active, all-electric homes and large families. The basic monthly fee of $142 includes up to 2,000 kWh per month with no time restrictions at no additional charge, and only 5.3¢/kWh thereafter.
Family Plan 2	Designed for smaller families. The basic monthly fee of $80 includes up to 1,150 kWh per month at no additional charge, and only 6.2¢/kWh, thereafter.
Apartment Owner Plan	Designed for apartment owners. The basic monthly fee of $41 includes up to 700 kWh per month at no additional charge, and only 7¢/kWh, thereafter.
EnergyMizer Plan	This plan gives a homeowner the lowest possible rate. The basic monthly fee of $100 per month, includes weatherization (caulking, duct-sealing, window heat mirror film, and weather-stripping where needed), and load controllers on water heaters, AC, and central heaters. Up to 1,300 kWh per month at no additional charge. Power is only 5.5¢/kWh thereafter.
Professional's Plan	Designed for DINK (double income, no kids) homeowners with considerable electric load, but low usage during normal business and peak hours. A basic monthly fee of $120 includes up to 600 kWh per month on peak and 1,000 kWh off peak at no additional charge, and only 7.8¢/kWh on peak, 4.3¢/kWh off peak, thereafter.
Hi-tech Plan	The basic monthly fee of $165 includes up to 1,200 kWh per month with no time restrictions at no additional charge, and only 5.3¢/kWh thereafter, a home automation system, on-site power and harmonics monitoring and priority trouble call, burglar, fire, and flood alarms, and a special computer access number so that the homeowner(s) can read their outputs, and a 220 watt two hour UPS for a home computer or TV-VCR combination.
Executive Plan	Designed for the busy executive with an active family. The basic monthly fee of $175 includes up to 2,000 kWh per month with no time restrictions at no additional charge, and only 5.3¢/KWH thereafter, on-site power monitoring and priority trouble call, and a 220 watt two hour UPS for a home computer or TV-VCR combination.
Leisure lifestyle	Designed for the weekend home or vacation site. Just $12/month includes basic electrical connectivity, and power line, burglar, and fire alarm services. Power is 11¢/KWH for the first 350 kWh, and 7¢/KWH, thereafter.

Buying and Energy Use Plans

In a manner analogous to how long-distance and cellular telephone companies market their products, Rescos will offer numerous tailored "energy-use plans." Table 14.3 shows several options aimed at the residential sector. Similar plans would target different types of businesses and industries. Often, these plans will correspond to the major theme or market distinction of the Resco, itself, but just as often a company will offer a wide range of plans designed to appeal to all segments of the market.

14.4 HOW DOES IT ALL WORK?

In a fully de-regulated, open access power industry, each consumer would receive offers from multiple sellers of electric power. These would come in the form of TV and newspaper ads, letters in the mail, and inevitably telephone calls from polite but determined sales people. Each retail company would tout its combination of price and services as unique. Every homeowner, and every business, would select a power supplier, and sign a contract with it, just as everyone has signed a contract for long distance services. For the sake of explanation here, the discussion will follow Mrs. Watts, a widowed grandmother who lives in Ampsville, and who has selected Big State Electric Services, a retail power company from out of her state, as her new power provider.

Mrs. Watts made her buying decision with the level of rationality and cool calculation that characterizes such decisions by many homeowners: Her son-in-law, Rick, works for Big State Generating Company. She was not aware that, since de-regulation, Big State Generation and Big State Electric Services have dis-aggregated into separate companies, and that Rick doesn't actually work for her new supplier. She never even considered anyone other than Big State, and she never read the fine print of their contract.

Big State would then make arrangements to sell power to Mrs. Watts, who is far outside the service territory Big State used to sell in, when they were regulated, by first arranging to communicate with her power meter. This might mean that they do nothing but schedule a local company to read her meter manually once a month, as in the "old days" before de-regulation, and have the information sent to them. But depending on the services they have promised to deliver, or the level of automation and cost reduction, Big State will perhaps install an electronic (smart) meter, which can communicate instantly, by radio or telephone line, with their central office.

Either way, by manually reading an old "dumb" meter, or by communicating to a smart meter electronically, they can now track when, if, and how much they are selling to Mrs. Watts. If they promised any advanced services, such as appliance monitoring, load management for conservation and lower bills, burglar and fire alarms, home automation, or whatever else requires the services of the smart meter or automated equipment, they would install it at this time, and make arrangements in their central office to program both the meter and other equipment needed to deliver these services, and to bill for them.

Big State now contacts the local wire company, Ampsville Power Delivery Company, the former distribution division of the former Ampsville Power Company, and informs it that it will be delivering to Mrs. Watts, and asks that the bill for power delivery to her location be sent to Big State. This wire company is regulated, and has a set price schedule for this service. Some wire companies will probably keep things simple and charge a flat rate – "$5 a month per home, every month," or they may charge a rate based on usage – "$2 per month plus 2.5 cents per kWh," in which case Big State has to arrange to transmit, electronically or otherwise, information on the amount of power Mrs. Watts consumes, so that Ampsville can bill them for delivery.

Big State also makes arrangements to move power, for Mrs. Watts and the other 37,893 customers it has in her area to Ampsville as a block, by contacting the local ISO in this part of the nation, and buying transmission capability so that it can move the power it has calculated it needs to Ampsville. The ISO has standard rates, and an "Internet" node to do this electronically. Big State is set up to do this in real time and for all of the region its sells to, not just for Ampsville.

Big State Electric Services now must find the power to sell to Mrs. Watts. It could buy it from Big State Generating Company, its former vertically integrated partner, but it has an entire trading department that plies the power exchange electronically 24-hours a day, picking up bargains on the spot market (short-term purchase market) for power, and it seldom buys from the same generator from one week to the next.

Thus, through what appears to be a complicated and circuitous set of business arrangements, Big State Electric Services obtains the power, arranges for its transmission over the grid, sets up its delivery to Mrs. Watts, provides her with any "high-tech" or advanced services by virtue of equipment it puts in her home, and bills her.

To Mrs. Watts, however, it's quite simple. She signed a contract with Big State something-or-other company. She has all the electric service she needs, and she gets one bill every month. The people who came to install the new meter were neat and friendly, and really seemed to care about her being happy with their service. They did the other things they had promised – caulking her windows and vents and installing new weather-sealing on her doors to cut down on energy usage. They also installed some sort of "control box" on her water heater, which they said would reduce its usage during peak periods so that she would get a discount, although it would still always have enough hot water to meet her needs. They put in a "lock sensor" and an automatic switch on her back door. As Big State's service man explained, it will know when she leaves home, locking her back door, and will then turn on a couple of her lights after dark if she's still out, so that burglars will think someone is home, all without her even having to think about it. Finally, Big State installed a burglar alarm hooked into their central office. Mrs. Watts is impressed with the security and the new features she gets.

She requested to be billed once a month for her actual usage, although Big State gave her several options, including quarterly, weekly, or fixed monthly payments. The bills come right on time, and they do seem a bit lower, although she can't actually tell for sure. As required by law and her state regulatory agency, the bill breaks out what she paid for power, for transmission, for distribution (local delivery), and it lists the discount she gets for water heater peak shaving, but she doesn't pay attention. The house is less drafty since the service men weather-sealed it, she still has plenty of hot water when she needs it, and she feels better with the burglar alarm integrated into her smart meter.

From the perspective of a consumer like Mrs. Watts, de-regulation is a good thing. The complexity is borne by the large companies competing for her business. But to her, there was more choice in services, and no real difficulty in making it work.

Regardless, Rescos will exist only in political jurisdictions which have fully de-regulated the power industry – where competition and open access exist at both the wholesale (transmission) and retail (distribution) levels. In a de-regulated industry, Rescos are the entity that most electric consumers think of as "their electric company," and will therefore put customer service and technological, business, or price advantage ahead of everything else in their bid to gain profitable market share.

In order to survive and in attempts to prevail, Rescos will be intensely competitive companies, with their focus on using electric energy sales as a lever to offer more profitable services and energy-related products. In some ways, they will be the most powerful players in the power industry, due to both their control of the "money flow" from the customer, and their involvement in all levels of the industry.

FOR FURTHER READING

H. L. Willis, *Spatial Electric Load Forecasting – Second Edition,* Marcel Dekker, New York, 2003.

P. Christensian, *Retail Wheeling Handbook,* Pennwell Books, 1996.

15

Service Reliability and Aging Infrastructures

15.1 INTRODUCTION

Modern utilities are under intense pressure to improve the quality of service for their customers, hold their prices low, yet still improve the bottom line for their stockholders. They must do more with less, a goal many long-time utility employees would point out is hardly new, but one that seems increasingly unrealistic in light of the efficiency increases, downsizings, and financial improvements that utilities have already made.

This chapter highlights two particularly thorny and intertwined issues at the core of power industry concerns. One is unavoidable and inexorable: aging T&D infrastructures. Every year transmission and distribution systems, many of them already quite old, grow older still. Much of the equipment in some utility systems is nearly worn out, and more is approaching that condition each day. The present situation, its trend, the challenges it will produce, and the possible solutions utilities could adopt are discussed in Section 15.2. Section 15.3 looks at reliability of service. Providing the 99.97% availability of power that modern consumers expect is a challenge under the best of circumstances, and one exacerbated by the higher failure rates and maintenance needs of aging infrastructures. The section presents basic reliability concepts, explains the challenges an operating delivery utility faces, and discusses the means that utilities employ to manage reliability, including asset management approaches to organizing a utility's decision-making.

15.2 AGING T&D INFRASTRUCTURES

Utility power systems are made up of incredibly robust equipment. Large power transformers routinely last fifty years in service, an incredible period considering the complexity of their design, the fact that intense magnetic and electrical forces are at work inside of them, as well as the number of intricate moving parts employed in their ancillary systems. Some utilities in the northeastern US have medium voltage (15,000 volt) circuit breakers that were manufactured just after World War I – not one or two but dozens still in service. Equipment that is simpler often lasts longer. Wooden utility poles are about as simple as equipment can get, and some last more than ninety years in service.

All of this equipment, still operating after decades of hard use, proves beyond a doubt that the power industry's engineers and manufacturers know how to design and build good machinery. Circuit breakers that last 90 years, transformers that last 75 years, and poles that last over a century are the exceptions. Most electrical equipment averages about 40 to 70 years in service before it fails or is so worn out that it has to be replaced. As is the case with almost all machinery, results vary in apparently random ways. Out of 1,000 service transformers built at the same factory in the same month to the same design, one will fail unexpectedly in the first year, one will still be in service 80 years later, and the majority will have lasted for something in between.

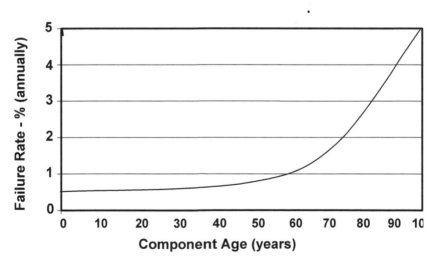

Figure 15.1 Electrical equipment fails at a faster rate as it ages. Shown here is the failure rate for a particular type of pad-mounted service transformer as a function of service age on one particular utility system.

Utilities certainly get their money's worth from such equipment. But there is a downside for the industry. A good portion of the poles in place at the time of this writing are approaching 60 years of age. A sizeable portion of the transformers in use in the industry are over 40 years of age. In some large utilities, the average circuit breaker is nearly 50 years old. As a whole, the electric utility industry is facing an aging problem. Some utilities, leaders in a category that no one wants to lead, are faced with entire systems that are or soon will be "aged."

Increasing Problems with Age

Aging infrastructures are a problem because as electrical equipment ages, it begins to fail at a higher rate, as shown in Figure 15.1. The data show that the utility can expect about 7 failures in an average group of 1,000 service transformers that are 15 years old, but about 35 failures (five times as many) in a group of 1,000 that has been in service for 35 years, or only three times as long.[1] Here, "failure" is used to indicate a malfunction so serious the equipment must be replaced: it cannot repaired.

Some types of electrical equipment, such as service transformers, normally malfunction in only a way that makes them unrepairable, or at least so expensive to repair that it is only sensible to just replace them. But other equipment, like voltage regulators, suffer breakdowns – malfunctions that are repairable. Generally, equipment with moving parts and the expensive equipment like large power transformers (which cost hundreds of thousands of dollars as opposed to a few hundred for a service transformer) tend to fall into this last category. Breakdown rate also increases with age, as does repair cost: spare parts for 50 year old equipment are not easy to find. One utility in the northeast US had to hire a local machine shop to custom build parts so it could repair a set of circuit breakers that had been in service since right after World War I.

How rate failure and breakdown rates increase with age depends on a host of factors. Sometimes the increase is linear – equipment twice as old fails twice as often. But usually it is exponential: failure rate may stay quite low for thirty years, then climb, slowly at first but faster with advancing age. Regardless, for utilities, the qualitative result is always the same: as equipment gets older, it has to be repaired more often, and more of it has to be replaced. Problems with reliability increase. Costs increase.

---

[1] Figure 15.1 is based upon the experience of one utility with one particular type of transformer used consistently to one loading standard. It is absolutely representative of the change in failure rate with age in most electrical equipment. But the absolute values are not general: readers should not assume that 28-year old transformers fail at a rate of 3.5%. Some situations might see a higher failure rate, others lower.

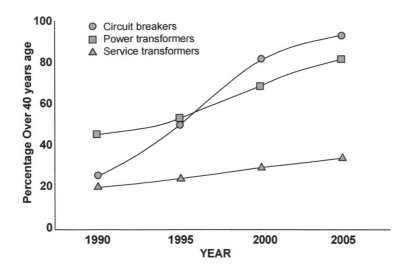

Figure 15.2 The percent of equipment that is "aged" (arbitrarily defined as over 50 years old) in the central region of a metropolitan utility in the US, plotted over time.

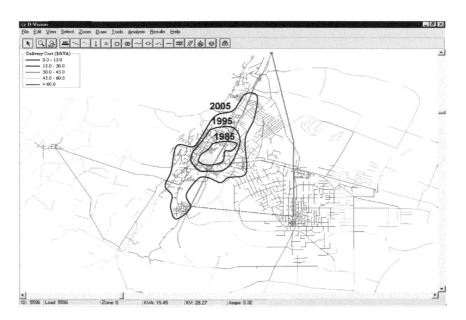

Figure 15.3 The geographic spread of "aged" equipment in a T&D system in the eastern US. Contours show years when a majority of equipment inside reaches 50 years of age.

Aging Equipment

There is no uniform agreement on exactly what "aged equipment" means. Some people insist that the actual chronological age of equipment is irrelevant: the important point is the equipment's condition, for that determines the equipment's breakdown rate and maintenance needs and failure rate. And most important to the utility (but usually hardest to estimate) is the remaining lifetime – when the utility can expect to have to replace an equipment unit because it has finally failed altogether.

Those are the factors that matter. But they are difficult to measure precisely. Breakdown rate accelerates gradually with age and it is difficult to determine when equipment crosses a threshold and is breaking down too often, etc. Therefore, for purposes of tracking and planning, the authors use a simple and effective definition.

> An aged power delivery infrastructure is any area of
> a utility system where the average age of equipment
> in service is greater than its nominal design lifetime.

Design lifetime is usually taken to mean 40 to 50 years. Electrical equipment is not designed to a specific service age: generally designers and engineers expect equipment to last more than 30 years. Thus, 40 years is a good rule of thumb, and many people believe good maintenance can push that another decade, to 50 years. Using this rule, an area is "aged" when a majority of equipment in it is over 40 years and not well cared for, or 50 years if well cared for. This definition is far from perfect, but its simplicity and explicit definition based on age permits tracking of aging infrastructure problems as shown in Figures 15.2 and 15.3, giving advantages in analyzing and understanding the problem, as well as in communicating it to decision makers at utilities and regulatory agencies.

Table 15.1 lists some typical characteristics of "aging infrastructure" areas in a power system. Usually, aging T&D infrastructure areas are just outside the core of major metropolitan areas, areas that were the "hot" growth areas 50 or more years ago, now themselves long-established areas of the city. Thus, many of the critical or important areas of a major city – those that one would like to have the very best quality of service – have electrical service infrastructure that are among the weakest.

Run to Failure Operating Policy

The equipment in most electric systems has reached the point of advancing age because of an equipment management practice utilities have traditionally followed: "run to failure" which means that equipment is left in place, even if it is giving signs of increasing age and stress, until it fails completely. Unless public or employee safety is jeopardized, electric utilities apply this approach to

Table 15.1 Characteristics of an Aging Infrastructure Area

- The area was originally developed, or last experienced a rebuilding boom, prior to the 1960s or 1970s. Most of the existing system structure (substation sites, rights of way) dates from before that time.

- The majority of equipment in the area is more than forty years old.

- The system is well engineered and planned, in the sense that a good deal of concern and attention has been devoted to it. It fully meets the utility's minimum engineering criteria.

- The area is seeing steady but perhaps slow growth in demand. It may be undergoing, or about to undergo, a rebuilding boom.

- The area is plagued by above-average equipment failure rates. Overtime for repairs in order to keep customer service intact is high.

- Reliability as seen by customers is not yet in jeopardy. Equipment outage rates began increasing several years earlier, but the problems they cause were "covered" by savvy operation and hard work. But that can't last forever.

- When a widespread interruption of service does occur, it is due to a series of events or a combination of failures that is *very* unusual; perhaps a "once in a lifetime" likelihood.

- With so much old equipment and so many breakdowns occurring, strange, bizarre, or "bad luck" combinations of events as in the bullet above occur frequently. Because they are bizarre, they are difficult to anticipate: the utility is often caught unprepared.

- Things go bad more and more as time passes.

just about everything in their system. Every year, aged equipment is repaired, patched, and coaxed along for another year or two until finally it is not repairable.

Run-to-failure is practiced because it makes sound economic sense, particularly from the short term: repairing equipment in order to get another year out of it avoids the major cost of replacing it. Thus, utilities compound the problem every year by stepping away from the obvious solution – begin replacing it. Before one judges them too harshly for this practice, one needs to recognize that they can not afford to replace it. This, and some subtle complexities behind run-to-failure, will be discussed later in this section.

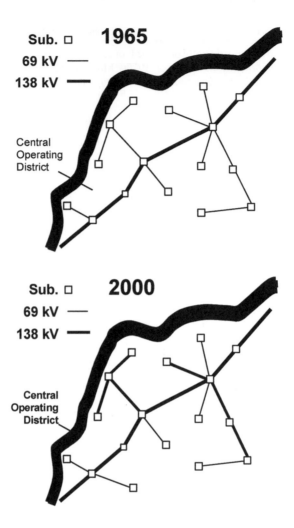

Figure 15.4 Top, in 1965, the central operating district of Metropolitan Power and Light used 16 substations averaging 128 MVA capacity each (total 2,048 MVA capacity) to serve a peak downtown load of 1,380 MVA at a 65% utilization factor, meaning it had a 52% (35/65) contingency margin. This diagram shows only the overall structure of the system. Many of the apparently radial lines feeding substations are in fact loops (two circuits) or double loops (four circuits). Bottom, 35 years of slow but steady load growth result in a peak demand of 3,058 MVA, more than double 1965's. In that time, Metropolitan Power and Light has been able to add only one substation site and expand four others in an area of increasing urban density. Total capacity now stands at 3,400 MVA, a 56% increase, for a utilization ratio of 90% and a contingency margin of 11%.

Replacement, Redesign, and Technology

Aging infrastructure areas sometimes face a second problem beyond aging of equipment and its tendency to break down more often and require higher maintenance costs. These areas have facilities and lines that were laid out and designed 40 to 50 years ago, when demand was perhaps only one-half to one-third of what it is today. The basic layout simply cannot provide the flexibility of operation and switching (contingency) capability needed for failure recovery at today's loading levels. Figure 15.4 shows a hypothetical example. Thus, in many utility systems, even if all aged facilities were updated with brand new equipment, the area would not be up to the performance level the utility would expect in those areas of its system it is now building (e.g., new growth areas, in the suburbs). But "expanding" the layout is often impossible – or as close to an impossible situation as an electric utility will ever face. The dearest commodity in an urban (aging infrastructure) area is land. This is the very thing that a utility needs to build new substations and add rights of way to "solve" this second problem – obsolescent layout. And it's the one thing it is least likely to get.

Technology is usually substituted for space and right of way. The example depicted in Figure 15.4 is quite close to the situation ComEd faced in Chicago in the summer of 1999, when both aspects of aging infrastructure came together to cause multiple blackouts of large portions of downtown. Replacing aged and failed equipment was not enough: upgrades in network capability were needed, even though no more land or rights of way were available. Part of the solution was a gas-insulated substation. This replaced a substation of nearly 80 years age with one of three times the capability.[2] GIS substations, automation, voltage upgrades, high-capacity cable, and, in extreme cases, super-conducting equipment are "technology solutions" that can be used in lieu of land and space.

Why Now?

The tremendous economic expansion in the US (and parts of Europe and Japan) in the two decades after World War II is what put much of the aging equipment – now somewhere between 60 and 40 years of age – in place. There is a very non-uniform distribution of equipment ages.

There is a secondary reason which has an element of truth to it although it is

---

[2] Normal, or "air insulated," substations use space (air) to insulate or avoid short-circuits between equipment: put it far enough apart and no sparks will fly, but substations can thus take up several acres if a lot of equipment is involved. GIS technology downsizes the equipment and then places it in steel "bottles" filled with sulfur hexafloride (SF_6), a gas through which electricity just will not pass. Whereas one needs many feet of air, one needs only a few inches of SF_6 to protect against short circuits. Equipment can be squeezed together. The substation is much smaller. Surprisingly, it is more reliable and durable too, because the equipment is "inside" a protective container.

sometimes exaggerated. In the 1950s and 1960s, utility costs and rates were dropping each year due to increasing economies of scale in generation and technological improvement in all levels of the electric system. Pressure to cut equipment cost was minimal: utilities bought quality and durability and were still able to lower or at least keep rates stable. But by the late 1970s costs had begun to rise and rate increases were the rule rather than the exception. Utilities began to push back on cost, and often put equipment manufacturers in a price bidding war to get their business. Manufacturers complied by working every bit of margin and over-design out of their equipment, so that it was "just good enough." For about forty years of service. Not more.

As a result, today some utilities find themselves the victim of an aging "double whammy": equipment installed in the 1950s had roughly a 50- to 60-year lifetime: that in the 1960s and 1970s had about 40 years expectation of service life. Both groups of equipment are now at the end of their lifetimes, so there is a glut of equipment ready to fail.

To some extent this perspective on aging is true: more than a few utilities do see a higher failure rate among power transformers installed in the 1970s than older units still left from the 1950s. But the contribution of this aspect to the overall problem is secondary: Aging infrastructures are a problem simply because so much of the equipment owned by utilities is old.

Aging Rate

An interesting measure of aging for all or an area of a utility system is to look at the "aging rate" of its equipment: how much the average equipment age increases in a year. For example, suppose that in one particular utility system equipment averages 32 years of age at the beginning of the year, and that during that year, no equipment fails, none is replaced because it needs to be upgraded, and no new equipment is added. Then at the end of the year the average age is 33 years. The system aged one year in a year.

On the other hand, suppose 1% of the equipment failed and was replaced or upgraded, and another 1% was added for growth. Then 98% of the system is 31 years old at the end of the year and 2% is new: average age is therefore 30.38 years: the system aged .38 years (about four and one half months) in the year. Most utility systems "age" about .5 to .7 years per year.

Run to Failure and Financial Pressures

T&D utilities have not replaced their aging equipment because they just can't afford to do so. Most can barely afford to replace equipment as it fails. One reason aging infrastructures are of such concern to so many industry insiders is that utilities may soon not be able to keep up. The logic is simple. Utilities can barely keep up now. As more and more of the equipment becomes aged, failure rate will increase and they will have to replace more and more equipment each year. They will not be able to keep up with the expense.

Can't utilities just pass on these quite legitimate costs of replacing equipment to their customers? After all, T&D utilities are regulated under a system that permits them to pass on costs. In theory, yes. But utilities, regulators, and customers alike have adjusted to the decades-long period in which infrastructure aging was not a problem – equipment was aging, but not yet aged. Replacement rates were low. Now, as they escalate, it will take time for all three stakeholders to adjust to the simple fact that spending will have to go up.

Beyond this, there is an element of common sense to "run to failure." Replacing equipment as it reaches a certain age, say 35 or 40 years, would cut off much of this problem, but it would also mean throwing away a good deal of lifetime in some equipment. Out of 1,000 units that are 40 years old, something on the order of 250 will last another 15 years. To replace all equipment at an arbitrary age of 40 would increase T&D cost in some systems by 25%. The rate increase would be unwelcome.

But there is something else at work, much more subtle, a characteristic in the thinking of utilities and regulators alike that the industry as a whole has yet to get past. The industry has never had to *manage* equipment lifetime. Aging was not a problem, and regard for it not even a need, until recently. It will take time for the industry to adjust to the fact that that must be done, and then to decide how to do it effectively.

Traditionally, and even into the 1990s, engineers, planners, and managers at utilities saw few failures – equipment just did not fail very often in the 1970s and 1980s. Failures were always viewed as anomalies ("That unit must have been bad from the start and we just didn't know it.") and unnatural – "That shouldn't have happened." What was missing from the industry's traditional culture was a regard for *failure as a natural consequence of the use of equipment.* Such a perspective makes failure something that needs to be considered and managed, a first step in tackling the aging infrastructure problem. This traditional attitude was compounded by two factors. First, equipment lifetimes were longer than the careers of most professionals involved in these decisions. Second, evaluating and predicting remaining lifetime in any equipment was (and still is, to some extent) a black art.[3]

The power industry has woken up to the fact that aging is a problem, and that equipment lifetimes must be managed. Methods and technology to address this well are still in their infancy. But technology is not the real problem: as will be discussed later, financing is the only real barrier.

---

[3] Methods for analyzing the condition of and predicting the remaining lifetime in major electrical equipment have improved immensely in the past ten years, but the authors doubt if anyone in the industry would argue that it is still the area of power engineering where the industry has the least confidence and experience.

Financial and Management Challenge

Equipment will continue to age. Failure rates will continue to creep upwards. Replacement and repair costs will continue to increase. Eventually the industry will be forced to adjust its prices for T&D so that replacement can be done and the "system" is sustainable. Today, in many parts of the US and the rest of the world, it is not: prices do not reflect a level of cost that will sustain the power system in good condition continuously into the future. This will be one of the major challenges for regulated T&D utilities in the first two decades of the 21st century: to develop sound, defensible methods to document the best policy for replacing equipment, and to convince regulators and customers alike that the cost is legitimate.

Asset Management

One way many utilities are stepping up to the challenge of addressing the aging infrastructure problem and the spending it requires is with a technique called asset management. Asset management is a management paradigm that orients the company's attention and efforts around the physical equipment (assets) and their "lifetime optimization," viewing them and all decisions about spending on a common business basis, seeking to minimize the risk of failing to achieve one's goals.

The term "asset management" and many of the concepts used in its implementation are borrowed from the investment community, where for decades "asset management" meant methods used to manage a portfolio of stock or financial investments and to minimize the risk associated with them. This risk basis fits the aging infrastructure situation particularly well. A utility never knows when a particular aged equipment unit will fail. It can only play the odds, minimizing the risk from both an operating and a business basis. Asset management provides a sound, procedural basis to do so and to manage spending in that way. Particularly when applied in concert with what are called reliability-based planning methods, asset management optimizes decisions about replacement versus repair, as well as how much and what type of maintenance to do.

Further, the decision making process in asset management permits a utility to compare dollars to equipment: Both are viewed as assets, and to a certain extent one can be exchanged for the other (money can be spent to buy equipment, equipment can be abused or neglected to save money). Exchanging money for equipment or vice versa makes sense if it leads to a greater overall total.

Asset management also represents a philosophical change, to one that says the system exists to achieve the company's *business* goals, not its engineering goals. Traditional utility cost management focused on dollars as dollars and simply minimized spending while keeping engineering criteria in check. Asset management views money spent on the system as an investment to improve business performance, literally a better "sell" from the standpoint of the business case. Ironically for many engineers, this non-engineering basis often results in more money, not less, being spent on the system.

Asset management has been adopted by many utilities for reasons other than *just* aging infrastructures. In particular, it helps predict and control costs with more precision and flexibility. But asset management *always* includes a much closer integration of capital and O&M prioritization than was traditionally the case, including, often, an explicit evaluation of expected lifetime and how that varies with use, maintenance (or lack of it), and other policies. It fits well into a culture that is looking at failure as a natural consequence of equipment use and that seeks to optimize the use of an equipment's lifetime, while minimizing the risk associated with no one knowing exactly what that lifetime is. Asset management cannot work miracles (old equipment still has to be replaced), but it is a sound, complete method that can handle the aging infrastructure problem well: minimizing financial impact, justifying the money spent, and managing the whole in a stable manner.

15.3 SUSTAINABLE-POINT ANALYSIS OF AGING INFRASTRUCTURES

Sustainable-point analysis is an approach to studying the aging of a T&D system (or for that matter, any other infrastructure) that uses as its foundation a fundamental fact that might seem counterintuitive:

> Any set of equipment in which units that fail are
> replaced with new units eventually reaches a stable point
> where its average age does not increase from thereon.

In other words, a T&D system does not age forever. At some point it reaches a point where its average age remains the same, the sustainable age: the average pole in the system is X years old this year. Next year, after failures and replacements are considered, new versus old will net to zero, so the average will still be X.

This *always* happens, with no exceptions, in any set of equipment where failures are replaced with new or old equipment is replaced according to some set of consistently applied rules. The process takes years to reach that sustainable point, but proceeds like this. For the sake of example, assume an electric utility installs 10,000 pressure-treated utility poles in year zero. When new and strong, these poles have a very low failure rate, only .20 percent per year. This means that in the first full year of service 20 fail and are replaced with new poles. Thus, at the end of the first year, the set of 10,000 poles consists of 9,980 poles that are 1 year old and 20 poles that are "new" (less than a year old). Average age of a pole in this set is .998 years.

In the next year, about 20 of the 9,980 original poles again fail (failure rate increases with age, but the failure rate at age 1 is indistinguishable from age zero). These failed units are replaced with 20 new poles. There could be a failure among the 20 poles that were replacements in the previous year, but in such a small set that is unlikely (the expectation is .04 pole – this is statistics, the authors realize that one cannot have four hundredths of a pole fail). From a practical standpoint none are expected to fail, but there is a slim chance. Thus at the end of two years, there are 9,960 poles that are 2 years old, 20 that are a year old, and 20 that are new. Average age is 1.996 years old. In these two years, the average age of the entire set of 10,000 poles has increased nearly by two full years, but not quite: it is .004 year – it is 35 hours short.

Over time, the original set of 10,000 poles is whittled down by failures, initially at a rate of 20 per year, gradually increasing with age as their failure rate increases: first to .25% at fifteen years (25 per year), then .5% at 25 years (50 per year), and then climbing faster as they continue to age. Every year, failed units are replaced with new poles. And the original replacement poles are growing older too: after 30 years that first-year set of 20 replacement poles is 29 years old, the second year's 20 are 28 years old, etc. Their failure rate is rising, too. Their failures are replaced with new poles, perhaps several times. After 30 years there is a good chance there is one pole somewhere in this set that has been replaced five times.

As time passes, this trend of the pole population growing older every year and its failure rate increasing continues. The set of 10,000 poles, now of very mixed ages from old to brand new, continues to grow a bit older, on average, each year, and to have an ever higher failure rate.

But this trend does not continue forever. Eventually a balance is reached, where failure rate is high enough that so many replacements are made each year, that these new units just counterbalance the aging of the remaining units. In this example, that *sustainable point* is reached when average age reaches 50 years. At that point, the composite failure rate (of the whole 10,000 pole set) is 2%. This means that 200 failed poles with an average age of 50 years fail and are replaced with new poles (age zero) each year. The remaining 98% don't fail – they last through the year and become, on average, 51 years of age going into the following year. Thus, the average age of the whole set is still 50 years (98% × 51years +.02 × 0 years). At this point, average age, and average failure rate, and with it the number of failed units that need to be replaced each year, remain stable thereafter.

Some readers might wonder if it matters that this example dealt only with averages. There will continue to be minor changes in the *distribution* of ages (the percent of poles that are this age and that age, etc) for a few more years, but

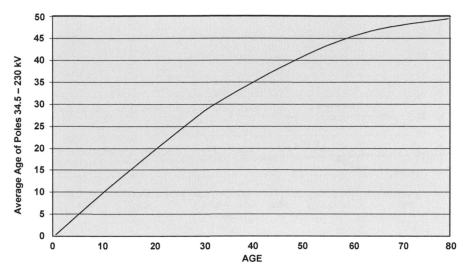

Figure 15.5 Average age of the 10,000 poles in the example discussed in the text. Eventually, they reach a sustainable point of age 50 years with a 2% failure rate. However, this takes more than 80 years.

with no major impact on average age and failure rate. And that does not affect the major point being made here: *every* equipment population eventually reaches a sustainable point, whether viewed with statistical averages, as here, or in more detail with analysis of its age distributions.

Characteristics of Sustainable Point Analysis Revealed

Every set of equipment where failed units are replaced with new or repaired units will eventually reach a sustainable point. Several points about how this occurs are worth noting:

Aging rate slows down as the population ages. In year one, the example population's average age increased by .998 year – effectively the population aged one year in that first year. In year 45, average age increase by only .59 year per year. Eventually, of course, rate of aging increases reaches zero, when the sustainable point is reached.

It takes a long time to reach the sustainable point. The population's average age reaches the sustainable point asymptotically (Figure 15.5). It takes 80 years for average age to reach 49.75, at which it is increasing by only .05 year/year. The final 20 years of aging, from 30 to 50 years average age, takes nearly 50 years in itself (from year 32 to year 80+).

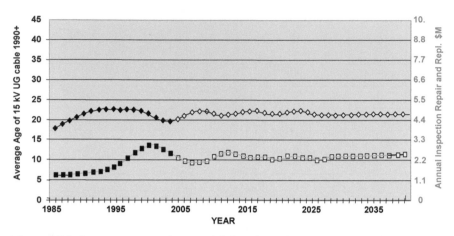

Figure 15.6 In rare cases, age (top trend, left scale) and costs (bottom trend, right scale) for equipment installed in the same year or over a short period will asymptotically oscillate above and below the sustainable point. This is data from a utility on the US eastern Atlantic coast, showing actual (solid) and projected ages and costs (open dots) for replacement of underground cable.

There can be a type of "oscillation" of age and replacement costs before sustainable age is reached, in cases where equipment has a highly exponential failure rate as a function of age, as some underground cable does (Figure 15.6). This occurs in only rare cases, but is frustrating for a utility to deal with if it happens, as it creates "cost spikes" every few years.

Everything covered above occurs in the same manner, quantitatively, even if the equipment is added gradually – say as 500 poles per year over 20 years, rather than as a single group in one year. The age distribution of the group is different from the beginning, but the statistics work out the same, the sustainable ages and failure rates work out the same, and eventually the set of 10,000 poles reaches the exact same statistical age distribution, average age, and failure rate, anyway.

Everything covered above occurs in the same manner, qualitatively, even if the utility adds a few new poles due to growth, not replacement, each year, too. The population will still tend to a sustainable average age and failure rate, but the steady supply of new poles due to growth each year will lower the sustainable age of the population by a small amount.

Sustainable point: a statistical concept

The sustainable point can be determined from statistical analysis of equipment purchase records, system operating history, failure rates, and maintenance records. The actual sustainable age of a population is often a moot point, of only academic and indirect value. Often, the average equipment in a utility system is nowhere near this sustainable age, so the figure is mostly an interesting statistical concept, far off in the future. The value is the analytical approach it permits engineers and planners to apply to ownership and operator decisions that a utility has to make in the course of managing its system. It forms the baseline point for an *asset base model* that projects present costs and failure rates into the short-term future (e.g., the next five to ten years) as in Figure 15.6.

Business Forecasting

Projections of remaining lifetime and failure rate trends can be made with the asset base model, to show how an aging equipment population will affect customer service reliability (the failures might cause outages) or utility costs (replacements cost money). Usually, one additional factor is included in any asset base model built for actual planning application: breakdowns. Breakdowns are malfunctions of the equipment that can be fixed by only repair, so they do not necessitate replacement. With poles, the example used above, there are few breakdowns: just about any failure or damage to a pole requires that it is replaced. But for many other types of utility equipment such as generators, transformers, voltage regulators, circuit breakers, and more, the majority of malfunctions can be repaired. Distinctions between breakdowns and failures and repair cost for breakdowns versus replacement cost for failures are usually included in a comprehensive sustainable asset base model.

Table 15.2 shows some results from one asset base model, developed and calibrated (adjusted to track all measured statistics and recent trends) for a utility in the central US. Results shown here are representative of the type of aging trends all electric utilities have, but the actual profile of aging and cost increases varies significantly among utilities in the US, and even more the world over.

The analysis summarized in Table 15.2 was done by first dividing the utility's asset base into various categories, fourteen in this case, represented in the table by rows, each a definite set of equipment. Usually, as here, it is necessary to sub-divide categories into sub-categories (not shown) for a variety of reasons of accuracy and detail. For example, the single row shown in the table as EHV/HV transformers was studied as six separate categories of different voltages and types of units that the utility owned.

Table 15.2 Projected Costs Now, "Eventually," and Ten Years into the Future, for a Utility in the Central US, Based on Sustainable Point Analysis of Its Asset Base

Level of System	Annual Repl. & Repair Cost	Current Avg. Age.	Sust. Age	Years 'til Sust. Pt.	Sust. Repl. & Repair Cost	Repl. & Repair Cost in 10 years
EHV/HV OH lines	3.9%	28	41	19	5.7%	5.0%
EHV/HV transf.	3.1%	29	47	21	5.0%	4.2%
EHV/HV breakers	3.8%	42	71	32	6.4%	4.8%
EHV/HV buswork	2.2%	35	74	40	3.7%	2.7%
Contr/prot.**	5.5%	22	45	25	9.2%	7.3%
Distr. subs. transf.	8.2%	36	45	23	11.6%	10.0%
MV breakers	5.4%	24	67	42	15.1%	8.2%
MV buswork	6.4%	29	74	46	16.3%	9.0%
MV OH lines	17.8%	24	29	11	21.5%	20.6%
MV UG lines	15.8%	23	25	12	17.2%	16.8%
OH service transf.	11.8%	25	29	13	13.7%	13.3%
UG service transf.	11.2%	26	28	15	12.1%	11.8%
OH LV lines	2.4%	23	28	9	2.9%	2.5%
UG LV lines	2.4%	27	31	8	2.8%	2.6%
	100%				143%	119%

The resulting model contains a lot of categories and sub-categories (78), each a set of data describing how much or what ages the utility owns and how failure and breakdown rates vary as a function of age, etc. Projected malfunctions, failures, replacement and breakdown costs, and reliability impacts are produced into the future, from which the utility can then plan.

Table 15.2, which presents only some of the results from the asset base model built for a utility in the central US, shows that eventually, when enough time has passed for this system to reach its sustainable points in all categories, its equipment operating costs will have risen by 43%. More significantly, however, they are projected to rise 19% in the next decade. This projection along with similar trends produced for impact on reliability, and labor force required to do the repairs,[4] provides both a cost trend for business planners, and a focal point for maintenance and service planners who want to study how to reduce costs, failure rates, or service problems due to equipment aging and failures.

Managing Sustainable Lifetime

The sustainable point is not set in stone. In addition to being a function of equipment type, weather, and time, the sustainable point and its trends are also a function of inspection, maintenance, service, and usage (loading) policies. Therefore, a utility can change the expected sustainable age, and the trends it sees in the future, by using the asset base model as a type of focal point for "what if studies."

Therefore, the projected 19% and 43% increases in costs in Table 15.2 may not be inevitable: the utility might be able to reduce these cost increases by changing how or when or where it does inspections, service work, or refurbishment on its equipment. Figure 15.7 shows details of one such change applied to the set of 10,000 poles used as the example earlier. This is an actual program of pole inspection, fungal treatment, and repair by injection of resins to selected poles on ten-year increments beginning at age 30, developed by the utility whose asset base model is diagrammed in Table 15.2. This program reduces failure rate dramatically in poles aged 40 years or more, which is about the age that the failure rate of the poles in this example begins to climb with further aging. Thus, the program effectively extends the sustainable age, and that reduces replacement costs and reliability problems in the foreseeable future.

---

[4] The cost of the labor for repairs and replacement is included in the costs shown. However, a separate breakout of projected labor needs permits the utility to estimate the number of skilled workforce it will need in the future, an important element of its planning, since, for example, it can take a minimum of five years to train a live-line worker.

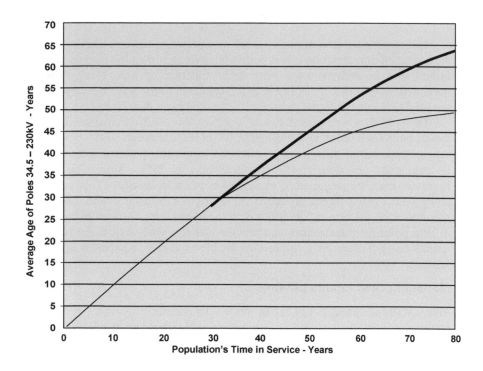

Figure 15.7 With an effective program of inspection and preventive service, sustainable age of the 10,000 pole example jumps to 68 years (bold line), a 36% increase over the original trend. Failure rate and failure costs drop by similar amounts – more than a third – saving much more than the program's costs. The utility used its asset base model and its sustainable point analysis to run various scenarios to determine that this particular program was the most cost effective possible.

A good asset base model, by which the authors mean one that looks at the distribution of ages of the equipment, not just its average age, and which is supported by accurate data, permits the utility to assess various responses to aging and failure rate increases and determine how to best deal with aging. Figure 15.7's program is in fact the optimum pole maintenance program for the utility. Its planners studied variations in when pole inspection should start (optimum was 30 years) and how often (optimum was 9-10 years). Starting sooner or doing maintenance more frequently reduces failures more than this program, but not enough to outweigh the added cost. The program considered a lot of details not discussed here, such as the fact that the utility has only a general idea about pole age in various areas it must inspect (i.e., most of the original poles in this area were installed between about 28 and 35 years ago). Overall, this program is quite effective. The best way to summarize its impact is to look at it as extending the expected lifetime of the average pole by 18 years, a 36% improvement on the value that the utility gets of its original investment.

Aging Infrastructures in the Power Industry: The Good and Bad

Most of the comprehensive studies of utility system aging and sustainable analysis that the authors have seen reveal a pattern broadly similar to that shown in Table 15.2: the various categories of equipment in the utility's system are not at their sustainable point and many are far from it. Generally, circuit breakers have very high sustainable ages (in rare cases, a century or more – such equipment can be rebuilt and refit almost indefinitely), while underground distribution cable and overhead distribution lines typically have low sustainable ages. In some utilities, distribution lines, both UG and OH, are quite close to their sustainable age (cable has a short equipment lifetime in many cases, overhead distribution is susceptible to storm and vehicle damage).

The fact that most systems are far from their sustainable point can be interpreted as good or bad news. It means that a good deal of equipment in the system has substantial useful life in it (good), but any system not yet at its sustainable point will see (usually slowly) escalating costs in the future, usually quite noticeable in the short term.

15.4 OUTAGES AND RELIABILITY

"Reliability" as normally applied by power distribution-utilities means continuity of service to their electric consumers. A 100% reliable power system provides power without interruption, power that is available all of the time regardless of if the consumer is using it at any given moment. In North America and Europe, almost anything less than perfection in this regard will garner customer complaints that the power supply is "unreliable." An availability of 99.9% might sound impressive, but it means eight and three quarter hours without electric service each year, a level of service interruption that nearly all electric consumers in North America, the UK, Europe, and much of the Pacific Rim would consider completely unsatisfactory. Typically, availability of utility power in first world countries is about 99.98%, or roughly for all but 1.75 hours per year.

Steady Trend of Increasing Emphasis on Reliability

Reliability of service – keeping the lights on – has always been a concern of electric utilities and their employees. Practices and guidelines developed during the first half of the electric century and honed into the 1970s produced generally satisfactory results. But beginning in the last quarter of the 20th century, reliability of service became increasingly important to all stakeholders in the electric power industry. Utilities saw a steady increase in the number and content of complaints about service interruptions.

Beginning in the 1970s, some state public utility commissions began demanding that utilities track and routinely report reliability of service. Within another decade, most state commissions were demanding reports on any "major events" (outages, blackouts) and annual surveys by the utility of its "worst performing" parts of systems along with documented plans for their improvement. Several state PUCs have gone further, to performance based rates (PBR), tying the earnings a utility will receive to the level of reliability it will provide its customers. Thus, much of the growing emphasis on reliability seen by utilities is in the form of increased regulatory scrutiny and emphasis, but it all springs from a growing demand by customers for continuity of service.

The digital economy and cultural dependence on electricity

One driving force behind the increasing need for reliability is the advent of computers and digital electronics. Before the widespread use of digital clocks and computerized equipment, short interruptions of the power supply were not troublesome and in fact were often not even noticed by most utility customers. For example, in the 1970s it was common practice for utilities to perform minor maintenance and switching operations "cold" – by shutting down parts of the distribution system for a few minute.

Electric utilities would routinely do so during the early morning (e.g., 3AM), when they could de-energize circuits for several minutes, interrupting power flow to customers, without creating any inconvenience or complaints. Few people noticed. Analog clocks fell a few minutes behind. But those sleeping, and even many awake, noticed no impact beyond that (if they noticed that).

But today, such a practice would lead to homeowners who wake up to a house full of blinking digital displays, and wake up later than they expected, because their alarm clocks did not operate. It would cause 24/7 computer systems at businesses to shut down, costing valuable production time. Utilities long ago gave up such practices and uniformly do "hot" maintenance and field switching, except in the rarest of cases. But this highlights an obvious fact: the use of computers and other digital equipment has made homeowners, business persons, and industrial plant managers all much more concerned about reliability of service, and particularly about brief interruptions of power supply.

But digital equipment alone cannot explain the growing societal emphasis on reliability of service. The real reason for the high level of emphasis on electric reliability is that both society in general and many individuals in particular have come to depend on electric power for just about every activity in their daily lives: people can't work and society can't function without it. Beyond that, our culture seems to expect continuing progress from every technology. New automobiles is expected to run smoother, require less service, and have better fuel economy than their predecessors. Every new generation of computers are expected to be faster, more capable, and have more features than the previous. And because the only two metrics that most people really notice about power are its reliability and its cost, people expect continual improvement in one or both, too.[5]

Outages Cause Interruptions

Although often used inter-changeably, the words "outage" and "interruption" have specific and different meanings with respect to electrical reliability. "Outage" means a failure of part of the power supply system – a line down, a transformer out of service, or a breaker that opens when it shouldn't. "Interruption" means a cessation of service to one or more customers. They had power a moment ago. They want power. They don't have it now, because of an outage of some critical element in their supply.

Interruptions are caused by outages.

Many people use the two terms interchangeably. In particular, more than a few people in and out of the industry use the term "a customer outage" when they really mean a "customer service interruption." For the most part they are understood when they talk to other people. However, to understand power supply reliability properly, and more importantly, to manage cost and reliability well, the distinction between outage (the cause of service problems) and interruption (the result) must be kept in mind at all times.

Frequency and Duration

Two different aspects of reliability receive attention in any type of power supply reliability analysis, whether of interruptions or of outages. These are the *frequency* (how often something occurs) and the *duration* (how long it lasts). With respect to customer interruptions, frequency refers to the number of times a customer's service is interrupted during a period of analysis – two times a year, five times a month, or every afternoon. Duration refers to the length of interruptions – some last only a few cycles, others for hours, even days. Usually, both are studied on an annual basis: number of interruptions per year and total duration (sum of the durations of all interruptions in a year).

---

[5] Even non-engineers understand that there are qualities like voltage and current associated with electricity that also can be measured. But somehow people think of that as "the electricity" itself and realize voltage doesn't need to be improved, a fact power engineers certainly appreciate. The truth of the matter is that the only two times people really notice electricity is when they pay for it and when it ceases to be available.

Extent

Frequency and duration are the two most important factors in reliability reporting: the bottom line, so to speak, on quality of customer service. However, there is a third factor that is important to the electric utility because it is the one over which it has considerable control and which it can use to manage reliability on its system. This is the *extent* of an outage's impact on service – how many customers or what portion of the plant's loads are interrupted by the outage of a particular unit of equipment.

The design of the electric power supply system – how its lines branch from one site to another and interconnect as they run up and down streets to deliver power to its customers – greatly influences just how much interruption occurs when an equipment outage occurs. A system designed with a certain type of circuit layout using an abundance of reclosers (a type of circuit breaker) and switches will have a low extent: outages may occur but each will interrupt service to only a relatively small number of customers. By contrast another system might have a higher extent; while no more equipment failures (outages) occur in this system, the average outage causes more customers to be affected. Turning that around, the average customer sees more interruptions because any outages in a bigger group of customers will affect his service.

Generally, reducing extent costs money, although clever design, attention to detail, and good record-keeping for field operations can reduce it by as much as 15% in some systems, without increasing cost. Beyond that, reliability-based planning methods optimize the cost spent to achieve systems that excel in this category of performance against other ways the utility can spend money. They are part of an asset management approach.

Reliability Indices

Reliability indices measure the reliability performance of an entire utility or parts of its system. Usually, such indices are evaluated on a monthly and yearly basis (Did we do better this year? Worse?). Their values, and details behind them, are used to manage reliability ("Why did this increase? Where did we have problems? Why?). The indices are also routinely reporting to the state regulatory agency, and perhaps to customers (on the utility's web site, in informational reports mailed with its bills).

The basic challenge facing anyone trying to measure reliability with a single number is how to relate the two quantities that everyone knows matter most: frequency and duration. Are two one-hour interruptions in a year "equivalently bad" to one two-hour interruption, or are two one-hour interruptions twice as inconvenient as the one two-hour interruption? Most people conclude that the correct answer lies somewhere in between, but few people will agree completely on exactly how frequency and duration "add up" in importance.

Opinions vary because the importance of frequency and duration vary tremendously from one electric consumer to another. There are some industries where a one-minute interruption of power will cause over ninety minutes of lost productivity: computer and robotic systems must be reset and restarted or a long production process must be cleared and bulk material reloaded before the plant can restart production. For those types of consumers, five one-minute interruptions may be much more serious than a single interruption, even if it is five hours duration. But other types of industries shrug off short outages as only a nuisance, being most sensitive only to long periods without power.

Popular Reliability Indices
Used by the Power Industry

The five most widely used indices for reliability reporting and analysis make are SAIDI and CTAIDI, which measure only duration, SAIFI and CAIFI, which measure only the frequency of interruptions considered long enough to have a duration, and MAIFI, which measures the frequency of events that were so short they had no "duration." This last category of events is usually defined arbitrarily, perhaps as events lasting less than one minute. A utility certainly has equipment that can measure duration to a precision less than one minute (so do many of its customers). But the point of tracking "short" events as a separate category is simply to gain a count of how many short events occur on a consistent basis: exactly how long they are is not critical to the utility of this effort.

To the utility, MAIFI matters a lot. Interruptions that take less than a minute are those that are restored (fixed, as far as the customers are concerned) quickly, by automatic and automated equipment. If their count (MAIFI) is going up over time, it indicates that the root cause of interruptions is increasing, but that automatic equipment is doing its job; if that equipment is not working well, the interruptions last longer and SAIFI, not MAIFI, increases. This is one of many clues that utilities use to constantly track reliability performance of their system so that they detect, as early as possible, deficiencies that need to be addressed. Either a minute, or five minutes, is a convenient cut off point for interruption duration in such analysis: interruptions lasting less than that time are counted, but considered to have no duration at all.

All five of the indices listed here count each customer interruption as a separate event: If service to the same home is interrupted three times in a year, that constitutes three customer interruptions. If one equipment outage causes simultaneous interruption of service of three customers, that too is three customer interruptions. In this example, the utility reporting period is a year, it has 1,000,000 customers, it defines momentary interruptions as those taking less than five minutes, and during this period it had 2,000,000 customer interruptions over five minutes in length which totaled 4,000,000 customer-hours out of service, and 2,400,000 interruptions that were less than five minutes long.

System Average Interruption Frequency Index (SAIFI) is the average number of interruptions, that are considered to have a duration, per utility customer, during the period of analysis.

$$\text{SAIFI} = \frac{\text{number of customer interruptions}}{\text{total customers in system}} \qquad (15.1)$$

Here, 2,000,000 of this category of customer interruptions occurred in a utility during a year, and if the utility has 1,000,000 customers, then SAIFI is 2.0

Customer Average Interruption Frequency Index (CAIFI) is the average number of interruptions considered to have a duration, experienced by customers who had at least one interruption during the period.

$$\text{CAIFI} = \frac{\text{number of customer interruptions}}{\text{number of customers who had at least one interruption}} \qquad (15.2)$$

The "S" in SAIFI means it averages the interruption statistic over the *entire* customer base (the system); the "C" in CAIFI means it averages over only customers who experienced at least one interruption. Customers who had uninterrupted service during the period are precluded from CAIFI. Perhaps of the 1,000,000 customers in the system, 500,000 saw no interruption in the year. Then all 2,000,000 interruptions were experienced by just 500,000 customers. CAIFI is therefore 4.0 although SAIFI of 2.0. The ratio of CAIFI to SAIFI gives an idea of how "spotty" reliability problems are throughout the system. The ratio in this example, 2.0, indicates a significant unevenness. A more typical value is about 1.15.

System Average Interruption Duration Index (SAIDI) is the average duration of all interruptions considered to have a duration, obtained by averaging the total time of customer interruptions experience over all of the utility customers – those who had outages and those who did not.

$$\text{SAIDI} = \frac{\text{sum of the durations of all customer interruptions}}{\text{total customers in system}} \qquad (15.3)$$

In this example, total customer hours out of service are 4,000,000 and there are 1,000,000 customers, so SAIDI is 4.0 hours, a level much worse than the average for North America (about two hours).

Customer Total Average Interruption Duration Index (CTAIDI) is the average total duration of all interruptions, but averaged only over the number of utility customers who had at least one outage.

$$CTAIDI = \frac{\text{sum of the durations of all customer interruptions}}{\text{number of customers who had at least one interruption}} \quad (15.4)$$

This is 8.0 in this example: Half the customers experienced all 4,000,000 hours of customer interruptions. Many of them would be quite upset because very few energy consumers in the US experience this poor level of service, even though it is better than 99.9% reliability of service.

Momentary Average Interruption Frequency Index (MAIFI) is the average number of momentary interruptions – those that are considered not to have a duration, per utility customer, during the period of analysis,

$$MAIFI = \frac{\text{number of customer momentary interruptions}}{\text{total customers in system}} \quad (15.5)$$

This is 2.4, from 2,400,000 momentary events divided by 1,000,000 customers. The total interruption index (from the consumer standpoint) is best estimated by the sum: SAIFI + MAIFI, which is 4.4. This is the average number of interruptions, including those thought both too short and long enough to measure duration, that an average customer experienced in the reporting period.

Analysis Using Reliability Indices

Reliability indices are used for a variety of purposes. For one thing, utility regulatory commissions require that SAIDI and SAIFI be reported, as a way of tracking and assuring the utility is performing well. Utilities also compare or "benchmark" themselves against one another to determine if, and why, they differ in reliability performance. But by and large the most useful application is to reveal trends and patterns, expose problems, and reveal how and where reliability can be improved.

Any of the reliability indices discussed above can be tracked over time to identify trends that indicate developing problems. Figure 15.8 shows frequency of customer interruption (SAIFI) for one suburban operating district of a metropolitan utility on an annual basis over a 35-year period. The operating district was open farmland (no tree-caused trouble) up to 1962. At that time suburban growth of a nearby metropolitan region first spilled into the area, with much construction and with it the typically high interruption rates characteristic of construction areas (trenchers digging into underground lines, early equipment failures of new distribution facilities, etc.).

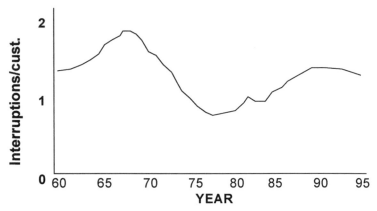

Figure 15.8 Electric service interruption rate for a region that grew from open farmland into a developed metropolitan suburb in the period from 1962 to 1970, during which time interruption rate rose due to the problems common to areas where new construction is present (digging into lines, initial above-average failure rates for newly installed equipment). Twenty years later, interruption rate began to rise again, a combination of the effects of aging underground cable in some residential areas and the effects of trees, planted when the area was new, finally reaching a height where they brushed against overhead conductors on occasion. The anomaly centered on 1982 was due to a single, bad storm.

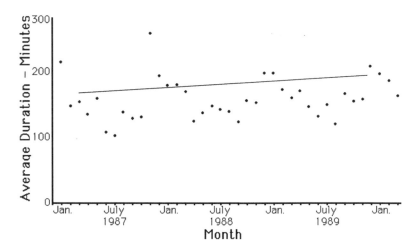

Figure 15.9 Average hours out of service per interruption on a monthly basis for a rural utility in the northern United States varies due to random fluctuations and changes in weather. It appears that duration of interruptions (dotted line) is slowly increasing over time, but further study is necessary before this conclusion can be confirmed.

Over the following ten years, as the area gradually filled in with housing and shopping centers and construction and system additions stabilized, frequency of interruption dropped. Twenty years after the area first began developing; interruption rates again began to rise because of two factors:

1. Gradually increasing cable failure rates due to age began to cause the outage rate on underground lines to increase.

2. Trees, planted by homeowners 20-30 years earlier, are now reaching a height and fullness where they interact with overhead lines.

An analysis over time like that shown in Figure 15.8 does not reveal the reasons (cable aging, trees). But it indicates a problem, which indicates that engineers and technicians should be assigned to determine the reason.

Figure 15.9 shows another tracking example using a reliability index, looking at the average duration of service interruptions in a utility over a period of several years. The increase in outage duration during each winter is clearly visible (winter storms stress repair resources and make travel to repair sites lengthier). While that might be expected, a long-term trend upward is discernible. Without additional study, it is not possible to determine the reason, but the utility can see in these statistics a reason for concern and investigation.

What Is Typical Utility System Reliability?

During the last quarter of the 20th century, reliability of electric service in the United States averaged a SAIDI of about two hours per year (i.e., the average energy consumer saw 2 hours in a year without power), a SAIFI of about 1.75 (slightly less than two "long" events a year), and a MAIFI of about three momentary interruptions annually. These are estimated values for performance across the industry in the period 1975-2000, and, as is typical of industry practice, excluding interruptions caused by "major storms or events." A utility that experiences ice storms every year cannot exclude interruptions from those; but the many, lengthy interruptions caused by a once-in-ten-year hurricane would be excludable. They would be tracked and reported separately, and might exceed the totals for the rest of the year. During the last decade, utilities nationwide have worked to improve reliability, and it now stands at an industry average SAIDI somewhere around 110 minutes.

Regardless, reliability varies a great deal from one utility to another. There are many reasons. First, some utility territories are much more difficult to serve reliably than others. Central Maine Power faces long, bitter winters with much snow and several ice storms every year. In addition, mountains and a hilly terrain make driving to repair sites slow going for its field crews during the winter. Not surprisingly, its service reliability is somewhat worse than average. Other utilities have milder climates and more benign terrain, and thus are more likely to have SAIDI and SAIFI values lower than average.

But beyond this, some utilities just do a better job of reliability. They design a better system, keep it in better condition, track down problems more quickly and accurately, and dispatch repair crews more effectively. Studies by the authors indicated that the best utilities in this regard greatly outperform the worst, by more almost two to one when adjusted for resources, difficulty of the terrain, etc.

But beyond that, service reliability varies from city to city and country to country because price/performance expectations are different. Traditionally, reliability of electric service in Europe was much better than in the US – but price was much higher, too. Within the US, people and regulatory commissions in some areas have decided they are willing to pay a bit more for better reliability: they get both better service quality, and higher prices for power. It is possible to deliver any level of power system reliability anywhere, if someone is willing to pay the price. In some developing countries, reliability of service is far less than in first-world countries, but that is accepted: In a country where some people do not have electric service, it is difficult for those who do to demand that what little money is available to expand the grid to new customers be instead spent to improve the quality of their service.

Generally reliability is better than the national average in dense cities, regardless of what country one looks at. In the US, SAIFI and SAIDI in major downtown areas are only about 1/3 of what they are in the surrounding suburbs. In other countries, although the national average might be higher or lower than in the US, a similar qualitative relationship applies: electric service is more reliable in the biggest cities. There are three reasons. First, urban populations seem more willing to pay higher prices for reliable power: people who ride on elevators every day and depend on traffic signals for a smooth commute usually value reliability of electric service a bit more than people who live in a rural area.

Second, problems that affect reliability are very much a function of the length of circuits in a system. Feeder circuits in a rural area might be twenty miles long, those in a city only two miles long. Customers in those two regions see a ten-to-one difference in the number of outages that potentially affect them.

Finally, there is an economy of scale to the power system a utility can build in a large city that permits it to build higher reliability designs (distribution networks) but still spread the cost over many customers, so that cost/customer is really not that much more expensive.

Table 15.3 gives reliability statistics for a dozen utilities around the world. These are values prepared by the authors based on data reported for a five-year period in the late 1990s and early 2000s.

Table 15.3 Average Annual SAIFI and SAIDI for Twelve Utilities 1988-1992

Country	Service Area Type	Climate	SAIFI	SAIDI(min.)
USA	Dense urban area	Hot summers, bitter winters	.13	16
USA	Urban/suburban	Hot nearly year round	1.2	68
USA	Suburban & rural	High isochronic (lightning)	2.0	97
USA	Urban & rural	Mild year round	2.1	122
USA	Agrarian/small towns	Hot summers, cold winters	1.1	168
USA	Rural, mountainous	Bitter, icy winters	2.8	380
Europe	Urban & suburban	Warm summer, bitter winters	.30	25
Europe	Metropolitan area	Warm summer, cold winters	.27	30
Europe	Agrarian/small towns	Hot summers, cold winters	1.2	120
Asia	Dense urban area	Near tropical year round	1.9	400
Africa	Large metro area	Tropical year-round	6	1200
Japan	Dense metro area	Hot summers, cold winters	.45	18

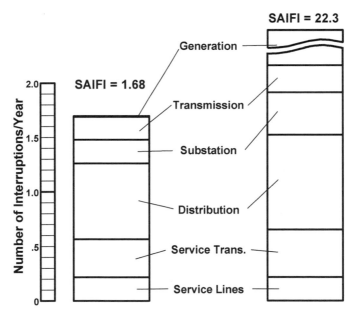

Figure 15.10. Left, breakdown of contributing causes of SAIFI for an urban/suburban utility in the US over a five-year period (average 1.68/year). Right, similar data from a utility in a third-world megalopolis, which had an annual SAIFI over 22, the majority caused by generation shortages.

What Causes Interruptions?

Figure 15.10 shows a breakdown of SAIFI contribution for a typical urban-suburban utility in the US, as well as the utility serving a very large metropolitan area in a third world country. Service problems due to T&D outages are roughly comparable in the two systems. However, power supply insufficiency – a lack of enough power to meet everyone's need – leads to rolling blackouts that overshadow everything else in the system depicted on the left.

In both utility systems, about 2/3 of all T&D-caused customer interruptions are due to equipment outages that occur on the distribution system (which consists of distribution lines, service transformers, and service lines, see Figure 1.1). This is the case in virtually all utilities. The majority of service problems occur on the distribution system because:

a) It is radial, not a network: outages are much more likely to lead to interruptions.[6]

b) There is much more of it.

c) It is built in easements rather than on rights of way that can be cleared of vegetation[7]

Managing Reliability

Utilities manage reliability through a combination of sound system *design,* good preventive *maintenance* to keep equipment in good working order, and *operational excellence* to restore service and repair equipment that fails unexpectedly or is damaged by storms or accidents.

Design

The generation and transmission systems (the wholesale level) of any first-

---

[6] "Radial" means that exactly one and only one line leads into every neighborhood, and that one extension and only one leads from that line to each home or business. In distribution systems, there are no backup routes, unlike in transmission networks (the word network means there is always more than one line leading everywhere). A network's higher cost is justifiable in transmission networks because they serve so many customers, but not for distribution where it serves only a few, or just one customer.

[7] Utilities build their major lines on rights of way they own. They can clear trees and vegetation that might interfere with the lines "aggressively" in order to prevent problems. Distribution lines are built in easements – in a space designated for utilities along the property line of home and business owners. The utility often trims branches away from lines but cannot "aggressively" handle the situation as it would on a right of way (cut the tree down so it does not fall into the line during a storm).

world utility system is designed with built-in margins of reliability that include an ability to operate without any interruption of service, and without any overloads, even if any one or two lines, generators, or large transformers fail. In addition, system planners and designers will have figured out ways that the system can continue to work, with slight overloads, for a brief period (large enough to make adjustments to other equipment) even if more equipment is out. The wholesale level is this robust for a number of reasons that will be covered in Chapter 16's discussion of blackouts.

The distribution level is generally not designed to be quite as robust, largely because it is the majority of T&D expense and to do so would nearly double system cost (transmission system costs are only about 1/6 of the total – "doubling up circuits" and other reliability measures at that level adds only a small increment to total cost). Instead, the distribution system is designed for quick service restoration through switching (if one line goes out, customers on that line are switched to a nearby circuit in a matter of seconds or minutes).

Equipment maintenance

Utilities inspect all equipment on a periodic basis. This includes visual inspections – from a car or truck or a helicopter – of lines, poles, and small equipment like switches and service transformers that may include the use of an infra-red scanner (overheating is a sure sign of impending failure, and overheated equipment shows up on such scanners). Larger equipment such as substation transformers and circuit breakers are often tested with quite elaborate diagnostic equipment that analyze the condition of internal parts. Preventive maintenance is done as deemed necessary. Repairs are carried out any time they are required.

Electrical equipment maintenance is expensive, and well-cared for electrical equipment so robust that it can be "abused" through neglect for several years before its condition deteriorates to the point that it gives more problems than normal. Thus, maintenance is one area where utilities that are hard pressed financially often cut back. There is little effect on operations for several years, then outages begin to be more frequent. Intense maintenance can restore the system to nearly its former condition, but some of the damage done through neglect will have been permanent and the equipment will have shorter lifetimes as a result.

Operation

The intent of power system planners, engineers, operators, and their management throughout a utility is that all of the equipment in the T&D system be in sound condition and operated on a continuous basis. Even when demand is far below its highest level, 100% of the T&D system is up and running, except for any equipment temporarily out of service while being serviced or repaired.

(This is in contrast to the generators that actually produce the power, which are started and stopped as needed depending on the number required to meet demand.) Thus, whenever any T&D equipment goes out of service, the utility operations department acts to put it back into service, quickly if it is badly needed, when convenient (i.e., scheduled so no overtime labor is needed).

Outages occur due to a variety of causes: equipment fails because it wears out, it is damaged by wind and weather, by vehicles hitting poles, etc., by trees falling on lines. Underground cable is sometimes gnawed through by moles, etc.

A utility monitors system condition and equipment status at its operations centers. There is usually one *system operations center* that monitors and controls generation and transmission (Figure 15.11). It is heavily computerized, monitoring transmission line flows and equipment status in "real time" through remote sensors at key substations. Computers in this operations center analyze data in a never-ending 15-minute ahead simulation of the system, in order to alert the utility operators if an overload or other undesirable situation is about to develop. If so, operators have time to make adjustments to avoid the situation.

The distribution system is controlled a bit differently. A large utility will have distribution operations centers located around its system, generally no more than one hours drive away from one another. Each center is where the

Figure 15.11 A system operator at a system control center for a large utility uses the twin-screen computer on the console in front of him to both monitor key readings on the power system and to simulate its operation over the next quarter hour, in order to see if problems are likely to develop. Several other operators sit at similar consoles spaced across the back of this system control room, each usually studying one particular aspect of the system's operation, its control, or plans for the following day. A very large computer-controlled map of the entire system, 60 feet wide and 22 feet high in this case, fills the front wall of the operations control room. It displays an up to the second report for all lines and equipment, color-coded by status (e.g., red means a severe problem). Photo courtesy of ABB Inc.

field crews for the region, the spare parts warehouse, and local engineering resources are located. Also in each will be a distribution dispatch center. This is a large room, with a road and circuit map of the distribution system for the region displayed along the front wall. Operators, called distribution dispatchers, sit at consoles or desks facing the map; each has a two-way radio to talk to field crews, and telephone lines connected to key utility offices and other dispatch centers. Status of the distribution system is displayed on the wall. Equipment outages or known problems are displayed on the wall map. Dispatchers analyze the problem and "dispatch" appropriate field crews and spare parts by radio.

Unlike all system operations centers, some distribution dispatch centers are not heavily computerized (although some are). Transmission systems have heavy bulk flows and large and very expensive-to-replace equipment that justifies the remote monitoring and computer simulation systems used. As a result, virtually all transmission control centers, in all utilities around the world, are very heavily computerized.

By contrast, distribution systems, which are made up of vastly more pieces all of which are far smaller equipment, are not nearly as heavily computerized – the cost of so much monitoring equipment would be prohibitive. Therefore, nearly all utilities get a large part of their information about problems on their distribution system from phone calls from their customers ("My lights are out."). Customers calling the utility talk to a service representative at a centralized call center, who take the information and relay it to the appropriate distribution dispatch center.

Many utilities, particularly the larger and more progressive ones, use a computerized *outage management system* in their distribution dispatch centers to analyze these trouble calls and to create computer-screen maps of their locations and the suspected problems that have caused them. Distribution dispatchers use that information to guide them in assigning field crews to do the work. The most advanced of these systems use mobile computing – the computers at the dispatch center send maps, design drawings and instructions directly to hardened laptop computers in each field truck. (Systems without mobile computing depend on the dispatchers to radio instructions to the field crews.) Outage management systems also track all service interruptions, maintaining a constantly updated set of SAIFI, SAIDI, and other operating statistics on every circuit and area of the system.

Many smaller utilities still run their distribution dispatch center the old-fashioned, non-computerized way. "Lights out" complaints and suspected trouble sites are indicated on the large wall map along the front of the dispatch center by sticking pins in the map or attached small stick-on notes. Dispatchers look at the map, analyze and prioritize repair needs, and radio instructions to each field repair vehicle. This gets the job done but is not nearly as effective as the computerized systems.

Premium Reliability

Most utilities will design and operate special circuits or equipment for customers who request, and are willing to pay for, above-average reliability of service. The most traditional of these situations is "roll-over" service to a critical site like a hospital. Not one, but two separate distribution circuits are run to the hospital, with a control system that automatically switches service from one to the other if the primary feed fails. Alternatively, an uninterruptible power supply (UPS, see Figures 8.8 and 8.9) or even backup generation might be installed at a customer's site. Sometimes the consumer owns this equipment, but many utilities will buy, own, and operate the equipment for their customers as a premium service.

FOR FURTHER READING

R. E. Brown, *Electric Distribution System Reliability,* Marcel Dekker, New York, 2002.

P. Christensian, *Retail Wheeling Handbook,* Pennwell Books, 1996.

H. L. Willis, *Spatial Electric Load Forecasting – Second Edition,* Marcel Dekker, New York, 2003.

H. L. Willis, G. V. Welch, and R. R. Schrieber, *Aging Power Delivery Infrastructures,* Marcel Dekker, New York, 2001.

16

System Blackouts and Operational Complexity

16.1 INTRODUCTION

Blackouts, and the dynamic power system behaviors that can lead to them, are among the most complicated phenomena that mankind has attempted to master, in many ways as difficult to engineer and control as a space probe on its way to Jupiter or a beam of quantum particles inside an atomic accelerator. This chapter covers power grid blackouts, discussing their salient points without the use of esoteric technical terms or equations. Despite the authors' best efforts, the reader with little or no technical background may be have to read through a section several times, and take some technical details on faith, before proceeding. In particular, Section 16.3 addresses several complicated concepts such as synchronized operation and phase angle differences that are difficult to grasp, even for engineers. However, the authors can think of no clearer way to explain these extremely important aspects of power system behavior.

This chapter is deliberately redundant, both because repetition is a useful tool in a self-tutorial book such as this, and because the authors anticipate that many readers will only select parts of this chapter to read. Section 16.2 provides a summary of blackouts and their causes, and compares them to other "lights out" situations, for the reader who wants the shortest possible discussion. Section 16.3 then kicks off a more protracted discussion with a look at the synchronized operation of electrical equipment, interconnected system security, and the types of events that trigger blackouts. It continues with a discussion of how grid operation, control systems, loading and voltages on major equipment, and operating problems interact with blackouts. Section 16.4 looks at the root causes of blackouts, and the ways the power industry tries to prevent them. Section 16.5 summarizes with a review of key points and some observations on the challenges the industry faces as it moves ahead.

16.2 BLACKOUTS: AN OVERVIEW

To a utility's customers, a "blackout" is when they and many other people nearby have their electrical service interrupted at the same time. This is certainly the chief feature of a blackout from the public standpoint: a widespread interruption of service, affecting more than just one neighborhood, perhaps an entire city, a state, or even a several-state region.

But to electrical engineers and power grid operators, the term blackout has a much more specific meaning: for some reason, a part of the power system consisting of dozens or perhaps hundreds or thousands of generators, lines, and substations disconnects itself and "de-energizes," in effect shutting itself down. This is the crucial distinction between a blackout and a widespread outage of power delivery caused by equipment failure or storms – the type of interruptions discussed in Chapter 15:

> In a blackout, there may be no equipment that has broken or failed, or the equipment that has failed may be such an inconsequential part of the whole that it has almost no importance whatsoever. What has failed is the ability of thousands of perfectly fine electrical units to work in harmony to form the "power system."

Equipment Failures: The Trigger But Not the Cause of Most Blackouts

The vast majority of service interruptions – situations when a utility's customers suffer a loss of electric service and "the lights go out" – are directly related to failures of electrical equipment (Chapter 15). A power line falls to the ground when a tree falls on it. Lightning hits a transformer and destroys it. A circuit breaker fails due to aging of its internal parts. A hurricane knocks down miles of distribution lines. In each case, equipment fails, and some or all of the customers "downstream" from its location(s) in the power system are without power until it is fixed or an alternate feed route can be established.

More than 98% of the service interruption hours that utility customers in the US experience are due to this type of cause –where full service can be restored only by fixing or replacing the failed equipment. Many of these are interruptions to a single home (the service drops leading to the home are damaged) or a small group of neighbors (the circuit serving their block fails). But occasionally equipment failures can cause quite widespread interruptions of service – the loss of a major transformer might halt power flow to several thousand homes and businesses in an area, a situation anyone living in the middle of that area would certainly identify as a "blackout."

Blackouts: A loss of system interconnection

In contrast to a widespread service interruption caused by a large equipment outage (e.g., failure of a major substation), a blackout can and usually does interrupt service to a much larger area. But again, the most important engineering or operating distinction is that during a blackout, while there may be some small amount of equipment that was damaged or failed, the vast majority of the equipment that is out of service simply "disconnected itself" from the power grid, so much equipment that "the grid" ceased to exist as an operating entity. The lights are out, but the equipment that constitutes the power system – the power lines, transformers, breakers, and substations – are for the most part all still in place and undamaged. But key switches have opened to disconnect one from the other, so that they are no longer connected together into a working system.

Blackouts are often triggered by an equipment failure *in company with* some other flaw or mistake, as will be discussed below: an equipment failure alone should never lead to a blackout. In what is perhaps the "classic" concept of a blackout, a key transmission line fails without warning (perhaps a tree fell on it or an aging insulator that somehow made it past inspection gives out). Regardless, protective equipment designed to prevent damage or unsafe operation switches the line out of service, leaving the rest of the grid to deal with all of the electrical demand that line was carrying at that moment. Electric power flow, moving at near the speed of light, instantly re-distributes itself onto nearby lines that are still working. Suddenly, they, too, are overloaded: they can't carry the electric load they were originally carrying plus the added burden from that failed line.

Now, before system operators can react to reduce demand or re-route power flow to unburden those lines, protective equipment designed to make certain that lines and transformers in the grid perform safely and without damage intervene, opening switches or circuit breakers throughout the grid to take those lines out of service. Power flow again instantly re-distributes itself onto other lines, which compounds the overloading problem for the rest of the grid even more. More lines and equipment shut down, too, almost instantly because the "overload" is now quite high for what equipment remains in the grid.

Such a "cascading outage pattern" of overloads and equipment switching itself out of the grid can ripple across five hundred miles of power system in less than ten seconds. Once started, there is virtually no chance that any human or intervention can prevent it. Therefore, since it is unrealistic to believe that unexpected failures will never occur, the only way to assure that blackouts never occur is to make certain that the grid is never operating in such a way that a failure could lead to the type of cascading outage of equipment described above. Instead, what should happen is that as that very first line fails and is removed from service, instantly, the demand re-distributes itself onto nearby lines, which *are* able to sustain the burden for at least the few minutes it would take for the human operators to take action to reduce loads or further re-distribute loading to mitigate any possible problems.

Nothing has to be repaired

And of course, since very little equipment is damaged, another distinction between a blackout and a service interruption due to equipment failure is that few *repairs* have to be made. To restore an interruption that is due to equipment failure, that failed equipment probably has to be repaired: a downed line put back up, a failed transformer replaced, etc. But to restore service after a blackout, the utility or grid operator has little or no repair to do: even if some unit failed and triggered the blackout, they may defer repair for the moment. Instead, they must "only" start and interconnect, through closing various switches, all the equipment that shut itself down when the blackout started. The authors stress the "only," because it can take hours, or even more than a day, to restart and reconnect the thousands of elements that might be involved in a major blackout.

Useful Analogies to Electric Blackouts

Numerous non-technical analogies have been used to help describe electric grid blackouts. Among those the authors believe convey useful points are:

> *A blackout can be likened to a torn sail on a sailing ship.* A single thread under tremendous stress weakens and parts during high winds. Deprived of the strength of that one strand, nearby strands pick up the strain, only to find it exceeds their strength. They also part, transferring the overload in a ripple through the fabric. In an instant the sail is torn apart.
>
> This analogy conveys well the concept of how a severe strain on a single part can quickly cascade to failure of the whole: just as in a sail, the electric load (equivalent to the strain on the sail) will re-distribute itself instantly, overloading nearby elements of the "fabric." This analogy is also good at conveying the relationship to loading: sails usually fail at times of severe wind loading, blackouts usually occur at or near peak demand, when major equipment is close to maximum loading.
>
> This analogy is less appropriate in one way. In the case of the sail, it is ruined: many of the strands besides that first failed thread are torn apart. In a power system blackout, most equipment is saved by automatic equipment that switches it out of service "just before it would be damaged."
>
> *A blackout can be likened to a marching band* that suddenly stops marching and playing in harmony. One moment eighty musicians are moving in lockstep and playing from the same songbook. Some event

disrupts their ability to follow instructions or coordinate, and each individual stops participating. (Perhaps the conductor trips and some band members lose the direction they need at a crucial point in their march, etc., or a player on the front of the row stumbles and those behind fall over him, the impact rippling through the ranks.) In an instant, the band changes from an organized entity to just a crowd of people each carrying a musical instrument and moving in his or her own direction and speed.

This analogy emphasizes several points about loss of synchronized operation and its impact on a system (band, in this case) – the players *were* marching in lockstep, just as the generators in a power system turn in absolute synchronization (a point which will come up in the more detailed discussions later in this chapter). It is also a good analogy for how it emphasizes that loss of synchronization means everything stops.

A blackout can be likened to a team of sled dogs, an analogy that is particularly apt for discussing the role of the generators versus that of the transmission grid in a blackout, which will be used later in this chapter

But regardless of whether the reader finds these analogies useful or not, the chief points to keep in mind about blackouts are:

Something disturbed the ability of the many units of equipment in the grid to work together in harmony, to the point that many tripped off line suddenly, so many that the remainder could not deal with the electric demand.

Most of the equipment that is out of service is undamaged. It did not fail, and as the system unraveled, it was saved from damage by being disconnected so quickly.

Two "bad things" had to happen at once: the failure or event that triggered the blackout, and some situation that permitted cascading of outages after that triggering event.

The root cause of the blackout is not the triggering failure event, but whatever happened to put the grid in a mode where a cascade could begin.

It can take quite a while to restore the system to full operation. The band conductor must pick himself up after tripping, dust himself off and restore as much dignity as he can, round up the now dispersing crowd, make certain each musician has his or her music, get everyone into place, communicate to all when and where they will begin again in both music and march, and re-start. All of that can take a long time.

When a blackout occurs, system operators must check each unit of equipment to make certain it is not damaged, then "re-assemble" the grid, piece by piece, switching one unit of equipment to another, and gradually adding more lines and facilities, always in a way that never overloads or violates any of the many operating criteria that would trip it out of service again, gradually "getting the lights back on." This can take hours, even days.

Blackouts Always Involve Many Separate Equipment Units Being Out of Service at the Same Time

The highest voltage line, or the largest power transformer, or the biggest power generator, is far too small to feed an entire city, much less a whole state. Thus, the only way that a whole city or state can be "blacked out" is if a lot of equipment goes out of service at the same time. Therefore, any blackout must involve many units of equipment going out of service at the same time.

As an example, the largest transmission line involved in the Northeast blackout on August 14, 2003 had a capacity that could support perhaps 1/10 of the power needs of Cleveland, just one of numerous cities that were without power for many hours during that event. Over 99.9% of the lines and equipment that did not work during that blackout were in perfectly sound condition, not damaged in any way. Blackouts occur because something disturbed equipment in a widespread area of the power system, consisting of thousands of units of equipment all designed to work in harmony, and it suddenly is unable to maintain that harmony.

How Can Blackouts Be Prevented?

There are several measures that must be taken to assure that a blackout does not happen. When a blackout occurs one or more of these conditions was violated (perhaps not intentionally)

1. Engineers and utility equipment owners must make certain that the grid has enough capacity and voltage margin to survive the sudden, unexpected loss of any line, generator, transformer, or switching point.

2. Engineers must assure that the power grid simultaneously satisfies hundreds of criteria for electrical stability and power flow requirements, and that it can stay within these parameters even if a large unit fails suddenly and power flow patterns shift instantly.

3. Engineers must make certain that control and monitoring systems can distinguish correctly between situations where they really should take action and remove equipment from service, and those situations where action is not essential and in fact might be harmful.

4. Grid operators must control the system to make certain it always operates so that items 1 through 3 are all satisfied at any moment.

5. Engineers and scientists in the power industry must make certain that the concepts and tools they apply to assure 1 – 4 above are always met are complete and sufficient to assure dependable operation.

All five of these requirements are more difficult to accomplish today than they were in the past. This is partly because the capacity margins (1 above) needed to comfortably survive major equipment losses are expensive and have shrunk: no one wants to spend more than necessary. Perhaps more important than the cost, large capacity margins mean additional "backup" lines and facilities, and irrespective of any regard for cost, there is considerable opposition to building new lines and facilities from an esthetic and environmental impact standpoint (see Figure 15.4 and accompanying discussion).

Second, in the search for more competitive de-regulated markets (see Chapters 11 and 12) and more efficiency in the use of equipment (see Chapter 3), power grids have become bigger – they cover more territory and include far more equipment than they did a few decades ago. The sheer size of these regional grids creates challenges of complexity and scale in analysis; the number of components becomes so great that it becomes practically impossible to check every combination of equipment settings in the whole grid to make certain that every contingency and situation is covered. Today, the best engineering and operating methods available can barely "see" or understand some of the complex ways a huge power system might slowly "back into" trouble.

Beyond this, the size of power grids creates another issue that has only recently been recognized. The distance across a multi-state grid might be so great that electricity, even though it moves at near the speed of light, takes a noticeable time to move from one end of the grid to another, often less time than a computer – for example those in protective control systems – takes to make a decision such as "This isn't serious," or "Switch this off, now!" Only since the 2003 blackout has the industry fully realized that dependable operation of a dispersed power grid requires very precise measurement of tiny differences in timing across the power grid, on the order of $1/10000^{th}$ of a second over distances on the order of a thousand miles. This is within the capability of modern control systems, but requires the very latest in equipment and software.

16.3 SYNCHRONIZED OPERATION OF POWER SYSTEMS

The phenomenon in which many disparate equipment units across an operating power system suddenly disconnect themselves so that the grid literally ceases to exist as an interconnected unit is caused by a sudden loss of what is called *synchronized operation.* This section presents a non-technical background summary on synchronized operation, and conditions that assure it can continue: *interconnected security.*

In an AC power system, the voltage fluctuates back and forth in a steady rhythm, 60 times a second (50 in European systems). The rotating generators create this fluctuation in the power system (see Chapter 6). It is *very* slight differences in the timing of these sinusoidal pulses, what are called *phase-angle differences,* that cause most of the power to move through a power grid, rather than, as many people assume, differences in voltage. In this respect the behavior of AC power and AC power systems is quite different than that of DC power and DC power systems. In an DC power system, voltage does not oscillate back and forth as in an AC system; power will only move from point A and B if the voltage at A is greater than the voltage at B. Furthermore, the amount of power moving between the points is proportional to the amount of voltage difference between them.

But in an AC power system, power flow is somewhat more complicated, because differences in voltage do matter, as they do in a DC system, but power moves and electrical behavior is also shaped by differences in the exact timing of the oscillation pulse at two points. If two points in the power system – a

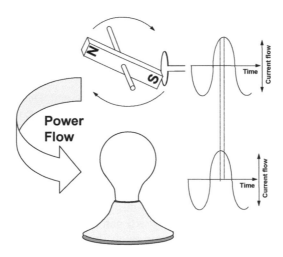

Figure 16.1 A generator and the load it powers, and the voltage at each location (from Figure 6.1). The generator produces sinusoidal power that is transmitted to the light bulb. Each pulse lasts 120th of a second (one back and one "forth" pulse together constitute a full cycle). Time is graphed with "already happened" to the right and "yet to happen" to the left, so the pulse at the generator began slightly before the pulse at the light bulb. Such small phase angle differences in the timing of the pulses (when they reach their peak) drive power from one location to the other in an AC power grid.

generator at one end of a transmission line and a factory consuming power at the other end – differ in the relative timing of their pulses, power will flow from the location "in the lead" (where the pulse reaches maximum amplitude first) to that which lags, as illustrated in Figure 16.1.

Complex Variable Voltages

More generally, slight phase angle differences at points throughout the power system are associated with how much and where power flows in the power grid. Voltage matters too – if the pulses at one location are bigger (higher voltage) than at another, that also makes power flow from high to low voltage.

Therefore, actually computing the power flows in a network is a complex undertaking, quite literally, for it involves complex variable mathematics, in which magnitude and voltage are combined, and all the numbers involved have two values, a magnitude and a phase angle. But conceptually, power flow and phase angle are easy to keep in mind for this reason: power systems are intended to function as *equi-voltage systems* – they are designed and planned so that differences in voltage from one location to another are very slight, only a few percent.

It is differences in phase-angle that are chiefly responsible for the movement of power in and around a power system. The difference in the timing of the pulses in Figure 16.2 creates an electrical force between the equipment shown. This force is just as real and palpable as if it were transmitted through a metal rod or a strong rope. This force *is* the power in a power system.

Synchronized Operation

In order for a power grid to operate in a stable and continuous manner, all of the generators must turn at exactly the same RPM rate. The required precision is phenomenal: over the course of a year they must turn within .0000001% – they cannot even vary, over the course of the year, by $1/10^{th}$ of a revolution in the total number of times they revolve. If this precision of all the generators in the grid is disturbed for any reason, unstable operation results, and a blackout almost surely follows within seconds.

Fortunately, the generators in a well-designed and well-operated power grid tend to be "self-synchronizing" – they will naturally fall into synchronization so that they all turn at precisely the same rate. This is a consequence of forces – energy – transmitted from one generator to another through the transmission grid. These forces tend to be considerable, but not absolute: any number of events can disturb them to the extent that the generators fall out of synchronization: a blackout follows within seconds.

Figure 16.2 illustrates this key aspect of utility operation, a fundamental concept behind understanding power system operation and blackouts. The transmission grid of a large power system interconnects all the generating plants throughout the system. Each generator turns as it generates power (see Chapter 7). They run at *exactly* the same speed. For stable, continuous operation, all must remain turning at precisely the same rate.

While the major purpose of transmission lines is to provide the means to move power to where it is needed, an important consequence of their interconnection of one generator to another is that each will sense how fast the other is turning, and they will fall into lock step with one another. Connected by their "electrical harness," each can sense the rotation of the other, at all times, as a type of electrical tug or push depending. If two generators are connected and the first tries to turn a bit faster than the other, the second exerts an electrical force through the transmission lines that says "Whoa there, slow down!" Similarly if the first slows, the second will try to pull it along.

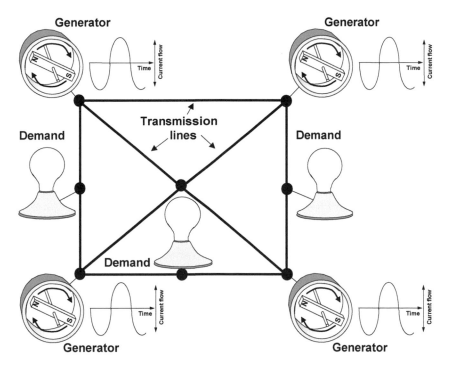

Figure 16.2   A power system consists of lines connecting generators and loads (demands). Here, four generators provide power for three load points. The generators are locked exactly in step with one another, their pulses *exactly* synchronized by magnetic – electrical forces traveling through the grid from one to the other.

In this regard, a group of generators and their transmission system is analogous to a team of sled dogs (the generators). The multi-dog team is held in check by a common harness (analogous to a transmission system) that transmits their energy back to the load (the sled).

Without a doubt, the major purpose of the harness is to transmit the energy from the dogs to the sled. But the harness also keeps all the dogs running at the same speed – it provides the strength to hold them together, as well as the "feedback" each dog need to participate as a member of the team: through the harness, each dog can sense how hard the rest of team is pulling and what is expected of it.

If any one dog wanted to go faster, it would have to shove against its harness harder and in some sense pull all the other dogs along with it, but that would not get it away from them: they'd still all be together. Thus, a particularly powerful dog cannot outrun its teammates in the harness no matter how hard it tries. But it *can* pull slightly ahead, relative to the others, by stretching its harness – say by half an inch – as it pulls that much harder. To stay that ½ inch out in front, it has to continue to put forth that larger effort.

The transmission grid in a large power grid links all the generators in a sort of "electrical harness." As a result, the generators in an AC power system such as depicted in Figure 16.2 do not turn at about the same speed as one another. They don't even turn at *precisely* the same speed as one another: when operating as they should, *they turn as if they are a single unit.* If one lags even a fraction of a revolution behind the others, its phase angle difference with the system opens, and that means it inputs less energy to the grid, so the work expected of it is reduced and it suddenly has a margin to catch up with the rest. Conversely, if a particular generator tries to speed up just a fraction compared to the others, it has to push ever harder as it tries to pull away, until it reaches a point where it can't provide any more power out.

In this way, all the generators run up to their output limit and stay pushed against their electrical harness. Each is turning at *exactly* the same rate, constantly pushing up against its electrical harness as all, cumulatively, tug the "load" (customer consumption of power in the case of the power grid) along with them.

A particularly good point about this analogy is that it can explain well how various capacities (sizes) of generator can all fit into the same system. The sled could be made to work with various size dogs: from small Chihuahuas to Huskies to Irish Wolfhounds. A power grid does work with all sizes of generators, from small to large. In both cases, each unit in the group contributes what it can, but all are moving along at the same rate: the dogs all run at the same net speed and the generators all spin the same RPM rate. At the end of the day the dogs have all covered the same distance and the generators have each produced just as many pulses of electric power.

Stable Operation

This "automatic" synchronization of generators is what permits a power system to work smoothly and reliably. To observe that such operation is routinely achieved is to ignore the many difficulties that must be mastered. The authors, no experts at mushing sled dogs, imagine that, like power system operation, it is not as easy to do as the best professional "drivers" make it appear. But the fact is that power systems are operated in this manner every day, all over the world. If reasonable care is taken in starting each generator, running it up to speed until it is at the same speed, then switching it into the system, it will push against the grid's "natural physical barrier" of phase angle and stay synchronized with the rest of the generators in the system.[1] The challenge to grid operators is that if it "stumbles" even a tenth of a turn from the pace of the others, it will trip off-line (break from its harness) and a blackout could begin.

Such operation, in which all the generators are working as one, is called *stable synchronized* operation. An AC power system cannot work at all if the generators are not both interconnected *and* synchronized in this manner. If one tried, power flows and voltages would fluctuate wildly in only a matter of seconds. Consider what would happen if one generator in Figure 16.2 were to turn at a slightly faster rate, say 3601 times a minute (60.0167 cycles per second) instead of 3600 times a minute (exactly 60 cycles per second). Over the course of that minute, the generator would go through one complete revolution compared to the rest of the system: its phase angle difference with the other generators would go from zero (perfectly in sync with the rest of the units) to 180 degrees out of sync, and back, all in one second. Because power flow is a function of this phase angle difference, in that one minute, power would first flow from the generator into the system, then change the other direction. Voltages would fluctuate wildly, between exactly what was intended, too much, then far too much, and then even more – enough to tear apart the generator and other nearby equipment.[2]

---

[1] It is worth noting that in a power system "sled dogs" are changed while the system is running. Generating plants are started, run up to speed, then switched into the system so that they synchronize themselves. Those shut down are first switched out and then shut down, all while the system is running. This is analogous to changing dogs while the sled is moving. One would have to mush a dog up to speed alongside the team, then clip its harness onto the larger harness. Similarly, taking a dog off would involve unclipping it from the team while it is still running, then letting it slow to a rest.

[2] This is purely a "theoretical" mental exercise. In the real world, this situation would never last the full minute, or even a major portion of it. Within a second or two, the fluctuations and voltage differences would be so great that protective equipment would switch the unit out of service, or electrical forces beyond the generator's design ratings would tear it apart.

The consequences of such fluctuating operation are why it is *never* permitted. Automatic equipment at every generator plant, and at many other points in the power system, is set to open circuit breakers and shut down generators in less than a tenth of a second if it sees any "off-frequency" operation – any difference in rotational speeds of operating frequency – even a very minor one such as the difference between 60 and 60.167 cycles.

System Stability and Security

Continuing a bit further with the sled dog analogy, suppose that a strap holding one of the dogs suddenly broke while all were pulling particularly hard, perhaps while going up a hill at high speed. One dog would break free, unable to pull any longer. At that instant, every other dog on the team would feel a sudden, heavier tug on its part of the harness. If the strap holding any of those dogs in place was too weak, it, too, would snap.

Thus, among other requirements, the harness has to be designed not just to be strong enough to handle the load of the sled and the pull of each dog, but to tolerate reasonable sudden transients and contingencies. Something like this is a key design goal for power grid engineers in every utility and regional grid operators. The transmission lines in any utility system are designed primarily from the standpoint of having enough capacity to move power from generation plants to where it will be consumed (i.e., the harness is strong enough to pull the sled). However, system design is checked for "stability requirements" so that the system will always have enough interconnection strength to take the shock of one line suddenly tripping out due to a problem (a criteria equivalent to the harness being able to take the shock of "taking up the slack" suddenly if one strap in the harness breaks without warning). Nearby lines must be strong enough (electrically) to tolerate the changes in flows and voltages that would result from the sudden loss of any nearby generator, transformer, or transmission line.

An additional aspect of generation-transmission system operation also has a sled dog analogy. Suppose that at the moment the strap breaks, one of the other dogs on the team is pulling as hard as it can, on the edge of slipping and with no strength left to give. The sudden "transient" – momentary shock transmitted through the harness – might cause it to trip or falter, which would exacerbate the situation to the point that other dogs are similarly tripped, etc. Instantly the whole team would fall out of step and tumble, quickly bringing the sled to a halt.

Therefore, the team musher needs to make certain that no matter how hard the dogs are pulling, at all times each has a slight margin left in case it must shoulder an unexpected transient shock. This could also happen if a strap broke due to the sudden transient snap of a dog slipping, etc. Thus, the harness has to have an extra margin of strength against any sudden unexpected event.

In a power system, the analogous situation would be a generator that is either so close to its limit that it cannot react with the additional power that might be needed if another big generator fails unexpectedly, or which cannot react fast enough during the milliseconds right after that failure. Either way, an unexpected event will cause it to slow its RPM rate. The instant that happens, because it is now turning too slowly, its automatic equipment would shut it down to prevent the wild voltage fluctuations that would cause, and to avoid damage to it.

Just as the sled driver must always control the team so that it never suffers a stability failure if a strap break, or a dog slips, so a power system operator must allow a small stability margin for every generator and transmission line. Operation must be *secure*.

Avoiding Blackouts

Thus, every dog in the sled team must always "operate" with a bit of "steady state" margin – an ability to shoulder a bit more burden in case one dog momentarily is lost (its strap breaks), and it must have "transient" margin against the sudden shock of a dog stumbling or a line parting without warning. Similarly the harness, too, must have equivalent capabilities.

In an electric grid, every generator and transmission line, and every substation bus and transformer, must have similar capabilities to handle both a short-term overload if any unit fails, and the "shock" of a sudden transient. Otherwise, an unexpected failure of a generator, line, or major transformer could lead to a quick cascade and a blackout. And since in the long run failures are inevitable, this means that the only way to avoid blackouts is to avoid ever violating these requirements: the grid must always be able to tolerate the loss of any component, without warning.

Again, if a blackout occurs, its root cause is not the unexpected failure or overload that was its trigger, but the mistake or control error in system operation that permitted the grid to operate in a mode where a cascading outage could happen if a failure did occur. The following two timelines outline the general way that many blackouts occur.

When automatic protective equipment works correctly

Time = 0.0 seconds. The system is near peak demand, but everything is operating well. Without warning a transmission line, generator, or key transformer in the grid malfunctions. Voltage in that line, generator, or transformer dips and current surges.

Time = 0.1 seconds. The "protective relays" monitoring the malfunctioning unit sense that something is wrong and open one or more circuit breakers (as many as required) to "isolate it" (de-energize it and remove it from the grid). This action is quick enough to avert major damage to the equipment, and more important from a system standpoint, before the malfunction can lead to intolerable voltage fluctuations. Instantly, the electric load (burden) on the surrounding system shifts to nearby lines and generators.

Time = .25 second. Unfortunately, a nearby generator is already loaded to its absolutely maximum, to the point that it cannot shoulder *any* additional load, or to the point that it cannot react to this event quickly. Either way, it begins to slow (its RPM drops relative to the rest of the system), perhaps by only a fraction of an RPM per minute. Its phase angle with the system opens as it begins to lag behind the rest of system, creating a situation called under-frequency (it is not keeping up with the system revolution, or oscillation rate). Voltages nearby begin to fluctuate, but before that can become a serious matter, automatic equipment called an under-frequency relay opens circuit breakers to switch it out of service.

This is the blackout's "root cause" – the first undamaged equipment withdrawn from service (there is nothing wrong with this unit). Regardless, it too is now offline. Electric burden shifts to other nearby parts of the grid.

Time = .40 second. The unexpected burden of losing the second unit, on top of the very recent shock of losing the first, causes other nearby generator(s) to slow, and their automatic equipment to cut them off line in a like manner.

Time = .60 second. Electric burden on the system has shifted rapidly to other generators and lines farther away. A lot of equipment throughout the system now sees the effects of this escalating series of events – fluctuating voltages and swinging transfers of load about the system, and some of it is slowing under too much burden, as the previous generators did. Protective systems for equipment throughout the system now are seeing conditions they don't like – a voltage too high here, a voltage too low there, an overload, under-frequency operation, etc.

Time = still less than one second. In a falling domino-process that takes at most a few seconds more, equipment throughout the system is almost simultaneously tripped off line by one piece of protective equipment or another, each seeing something it doesn't like. The system literally comes apart, with lines and generators switched off from the system. Interconnection is lost.

In this scenario, protective equipment worked correctly. The blackout occurred because the system was in a condition, prior to the blackout, where it could not tolerate the triggering event. The "mistake" that led to the blackout was letting the system get into that state.

Automatic equipment does not work correctly

If automatic equipment does not work either because it fails to do its job, or because someone thought that a blackout could be avoided by setting it to react more slowly, things are get much worse, much sooner.

In this scenario, the original, first malfunctioning generator or line fails, as before, but now, for whatever reason, it is not instantly withdrawn from service by its protective equipment: something goes wrong. As it fails, voltages begin to fluctuate and nearly equipment overloads, perhaps for only milliseconds, as power flows fluctuate wildly. Depending on the specifics, any number of rather remarkable events happen in quick succession, none of them good. For the sake of example, assume the malfunctioning generator explodes in a hail of shrapnel and electrical fire. The electrical shock of this catastrophic failure ripples into nearby transmission lines; fluctuating voltages reach nearby (electrically nearby – this might mean dozens of miles) generators. Their protection equipment instantly takes them off line. A process of system disintegration that took a few seconds when everything worked correctly is over in less than a second. Beyond that, one generator is ruined beyond repair, perhaps more. A lot of minor equipment nearby is also ruined or stressed to a point it is suspect and must be carefully tested before being put back in service.

Finally, a third alternative occasionally occurs. The protective system "failure" could have been the opposite; there was no problem with the generator but the sensors thought there was, and the protection system opened switches and just as if there were – and from there the same disintegration of the system began. There have been blackouts triggered by problems with protective systems making such mistakes. This is why utility relay engineers, the people responsible for specifying, setting, and checking protection systems, are meticulously careful about all protective settings and their coordination.

But the point of all the foregoing is that, somehow, a single event triggered by a failure – of a generator or a protective/control system – triggers an event that cascades through the system, leading to widespread disconnection of equipment that is, for all intents and purposes, in perfectly good, working condition. *The real flaw that leads to the blackout is not the failure that triggers the event, but the fact that the rest of the system is operating, at that moment, in a state where it can not tolerate this event.*

System Security

Failures, accidents, and unexpected events cannot be completely avoided. Thus, the only way to assure that a blackout never happens is to maintain *interconnected system security* at all times – have enough lines in service and enough generators on line with enough margin that the system can stand *any* shock it is likely to see without disturbing the synchronization (spinning rate) of any generation in the system or otherwise overload of trip off other equipment. That is the *chief* concern of utility system operators. Although they are committed to keeping the lights on, and to achieving the maximum efficiency they can, they will sacrifice either or both of those goals in order to keep the power system in a secure state, if necessary.

There are two reasons for this high priority on operational security. First, as outlined earlier, the system cannot work at all in any other mode: without interconnection and synchronization it will develop voltage and power flow fluctuations within seconds and become inoperable. Second, it is not easy to get the system back into interconnected, synchronized mode again.

After a blackout, generation operators must do a "black start." They have to begin with a generator that has "black start capability" (equipment and facilities so it can be started when there is no nearby source of electric power. Not all generators as so fitted). They must run it up to the exact speed of system RPM (equivalent to 60 cycles), switch it to a transmission line, then use power from that generator to run a second up to speed, check that it is running at the identical speed and switch it onto the line. The two generators will fall into synchronization themselves, instantly, and the operators now have an elemental system. Then, one by one, other generators are run up to speed and other lines are switched into this mix until all are "back on line." This can take days in particularly bad situations.

Maintaining System Security

In order to avoid ever getting into a mode where the system is not secure, operators in the system control center (see Figure 15.11) run continuous computer simulations of *system security*. These systems monitor key points in the grid, then use that data to cycle through a "what if this happened?" analysis of the consequences of every possible failure that could occur in the next 15 minutes. For each, they determine if that failure could be tolerated. If so, they move on to the next possibility, if not, they alert the system operator, and usually then run a subsidiary program to identify the best solution ("If you lose generator A you will lose B within five seconds and start a cascading blackout. Start generator C now so it can share the burden, and there will be no problem if you lose A.") Electrical forces act so quickly that this is the *only* way blackouts can be avoided. One cannot wait until a failure occurs to act: even the fastest devices will fail to react quickly enough. The utility must avoid ever having the system operate in a mode where a cascading outage could begin.

If the system simulation computer detects insecurity, operators will do whatever they can to avoid it. They will change the generation dispatch (which generators are producing what amounts of power), adjust flows and settings on transformers and substation buses and other equipment to re-distribute flow through the grid, and, if necessary, shed load – deliberately disconnect some customers from the grid in order to lighten the burden on the system.

Impact of Loading

For a number of reasons, blackouts usually occur when the system is heavily loaded: high loads create situations where interconnection capability is squeezed and margins are small: the harness isn't likely to break and a dog is not likely to slip when the team is just ambling along on a smooth surface.

16.4 WHY DO BLACKOUTS OCCUR?

Blackouts have always been with the power industry – a major one seems to occur in North America roughly every ten to twenty years (Figure 16.3) and may very well be with the industry forever. The fundamental reason that blackouts occur is not technical, but cultural: people want to build ever-larger power grids and operate them ever closer to their limits. From any practical standpoint, blackouts *can* be avoided, or at least made to be so unlikely that they are literally not even a once in a lifetime likelihood: Sections 16.2 and 16.3 discussed some of the reasons blackouts occur and summarized how they can be avoided. But the history of the power industry, and its blackouts, has shown that new technologies are eventually used to extend performance, not reliability of grid operation.

As mankind learns more about electrical systems, and develops better technology, it uses that capability not to assure that its existing power grids are more dependable, but to permit it to reconfigure and build bigger grids and to operate them with thinner-than-ever capacity margins. Thus, "power grid reliability," defined as the proven ability to avoid blackouts, has not really improved in the last thirty years: In North America a major blackout has always been likely to happen once every ten years of so. There is really no reason to think this situation will change in the future.

Major blackouts are usually a "learning experience" for the industry. When a widespread, lengthy blackout occurs, a thorough investigation by the power industry's own organizations (i.e., NERC, IEEE) and/or the government (FERC,

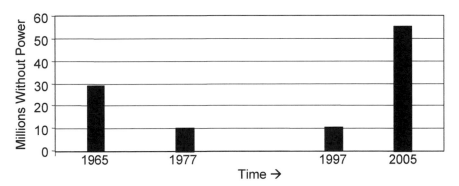

Figure 16.3 Blackouts affecting ten million or more energy consumers in North America. Events depicted here are blackouts in the Northeast US (1965), Northeast US (1967), Western US (1977) and NE US and Canada (2003). Despite industry efforts, one such blackout seems to occur every ten years or so.

DOE) is carried out (see For Further Reading). This identifies or verifies a new technical concept that needs to be considered in order to avoid blackouts (e.g., voltage collapse or the need for wide-area measurement). New rules and guidelines are established. New technologies are deployed to address it. Power grids are now "immune" to blackouts from this cause. But this capability permits operations of bigger grids, or grids operated closer to the limit, either of which will save money and improve efficiency (i.e., make money). Inevitably, the grid is pushed in this direction. Calculation and guidelines show that, even so, because of the new capabilities, the grid is still reliable.

But inevitably, this new knowledge and technical capability is applied to permit the interconnection and operation of more complicated (bigger, more widespread) grids, operated closer to the known limits of secure operation (with high loading, and less margin). Grids are again operated right at the edge – and somewhere, sometime, someone pushes past that edge, and another widespread blackout occurs. This cycle tends to take about 10 – 20 years.

Steady Increase in Complexity

The basic causes of power grid blackouts are easy to articulate ("never operate in an insecure mode" and "always have sufficient margin for any contingency"). But application of those concepts to an actual grid can require a good deal of intricate and precise calculations, followed by very knowledgeable interpretation of the analytical results, before one can determine if a particular power grid is operating within these criteria, and if not, why and how that situation can be corrected. The bigger the grid, or the more closely to its limits it is operated, the more difficult it becomes to do this analysis correctly.

In 19th and even into the early 20th century, the power industry operated few interconnected grids larger than the generator and the distribution system required to serve a small town. Cities like New York and Cleveland were served by several separate, non-interconnected power grids, each a very local assemblage of one or two generators and lines leading to nearby homes and businesses, even if these "mini-grids were all owned and operated by the same utility company.

As engineering methods improved more, and ever-newer and more improved equipment and systems technologies were developed, interconnection was carried out over wider areas. Over time, the small community systems were integrated (connected one to the other) into city-size, county-wide, and in some cases state-wide systems. This resulted in much improved reliability and economy of operation.

By the early 1950s, with few significant exceptions anywhere in the world, every electric utility, even the largest ones, had a single integrated power grid. In other words, each utility operated only one system that covered all the territory it served, even if that was an entire state, and the only boundaries between systems were at the boundaries of separate utility service territories.

Beginning in the mid 20ᵗʰ century, for mutual support, groups of neighboring utilities began to enter into "power pool" arrangements in which they connected their systems one to the other, in order to provide for mutual support – if one lost a generator suddenly, a neighbor would "loan" them power until they could get another up and running, etc. Over time, these power pools grew stronger (initial very weak links were reinforced into major power arteries) and wider (more utilities joined the group). It was the existence of these regional power pools, as they had become by the 1980s, that formed the basis for the large, regional, interconnected transmission "markets" that are the foundation of de-regulation in the wholesale power industry (see Chapter 2 for the reasons de-regulation led to larger grids. For more history on the growth of the industry and power pools, in general, see Chapter 3).

One can measure the complexity of power grids in terms of geographic size, number of components, amount of demand served, or the number of customers connected, and the conclusion is always the same: the average size of grids in North America and elsewhere slowly but steadily got bigger, as utilities merged, or as neighboring utilities joined ever larger power pools. Thus, over time, while there were fewer and fewer separate power grids, the average size of which steadily if slowly increased.

As the number of elements in a grid increases, the complexity of analysis of its performance increases and operation of it on a daily basis becomes more demanding. A small power grid with ten generators will typically have many dozens of operating modes (ways that combinations of generators and switched lines can be run). Each must be checked carefully from the standpoint of numerous operating criteria related to engineering and interconnected operation, including all of the stability and interconnected security issues alluded to in Sections 16.2 and 16.3, before one can determine with certainty how to operate it so it can avoid blackouts. Similarly, the operator will have to understand the characteristics of each of the several dozen operating modes, and what they mean with respect to sudden losses of equipment, demand level shifts, or other sudden changes, so that system operation can be performed well despite sudden unexpected events.

Poring over dozens of operating modes is not a challenge for modern engineering analysis methods. In fact that has been within the capability of the electric engineering fraternity for nearly a century. But the number of operating modes increases exponentially with system size. A system with only a hundred generators would have far more than ten times as many modes to check and verify – probably something closer to one hundred times as many. Furthermore, each of those modes would involve more equipment and more equipment interactions, and be that much more difficult to analyze.

Analytical resolution and new factors

Increasing grid size adds two further challenges. First, in addition to creating more operating modes to check, and making each of them more involved in numbers of items that need to be analyzed, increasing the size of the grid means that more precision needs to be applied in certain steps of all that analysis and operation. The truly massive, multi-regional grids of the 21st century require incredible precision in the analysis of certain power flows and equipment status: a very slight change in one place can make a significant change somewhere else. Thus, requirements for measuring equipment at key stations in the grid, and computation systems that monitor those measurements increase further.

Second, as grid size increases, the power industry occasionally encounters additional phenomena, for the most part not entirely anticipated, that have to be considered. Several times during the history of the power industry, the creation of ever-larger grids that spanned much larger distances, included more equipment than ever, or worked closer to their capacity edge, revealed new physical principles that had to be included in engineering and operating analysis if one wanted to keep their operation reliable. Details of these phenomena go far beyond the scope of this book. They include things like sub-synchronous resonance, in which voltage and current will oscillate back and forth in an unwanted, growing, and ultimately damaging manner, and voltage collapse, which as the name implies means that voltage can suddenly disappear within a split second, as well a number of other rather esoteric and complicated technical phenomena. In each situation in the past, these somewhat-understood phenomena were involved in a large blackout, which led to a much better understanding of it, along with new methods and new rules for grid operation. Such occasional learning experiences seem unavoidable.

The modern complexity challenge

In almost all cases in the past, these new phenomena had been partly anticipated through theoretical analysis and research, but not completely understood as to if and how they might really interact with operation. That is the case with modern power grids, and a new challenge facing the power industry. Modern power grids have reached the geographic size where the time it takes electricity – and more importantly the impact of an event such as an equipment failure or short-circuit – to travel across the grid is noticeable compared to other time spans needed in analysis and operation.

For example, it takes only 1/465th of a second for electricity to travel 400 miles, about the distance between Cleveland and New York, both of which were without power during the blackout on August 14, 2003. Yet 1/465th of a second is 1/8th of a full cycle (see Figure 16.1). It represents more than 45 degrees of phase shift. In an industry where 1 or 2 degrees of phase shift over the span of a second can make the difference between secure synchronous operation and a blackout, this is a huge amount of time.

Modern grid operators can measure the timing of any aspect of their systems to within $1/465^{th}$ of a second: they can measure times as small as $1/1000^{th}$ of a second quite well. But what they cannot do is measure timing that precisely on different equipment separately *over long distances,* such as the distance between Cleveland and New York. They can measure an event in one part of the grid and determine that it took $1/400^{th}$ of a second. And they can measure an event 400 miles away of the system and determine that it took $1/200^{th}$ of a second. But generally, grid operators cannot determine that these two events, separated by 400 miles, occurred exactly $1/465^{th}$ second apart, or $1/500^{th}$ second, or $1/200^{th}$ second, or just how far apart in time they were, with any precision approaching what seems to be needed.

This inability is not the cause of the August 14, 2003 blackout: had the operators of the power grids involved had this capability, the blackout would very likely still have happened. But a lack of this timing analysis capability hampered the industry in post-blackout study, to determine what really happened and how it can be avoided.

Furthermore, this ability to coordinate accurate timing measurements over wide distances is a part of the solution to such blackouts. A wide-area measurement system (WAMS) that would track timing over a 1000+ mile wide grid with precision to $1/10000^{th}$ of a second would permit coordination of automatic protective control equipment throughout the grid such that they could be programmed to avoid blackouts cascading from one area to another. Had such a system been fully deployed on August 14, it would have terminated the blackout before it spread widely – probably limiting it to only the area in northern Ohio, where the triggering event (a series of high voltage lines that tripped out due to overloads and short circuits) occurred.

De-Regulation Magnifies Complexity

To recap, there are various very compelling reasons to build ever-larger power grids. For a host of reasons, people in all walks of life, including utility owners, grid operators, politicians, regulators, and customers, demand that these grids be pushed to the limit.[3] Occasionally things don't work as well as expected and a

---

[3] It is unrealistic to blame utility owners or operators alone for the omnipresent pressures to "push" the grid to its limits. Customers want more power, at low cost. Politicians and regulators want them to have it, and don't want to approve rate increases or construction of additional lines to case burden on the grid (new lines have environmental and esthetic impacts). Everyone wants a low-cost grid that nonetheless performs to a high standard. Sometimes the cumulative effect of those pressures is a grid that is operated up to its limit, so that infrequently, but occasionally, a blackout occurs.

blackout occurs. Often a large blackout reveals a new phenomenon or drives home a lesson to the industry that results in new guidelines and technology, so that that type of blackout never happens again. But grids continue to grow and something new surfaces to add more complexity, and cause the next blackout.

De-regulation accelerates the trend in grid size growth

As power grids grew in size and complexity through the 20th century, for the most part the industry's technology and skills were able to keep pace because *the power industry let its developing capabilities constrain the growth of its power grids*. It approached the operation of ever more complex grids incrementally, being careful (some might say overly careful) to not overextend its capabilities. In the 1970s, one of the authors (Willis) was co-chairman of reserve planning for ERCOT (Electric Reliability Council of Texas – a group of utilities in Texas that operated in a pool for mutual support and economy). From personal observation, it was abundantly clear that ERCOT's plans, like those of many other power pools, were dictated by a deliberately prudent policy to not overly extend the proven capabilities of available analytical and control technologies. The rules were: "If we can't analyze it well, we won't build it. If we can't monitor and control it, we won't try to operate it."). There is no one in the industry that would credibly argue against such a policy then, or today.

De-regulation pushed the power industry to interconnect its large power grids into ever-larger regional and multi-regional grids. In some cases, it mandated that larger grids be created. The reasons are covered elsewhere (see Chapters 2, 10-12). This accelerated grid growth and a consequent increase in complexity beyond that which would have occurred had the industry been left to its traditional "technology constrained" growth limitations.

Many industry experts were concerned about the operation and control of the very large grids that de-regulation had created prior to August 14, 2003. Many still are. This led, among other things, to some conservatism in operating the grids and policies often interpreted by proponents of de-regulation as opposition to de-regulation. It also led to development of new engineering methods and technologies. Among these is WAMS, the precise timing systems described earlier, which are needed for large-area grid operation. While the basic technology was developed just prior to the August 13 blackout, at that time WAMS was not widely deployed. In fact, at the time of this writing, nearly two years later, they still aren't.

Figure 16.4 is the authors' estimate of how grid complexity grew during the past century. "Grid complexity" is computed as the number of major elements in the grid (generating plants, key transmission lines, major high voltage buses) times the number of simultaneous goals the system operator is trying to achieve:

Complexity = (number of interconnected components) x (number of goals)

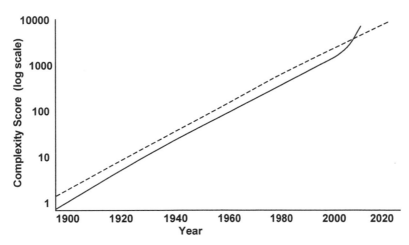

Figure 16.4 Complexity facing the operators of interconnected power systems increased steadily during the 20th century and into the 21st (solid line). This was always kept within the rate of growth of technology to analyze and control grids (dotted line.) De-regulation accelerated that trend to larger grids starting in the late 1990s, by adding several new criteria to operations requirements (see Chapter 12, as well as the discussion earlier in this chapter, and Table 16.1) to where grid operating complexity perhaps exceeded modern control capabilities. The industry is working hard to both accelerate technology and deploy modern systems based upon new concepts, but it takes time. This figure shows that it could well take the industry perhaps a decade to catch up.

Whether this is the best possible measure of complexity is certainly open to question, but this measure is representative and relatively easy to both compute and communicate. The authors applied this to what they considered to be the average size "grid" operating in the US in each decade. "Average size" has been estimated by examination of historical ownership records of selected utilities and total US demand. The authors looked at twelve utilities scattered across the US, each known to be the result of many mergers and system aggregations going back into the early 20th century (such as Niagara Mohawk, now the western operating division of National Grid USA, which was at one time many separate utilities).[4] The resulting trend gives an indication of the steady growth in operating difficulty that faced power grid operators.

The measure the authors have used is based on the number of elements in the grid, and tends to give a lower estimate of grid growth than measures based on

---

[4] See Chapter 3 for a discussion of mergers as far back to the early 20th century.

number of customers, amount of load, or geographic area would. This is because during most of the 20th century time the size of electric equipment grew significantly. Generators in 1980 were five to ten times the size of generators in 1910. Therefore, a grid in 1950 might cover ten times the area and customer base of one in 1920, but contain only twice the number of elements.

But a measure based on *number* of components and goals, rather than customers or load, is most appropriate for assessing operating challenge and blackout possibilities due to complexity: a big generator is not much more difficult to control than a small one, but the ten small ones it would take to equal its output are much more difficult to control, as a group, than that one large unit.

Impact of the increase in the number of elements

Thus, whether individually large or small, it is the *number* of separate, interconnected elements that contributes most to complexity of operation. Three aspects of this count are worth comment with respect to today's operating challenges.

Smaller generators made something of a comeback in the last two decades of the 20th century. Until then the trend had been toward ever-larger generating units (up to 1,200 MW, such as either of the two units at the South Texas Nuclear Plant). But certain technological advances in gas turbines, the PURPA act and de-regulation, and low gas prices, as well as changes in generation costs and market structure, created business advantages for medium size units (150-300 MW) during the period 1985-2000. The number of generating units connected to some power grids grew quite fast in the late 1990s and early 21st century for this reason alone. This slightly exacerbated the already-accelerated trend of large-grid creation.

De-regulation took the very large, interconnected regional power pools that existed prior to de-regulation and institutionalized them as the basis for modern, de-regulated regional wholesale marketplaces. In many cases, boundaries of these grids were flexibility (utilities in the central US could choose from a number of grids, north, west, or south of them, to join). But overall, they became larger and the complexity involved increased, particularly since there were many interconnected one to the other. (The northeast blackout on August 14, 2003 involved five regional grids and nearly affected two more).

De-regulation took control of the generation that vertically integrated utilities had had from 1900 into the 1990s and made it slightly less absolute. Theoretically, ISOs (independent system operators) have sufficient control of the generation on the grid to be able to control it. But their control is more limited than was the control of power pool operations prior to de-regulation. Then, there was no doubt who was in charge ("My boss is your boss. I need you to run your generator like this, now!"). Under de-regulation, generation owners have a contractual obligation ("We have a contract. Please change your output now, even if it inconveniences you."). The barrier to good control here is less the actual situation than the learning required for the industry to adapt new rules and procedures that fully accommodate it.

Increase in the number of goals

De-regulation also increased the complexity of grid operations by increasing the number of goals for system operation (Table 16.1). Of course, *interconnected security* was and is and always will be the primary goal: if the grid can't be held together, no other goals can be sustained.

The next most important goal is service reliability: keep the lights on to all customers. The earliest power pool arrangements – regional grids in their infancy – were established for one reason: *operating reliability.* If one utility had a generator fail, it could borrow power from its neighbors until repairs could be made or another generator started and brought on line.

Beyond security and reliability, utilities also traditionally used power pools for improvement in *economy:* participants pooled their generators into one large set and ran them in a way that minimized fuel needs and wear and tear, settling who owed who off-line, at a later time. Under de-regulation this function changed to what can be called market accommodation – *business, rather than economic factors, drove the optimization of generation pattern.*

But to system operators, either way – regulated or de-regulated – the system must have enough generation on line and running to meet demand and allow margin for contingencies while serving the load and meeting security and reliability goals. Under regulation there had been a "dispatch merit list" of generators to be used as needed and the goal was to achieve "economy" in using them. Under de-regulation, there is a list of generators that has been developed of people who quoted the lowest prices and made deals with buyers.

Table 16.1 Goals for Interconnected System Operation

Early 20th Century	Late 20th Century	Early 21st Century
Security	Security	Security
Reliability	Reliability	Reliability
	Economy	Business "Economy"
		Open Access/Transparency

To these three goals, interconnected security, reliability, and "economy," de-regulation added a fourth: *open access*. The grid has to be operated in a way that provides equal opportunity for its use to many players, not just the owners of the various parts of the grid (as was the case with power pools). This is a goal mandated by de-regulation, basically because without it, de-regulated market operation will not work well.

Thus de-regulation increased the number of goals that system operators had to address by 33%. A more relevant guide to how burdensome this increase was might be to look at the number of interactions operators had to keep in mind. With three goals, there are three interactions (how security interacts with reliability; how reliability interacts with economy; how economy interacts with security). With four goals, there are *six* interaction pairs, twice as many to consider and balance as before de-regulation. In some sense, operation is probably twice as difficult. But in spite of this, the author's index uses only the number of goals, for a computed 33% rather than 100% increase.

A jump in complexity

Figure 16.4 shows that the cumulative effects of de-regulation created a surge in complexity, which by 2005 was roughly equal to more than ten years of complexity increase at the traditional rate of growth. As indicated by the dotted line, the traditional pace of technology progress in the industry would require perhaps a decade or more to catch up. Effort and focus can no doubt shorten the period it will take for its control capabilities to catch up. But it will take some time and, in the interim, system operators throughout North America, and elsewhere, will operate a bit more conservatively, and worry a lot more.

16.5 SUMMARY AND CONCLUDING COMMENTS

Modern power systems consist of hundreds of thousands of generators, transmission lines, transformers, and switching equipment, all interconnected electrically and all working in concert with one another on a milli-second to milli-second basis. Blackouts occur when stable interconnection of these myriad parts is disturbed in a way that causes electrical equipment in some part of the grid to be unable to keep up with the rest, or when that disturbance is sensed by automatic equipment as a potentially unsafe or damaging situation. Regardless, interconnection of a small part of the grid is broken, and for some reason the system operators are unable to stop that cascading into overloads of other equipment and a subsequent, and very rapid ripple effect as the protection systems of other major equipment, sensing something is not normal, all disconnect to avoid conditions that could be potentially damaging to them.

In a blackout, within a few seconds, a multi-state wide power grid can go from hundreds of thousands of units all connected together, to being thousands of separate elements, electrically isolated from one another, and all "shut down."

Not only does power cease to flow to hundreds of thousands, millions, or perhaps even tens of millions of homes and businesses, but it takes hours to "put the system back together again," by starting and switching equipment together in a tedious interconnection by interconnection process. Table 16.2 is a list of characteristics of a blackout.

The only way to avoid blackouts is to operate a large power grid so that it always has interconnected security – a margin of capability so that it can such tolerate any sudden failure or unexpected event without serious overloads or undervoltages. In a very real sense, blackouts start not when an unexpected event happens and ripples through the system, but when the power system first begins operating in a mode where a blackout could occur if something unexpected did happen. For this reason, system operators never knowingly permit a power system to stray out of secure, stable operation.

"Knowingly" is the issue. Power grids have become so complex due to the regional size needed to achieve good "market efficiency" under de-regulation (see Chapters 2, 11, and 12) that they come close to exceeding the power industry's ability to understand them, anticipate insecure conditions, and control them so that blackouts do not occur. This is certainly not an industry "crisis." A few more blackouts may occur, and will be frustrating if they happen, but will not destroy the industry. Much more likely is that system operators throughout the industry will adopt more conservative operating rules, providing a wide margin for "what they don't know," and this will provide the needed margin of security for power systems. If not, the margin will widen with the next blackout.

Table 16.2 Characteristics of a Power System Blackout

- Service is interrupted to many electric customers over an extensive area.
- Most of large elements of the power system are "de-energized" (disconnected and without power).
- Little or no equipment is damaged.
- The blackout occurred almost instantly as automatic equipment took over to break interconnections because it sensed operating conditions that could lead to major equipment or customer appliance damage.
- Some area of the power system, but usually only a very small part of the area blacked out, was under tremendous stress – overloads or low voltages – just prior to the blackout.
- Operators will spend a good deal of time – perhaps as much as a day – restarting and re-interconnecting the grid.

What Can Be Done to Prevent Blackouts?

The power industry in general, and several pro-active regional grid operators in particular, certainly recognized early in the de-regulation process that increases in complexity would require improvement in analysis and control technologies. Sections 16.2 and 16.3 summarized several measures that must be taken to assure that a blackout does not happen. Looking at these from the standpoint of modern needs reveals pluses and minuses for the industry.

1. *Engineers and utility equipment owners must make certain that the grid has enough capacity to survive the sudden, unexpected loss of any line, generator, transformer, or switching point.* This is a very firm rule in the mind of all power system engineers and operators, but modern systems have "pushed" more than might be expected. Since de-regulation, few new transmission lines have been built – the cause of this hiatus on construction is political, not technical. Yet demand has increased. The result: grids are operated "closer to the limit."

2. *Engineers must assure that the power grid simultaneously satisfies hundreds of criteria* for electrical stability and power flow requirements, and so it can stay within these parameters even if a large unit fails suddenly and power flow patterns shift instantly. Grid complexity increases, and the need for new monitoring equipment and control center computers has increased. For the most part, the industry has just barely been able to keep up with this demand.

3. *Engineers must make certain that control and monitoring systems can distinguish correctly between situations where they really should take action and remove equipment from service, and those situations where action is not essential and in fact might be harmful.* This is where new technology, specifically WAMS among other solutions, must be deployed. The industry is working on this, but it takes time to deploy so much new equipment, so much of it very advanced and requiring new technical skills that are in limited availability.

4. *Grid operators must control the system to make certain it always operates so that items 1 through 3 are all satisfied at any moment.* Operators want to do this, but the cumulative effect of the challenges outlined above makes this difficult during peak periods.

5. *Engineers and scientists in the power industry must make certain that the concepts and tools they apply to assure 1-4 above are always met are complete and sufficient to assure dependable operation.* This is the current technical challenge, and a source of considerable urgency in the minds of many experts in the industry.

All five of these requirements are more difficult to accomplish today than they were in the past. But what is clear at the time of this writing is that the industry is challenged by the size, span, and complexity of modern power grids and the needs of de-regulation. Until analytical models, security assessment routines, and operating control center capabilities improve, the industry cannot fully realize its ambitions for de-regulation and growth.

And the challenge will not end with the current level of grid size and complexity: grids will continue to grow in the future. Already there is talk of connecting power grids throughout the US into a grid spanning the continent. South American utilities have long-range plans to interconnect hydro generation in the Andes to the huge demand centers along the Atlantic coast, such as Rio de Janeiro and Buenos Aires. Both of these long-range concepts would require grids that would span the distance that electricity travels in $1/60^{th}$ of a second – a full cycle at 60 Hz – raising several new and interesting challenges and, in addition, creating some unusual technical opportunities (lines where electricity takes exactly $1/60^{th}$ of a second to traverse them can be set up so they do not need transformers at either end). In addition, there is a recognition that even larger, multi-continent grids could bring about significant improvements in environmental and economic impacts of power usage. Therefore, the century-long trend of gradually escalating power grid size and complexity shown in Figure 16.4 is almost certain to continue, along with a continuation in the growth of technical and operational challenges for the electric power industry.

FOR FURTHER READING

P. Fairley, "Unruly Power Grid," *IEEE Spectrum,* pp. 22–27, August 2004.

S.H. Horowitz and A.G. Phadke, "Boosting Immunity to Blackouts," *Power and Energy Magazine,* September/October 2003.

V. Madani and D. Novosel, "Taming the Power Grid," *IEEE Spectrum,* August 2005.

NERC Recommendations to August 14, 2003 Blackout — Prevent and Mitigate the Impacts of Future Cascading Blackouts; www.NERC.com

D. Novosel, M. Begovic, V. Madani, "Shedding Light on Blackouts," *Power and Energy Magazine,* January/February 2004.

M. Shahidepour and M. Alomoush, *Restructured Electrical Power Systems – Operation, Trading, and Volatility,* Marcel Dekker, New York, 2001.

Western Systems Coordinating Council Disturbance Summary Reports for Power System Outages Occurred in December 1994, July 1996, and August 1996, respectively; www.WECC.biz.

Glossary

Administration. This word has many meanings outside the power industry, as for example with respect to the federal government's executive branch (the "Administration"), or the main office building at a university ("go over to administration."). However, within the power industry it refers to a type of federal or state authorized operation of a power generation and transmission system in a region. See *Power Administrations, Power Agencies, and Power Authorities* for more specific details.

Agency. See *Power Administrations, Power Agencies, and Power Authorities.*

Alternating current (AC). The type of electric power provided by all electric utilities worldwide, both current and voltage oscillates either 60 or 50 times per second, respectively, in American or European type power systems.

AM/FM systems. Automated mapping and facilities management systems are computer systems that maintain map drawings and equipment records in a computer database, using a coordinated relationship between maps and equipment files. See also *Geographic Information System.*

American Public Power Association (APPA). An electric industry trade organization representing the municipal and public power district electric utilities in the United States. It sponsors technical and managerial information exchange meetings among its members, and acts as a lobbying and public information organization on behalf of its membership.

Amp. See *Current.*

AMR. See Automated Meter Reading

Ancillary services. Those sold in addition to a primary service. With respect to the primary service of *wheeling* power, ancillary services refer to various aspects of wheeling *support*, such as voltage regulation, reserves, etc., which must be unbundled , i.e., offered for sale as separately priced options in addition to the basic service of moving power.

Asset management. An organizational and priority structure that orients an electric utility's managerial and objectives efforts around the use of its physical equipment (assets) and their "lifetime optimization," viewing them and all decisions about spending on a common business basis, seeking to minimize the risk of failing to achieve one's goals.

Authority. See *Power Administrations, Power Agencies, and Power Authorities.*

Automated Meter Reading (AMR). A computerized system the utility uses to read energy meters remotely. There are two-way systems in which the central utility billing computers "talk" to meters at homes and businesses, and "drive by" systems in which meters can only "talk" to a computerized van as it drives by once a month.

Automation. Equipment and systems which can be monitored and controlled remotely, and in a coordinated manner. Automation generally requires that equipment have variable features which can be altered (switches or control status that can be changed); remote measurement of key variables such as voltage; centralized analysis and remote control capability. See: *Distribution Automation, Energy Management Systems,* and *SCADA.*

Autotransformer. A type of power transformer that has only one winding which magnetic forces in the transformer core force to act like the two in a "normal" transformer. They have operating and cost advantages that make them preferable in some types of power applications, and are widely used.

Blackout. A blackout is the sudden loss of power supply service to a widespread region, affecting much more than one major equipment unit of the system, that occurs due to loss of interconnected security and stability at the transmission-level.

Brush, Charles. An early inventor/engineer/proponent of electric power, and a rival of Edison and Westinghouse. He founded a number of electric utilities, including the Cleveland Electric Illuminating Company, still doing business as an operating company of First Energy.

Bundling. Bundling refers to grouping a number of services or products into one package at one price. See *Ancillary services* and *unbundling*.

Cascading Outage. A cascading outage is a series of events in which the failure or de-energization of one unit of equipment leads immediately to the failure or de-energization of another, and so forth, until a large set of equipment is out of service. The most typical cause of a blackout.

CBM. See Condition-Based Maintenance.

Co-generation plant. Originally, a facility that produced both electricity from a steam turbine generator, and steam/hot water for industrial uses from a single boiler system. Now, often used to designate only the portion of privately owned power that is sold to the local electric utility.

Co-generator. A non-utility company or individual who owns a generator and sells power to the local regulated electric utility.

Competitive energy service companies. Retail sellers in the competitive electric marketplace. In many de-regulated electric industry structures, there are half a dozen or more companies vying to sell electric power and services to consumers. Many of these companies offer other forms of energy, e.g., natural gas, propane, fuel oil, and such energy services as weather sealing and other conservation measures. See *Retail Energy Services Company*.

Competitive power generators. Wholesale sellers in the competitive electric marketplace. Usually, in any de-regulated electric industry structures, there are many (dozens) of companies vying for the sale of electric power. See *Generating company*.

Condition-Based Maintenance. A method of allocating money and resources to equipment maintenance based on evaluating the condition of various equipments and giving the highest priority to equipment with the worst condition ratings.

Contract for differences. A financial agreement between two or more users of a power transmission grid that promises to compensate one or more or them if, due to the system operator's decisions, his or her electrical costs rise above those he originally expected when ordering power or transmission service from the system operator, as the case may be.

Criteria. A criterion is a factor or aspect desired in the solution of a problem, for which a value at desired minimum, maximum, or with a certain range will be sufficient, and for which no additional benefit derives beyond that being met. An example is voltage level, which only has to be maintained with range on a power system (e.g., 113-126 volts at the service level). This contrasts to an *attribute*, which is a factor one wants as much of as possible (e.g., cost reduction).

Current. The actual electrical flow, measured in amps, an arbitrary quantity of measurement. See *Voltage* and *Power.*

Delta, as in a delta circuit or delta configuration. A basic type of electric line configuration in which there are only three phase conductors with no "neutral" or ground wire (a delta circuit might have a total four or five wires on the towers or poles, two additional one or two being *shield wires,* but these are not part of the circuit. They are considered instead part of the structures.). Delta configuration is the predominant approach to transmission line design, but not as widely used in distribution. See also *Wye* circuit or configuration.

Demand. As used in the power industry refers to the demand for electric power. The standardized meaning refers to the amount of power needed in a specific interval of time, usually a quarter hour or an hour unless otherwise specified. See *Load.*

Demand-Side Management (DSM). The traditional, regulated power industry terminology for effective management of electric energy usage on the consumer side of the electric meter: balancing conservation, building heat-loss efficiency, appliance and equipment efficiency, environmental concerns, and load control and automation against utility resources, in an effort to minimize cost and/or societal impact. Often called demand response in a de-regulated industry.

DG. See Distributed Generation.

DR. See Distributed Resources.

DS. See Distributed Storage.

Direct current. A type of electric flow in which current and voltage are constant and do not oscillate. See *Alternating current.*

Disconnect. (1) as a verb, it means to terminate power delivery by opening a switch or otherwise opening a circuit, (2) as a noun, it can mean a customer- or work-order (authorization paperwork) to disconnect a customer from the system, either because they requested termination of service (they have moved, etc.), or for non-payment, (3) as a noun, it can also mean a robust if simple switch/fuse device used on medium voltage ($\approx$15kV) overhead distribution, so called because it can be used to disconnect a circuit or equipment when needed for repairs, etc.

Distributed generation (DG). The use of small (5 kW – 5,000 kW) generators located at or close to customers and interconnected to the distribution system to provide some or all of the power for a utility system.

Distributed resources (DR). Distributed resources include distributed generation, distributed storage, and various DSM measures such as load control and RTP.

Distributed storage (DS). Distributed storage means the use of small (5 kWh-25,000 kWh) energy storage devices, essentially large UPS units, located at or close to customers and interconnected to the distribution system to provide both peaking capability and reliability ride-through during power supply outages, thereby improving reliability.

Distribution. (1) The power system function of delivering electric power to the end consumers (homes and businesses), (2) that portion of the electric power system that performs distribution: normally poles, wires, transformers, and control equipment operating at "primary voltage" (between 34.5 kV and 2 kV) and at "utilization voltage" (230-250 volts in Europe, 110-125 volts in the US, 100-110 volts in Japan).

Distribution automation. Includes a wide variety of remote monitoring and control capabilities installed on the distribution system, including all or any of: remote switching, automated volt-VAR (capacitor and regulator) control, automated meter reading, power quality monitoring, control of distribution generation and storage.

Distribution company (Disco). A regulated company that in a de-regulated power industry has a monopoly franchise on the local power distribution system. It owns and maintains the lines that route power to consumers in its service territory, and charges a regulated rate for its use.

Distribution Management System (DMS). A DMS is a computerized control system for distribution operations functions. It is somewhat less specific as to application than terms such as OMS or D-SCADA, but usually includes all those functions plus field force tracking and work order or activity management.

D-SCADA (Distribution Supervisory Control and Data Acquisition). SCADA is a rather old term for power transmission control system. D-SCADA refers to control systems, usually based around equipment and communications systems installed at substations, which monitors and controls the distribution feeder system.

Easement. Land, usually public land along roads or railroad tracks, which has been dedicated for use by public utilities and on which the electric utility can build power delivery lines. See also *Right of Way.*

Earth return circuit. A power line that uses the earth as the return conductor. An earth return circuit has only one conductor (a single wire) and is therefore quite inexpensive compared to any other circuit option. Soil and rock do not conduct electricity well, but there is a lot of it (a whole planet full) so an earth-term circuit can be made to work – not well, but good enough for some needs.

Edison Electric Institute (EEI). An electric industry trade organization representing the investor-owned electric utilities in the United States. It sponsors technical and managerial information exchange meetings among its members, and acts as a lobbying and public information organization on behalf of its membership. Originally called the National Electric Light Association, it was established in 1898, by Samuel Insull, Thomas Edison's chief assistant.

Edison, Thomas. Perhaps the most famous early inventor-engineer associated with electric power. He invented the light bulb, many other electrical gadgets, e.g., the phonograph, and many types of electric equipment, most of which were superseded by designs fostered by *Westinghouse*.

Electric load. The demand for electric power, by consumers, is seen by the power system as a "load' or "burden" that must to be electrically supported by its equipment. See *Load.*

Electric Membership Co-operatives (EMCs). Electric utilities owned by their customers. When the electric industry was just forming, local farmers and businesses in many rural areas of the United States pooled their resources to build a jointly-owned rural electric system, obtaining financial support from the US government.

Electric Power Research Institute (EPRI). A non-profit research and development organization formed by electric utilities in the United States in the 1970s. Since electric utilities at that time did not compete against one another, they thought it best to combine their R&D funds in a cooperative pool, sharing their experience.

Electric Power Supply Association (EPSA). An electric industry trade organization representing competitive power suppliers active in the US and global power markets. It sponsors technical and managerial information exchange meetings among its members, and acts as a lobbying and public information organization on behalf of its membership. It was founded in December, 1996, as a result of the merger between the National Independent Energy Producers and the Electric Generation Association.

ELSCI (Enterprise Level System Control Integration) or sometimes ELSSI (Enterprise Level Substation Systems Integration). Combination of coordinated substation automation and control systems that communicate with a central computer system and a data archive system for historical and equipment information gathered from that system, used to analyze operation, maintenance, and management of substation and feeder equipment.

Energy Management System (EMS). The computer control system used by a large electric utility, or Genco, to operate and coordinate all the operations of many power generation units located at widely spaced power plant sites.

Energy Services Companies. See *Retail Energy Services Company.*

Enterprise or enterprise-wide or level. The enterprise is the entire organization or business. An enterprise-wide system extends over the whole business. Thus, an enterprise-wide data system would be a database covering all data used by the utility and/or available to all functions in the utility. See *ELSSI* for example.

Esco. See *Retail Energy Services Company.*

Exempt Wholesale Generator (EWG). An independent power producer that generates power and sells it on the wholesale market, and is exempt from restrictions normally imposed on electricity providers by the Public Utility Holding Company Act (PUHCA). This class of company was created by the Energy Policy Act (EPAct) of 1992 to expand competition in wholesale electrical generation.

Fault. A short circuit, a situation when electric flow is diverted from the path intended by engineers and operators to anywhere where its flow is not wanted, is called a fault.

Federal Energy Regulatory Commission (FERC). An independent agency of the Department of Energy, composed of five commissioners appointed by the President of the United States, and confirmed by the Senate, who serve staggered five-year terms. The commissioners have a large staff of attorneys, economists, engineers, and policy analysts.

FERC Orders 888 and 889. These rulings by the U.S. Federal Energy Regulatory Commission ordered open transmission access at the wholesale level, thus creating a competitive generation marketplace (order 888), and mandated how communication of transmission capacity availability and reservation of service would be communicated to users of the system (order 889).

Fuel cell. A device that converts natural gas, methane, or other combustible fuels directly to electricity through a electric-chemical process, without moving parts and rather quietly.

Gas Insulated Switchgear (GIS). High-voltage substation equipment which is placed inside pressurized containers filled with a special type of gas, which acts as a very strong insulator. Inside the gas, electrical separation distances (as between equipment of opposite voltage polarity) that require several feet of separation in open air can be reduced to inches, and as a result the substation can be reduced in size, to fit in tight spaces, such as in building basements or in special vaults (underground rooms) under the street.

Genco. See *Generation companies.*

Generation. Power production that involves actually manufacturing the electric power. This is done with electric generators (see Chapters 6 and 7).

Generation and Transmission (G&T) utilities. Owners of generation and transmission facilities that sell power on the wholesale market, but do not own distribution or sell power to consumers. Under the traditional regulated industry structure, several dozen EMCs or municipal utilities in a region would pool their resources to jointly own a G&T that provided them with power.

Generation companies (Gencos). In the de-regulated power industry, they are electric power manufacturers. They own generation units and produce electric power, which they sell "at their site," in the same manner that a coal mine might sell coal in bulk at its railhead. In each case, some transportation mechanism (transmission line, railroad) exists to move the commodity to the point of consumption, but the business concept is production and sale *at the site.* Electric generating companies produce power, fed into an electric power system *owned by someone else,* then moved to the point of consumption over the electric lines.

Generator. A machine for producing electric power by transforming a fossil fuel, nuclear fuel, or solar, wind, geothermal or hydro energy into electric power flow. Also, increasingly used for a company in the business of owning generators and producing and selling power.

Geographic Information System (GIS). A "smart" computerized mapping system, used by utilities to track all of their disparate distributed equipment, customer locations, etc. It is similar to an AM/FM system, differing only in technical details of implementation.

GIS is an acronym with three different meanings in the power industry. See *Gas Insulated Switchgear, Geographic Information System,* or *Government Instituted Structure.*

Global Energy Network International (GENI). A non-profit organization based in San Diego, California, dedicated to promoting Buckminster Fuller's vision of a global energy grid, to connect renewable resources (predominantly hydropower in remote mountainous areas such as the Andes, Kamchacta, and Siberia) to the population centers of the world, thereby reducing mankind's dependence on fossil fueled generation. Although Fuller's vision of a vast global energy grid may not be attainable in our lifetimes, GENI promotes responsible energy use and wise management of the earth's energy resources on an international scale.

Government Instituted Structure (GIS). Government institutionalized structure, refers to a power industry structure mandated by government regulation, even if the industry itself is partially or completely "de-regulated."

Ground. The common, zero-voltage location or reference for all equipment and reference in the power system. So called because the earth (ground) is supposed to be at this level of voltage (i.e., at zero voltage).

Heat rate. A "fuel economy" rating for generating plants, the number of BTUs (British Thermal Units, a measure of heat energy) required to produce one kilowatt hour. The lower the value, the more efficient the generator. The average generator in use worldwide probably has a heat rating of about 11,200 BTU/kWhr; an acceptable rating for a good new generator is anything below 10,000, and the best that can be done using proven technology is about 9,000. Like automobile fuel ratings ("My car is rated at 25 mpg but with all my city driving I only get 22") there is often a difference between the rating and actual economy ("Our units are rated at 10,200 BTU/hour, but cycling their output up and down on a daily basis to follow peak and minimum demand periods worsens that, so we average 10,700.").

Hydroelectric power. Electric power produced by using falling water, usually from the upstream side of a dam, to turn an electric generator. Hydro-power plants cost slightly more to build and operate than other types of power plants, but they require no fuel, and they are relatively (but not completely) environmentally benign.

IED. See *Intelligent Electronic Device.*

IEEE. See *Institute of Electrical and Electronics Engineers.*

Impedance. The quality of any material to oppose the flow of electric current through it. Impednance is a complex variable quality consisting of two parts, *resistance* and *reactance.*

Independent Power Producers (IPPs) or **Non-Utility Generators (NUGs).** IPPs and NUGs are *private* companies owning generators and producing electric power. The Public Utility Regulatory Practices Act (PURPA), of 1978, required by law that electric utilities buy power from an independent power producer (IPP), if it is willing to sell power for less than the utility can produce it itself. Under de-regulation, these two become Generating Companies (Gencos).

Independent System Operator (ISO). Although called an operator, the ISO, like the air traffic controller, is actually an entire organization and infrastructure. An ISO has operational control of the transmission system over a wide region, e.g., California. The ISO runs the interconnected power system, provides open access to transmission facilities, tracks usage, settles bills, and disburses money.

Intelligent Electronic Device (IED). Any monitoring or control equipment installed on the power system that has both: 1) an internal processor of some sort, and 2) two-way communications capability from and to a central control hub (whether at the utility or a local substation). Examples: an automated meter in a 2-way AMR system; a remote control switch on a feeder.

Institute of Electrical and Electronics Engineers (IEEE). The largest professional organization in the world, consisting of over 250,000 member electrical engineers. Its various committees help set standards and guidelines for efficient and safe use of electric power and power equipment.

Integrated Resource Planning (IRP). Involves the electric utility's including the utility, DSM, and conservation resources in its analysis of "what is best" when planning expansion to meet future energy needs.

Interconnected Security. The qualities of margin and stable operation in a power system that assures all generators and equipment will continue to work together in harmony.

Interruption. An interruption is the cessation of power delivery to a customer or area of the system. It is caused by an equipment *outage* or a system *blackout*. The event means that one or more customers had their power supply discontinued for a period of time due to a failure, mis-operation, or other problem in the power system or its control systems.

Investor-owned utilities (IOUs). In the traditional regulated industry structure, many monopoly franchises for electric service in the United States and many other nations were held by profit-motivated companies that raised money for their operations and investment by selling stock in their company, attracting stockholders with the promise of dividends on their investment. Although essentially the same types of business as General Motors, IBM, and Motorola, IOUs were considered lower yield, but safer investments.

Load. In power engineering, the demand for electric power. Consumers create a "load" on the power system, which it must meet.

Load aggregator. An organization or company that pools together many consumers into a large block of demand for power, of sufficient size to buy on the wholesale market.

Local distribution companies (LDCs). Local monopoly-franchise electric utilities that own *only* the local distribution system, and provide retail sales and services. Traditionally, LDCs have distributed and sold electric power to all the customers in their service territory, and acted as "the local electric utility." Examples of such companies are many municipal electric departments in smaller communities, which own no generation equipment, and many rural electric cooperative utilities.

Locationally Based Marginal Costing (LBMC). A method of transmission pricing in which the actual (technical) cost of power at every location in the power grid is computed using some mutually agreed upon method, and the price for power transmission between any two points in the grid is then defined as the difference in the computed local prices.

Lock out. The result of a recloser operating through its programmed cycle of several attempts to reconnect a circuit, only to fail because a fault or other serious problem was detected each time it tried. It "locks out" with the circuit denergized (lights out, power off for safety).

MAIFI. Momentary Average Interruption Frequency Index – the average amount of times in a period (usually a year) that a customer in the utility system experienced a sudden cessation of power availability that lasted only a few moments. (See MAIFI limit).

MAIFI limit. The amount of time that an interruption can occur before it is considered to have a duration that is counted into SAIDI statistics. Definitions vary from only a few seconds to up to five minutes depending on the utility. IEEE standard 1366 recommends 5 minutes.

Merchant generator. A company or investor group building a generator(s) purely to sell power into the de-regulated electric power market. See *Generating company*.

Merchant transmission company. A private company operating outside of normal regulated structure that owns and operates one or more "toll-transmission lines."

Meters. (1) Used when counting the number of delivery points in a utility system. The utility will report "We have 1,234,567 connected meters in the system." (2) The term "meter" is used for any device that measures power flow. Usually one is installed at every customer for billing purposes and many others located throughout the system to measure flows for monitoring purposes.

Micro gas turbine (MGT). A very small gas turbine generator unit, usually of less than one megawatt power output, and often only 25 kilowatt output.

Municipal utilities. Electric utilities owned and operated by a town, city, or metropolitan area government. About 800 communities own and operate their own electric utility system. The largest one in the United States is the city of Los Angeles.

National Rural Electrification Cooperative Association. An electric industry trade organization representing the rural electric membership cooperatives in the United States. NRECA was formed in 1940. It sponsors technical and managerial information exchange meetings among its members, and acts as a lobbying and public information organization on behalf of its membership.

National utilities. A number of nations own and operate their electric utility on a national basis, either as a governmental department or as a single, government-owned company, which, while legally separate, has a symbiotic relationship with the government and is closely constrained by government policy. An example is Electricité de France, which serves all of France.

Neutral conductor or neutral wire. The fourth wire in a wye-connected circuit. It carries the imbalance among the three phase conductors and is often grounded.

Non-Utility Generators (NUGs). See *Independent Power Producers.*

North American Electric Reliability Council (NERC). NERC is a non-profit organization formed in 1968 by the electric utility industry to ensure reliable, adequate power supply in North America. It is made up of ten regional councils comprised of electric utilities in the United States, Canada, and Mexico. NERC defines standards, rules, and recommended forms of cooperative interaction among utilities and electric users that foster system reliability.

Nuclear Regulatory Commission (NRC). The final arbiter of regulatory issues on all matters relating to nuclear power plant construction, licensing, and operation. Formerly called the Atomic Energy Commission (AEC), it was established in 1946 by the Atomic Energy Act to manage the military nuclear uses of atomic energy. Its mission expanded to a focus on *peaceful* uses in 1954. It was metamorphosed into the Nuclear Regulatory Commission (NRC) in 1975.

OMS. See *Outage Management System.* Also see *Trouble Call System* and *Distribution Management System.*

Open access. Non-discriminatory access to the use of facilities, in this case transmission and distribution systems, by all potential users.

Outage. The cessation of function of a unit of equipment in the power system is an outage. It may cause an *interruption* of service to a customer.

Outage Management System (OMS). A computerized system used by a distribution utility to track customer "lights out" complaints and, through pattern analysis and other computational means, determine where equipment is outage, and when and where to dispatch field forces to repair it.

Performance-based rate (PBR). A price or rate system in which an electric supplier or electric utility is paid based on performance, not cost. For example, a utility might be given a performance-based rate that meant it would earn its typical return on investment if it produced with its typical degree of reliability, e.g., perhaps 90 minutes average out of service per customer per year. It might earn an extra million dollars per minute for every minute it could reduce that, but have to pay a million out of profits for every minute over that if its performance fell.

Phase-to-phase. A term for measurement of a quantity (e.g., voltage) between two phases of a three-phase power system. This is the normal way voltages and other voltage related factors are measured and recorded in power systems. Phase-to-phase voltage is about 73% higher than phase-to-ground voltage, which is the way a homeowner would measure voltage in an appliance or home wiring system: 138,000 volts phase-to-phase is actually about 80,000 volts phase to ground (which is the voltage one would feel – for just an instant – if one were to grab hold of the "138,000 volt" power line). The distinction is made both for engineering purposes and because usually electric demands are connected across two phases, so the effective voltage they see from a 138,000 volt line is 138,000 volts, not 80,000 volts.

Poolco. An independent company or government-run organization that both operates the electric transmission grid and takes bids from competitive electric suppliers to provide power, dispatching generating plants by buying and selling wholesale power. A Poolco operator basically provides the functions of both an Independent System Operator (ISO) and a Power Exchange (PX).

Postage stamp pricing. A method of pricing power transmission services in which location and distance do not matter: as with mail, everyone pays the same unit price regardless of their location or the distance they wish to move power.

Power is the ability to do work. Electrical power is measured in *watts,* one watt being an amp of current pushed by a volt of voltage. Since a watt is a small amount of power, equal to 1/746 of a horsepower, electrical power is usually measured in terms of kilowatts (thousands of watts) or megawatts (millions) or gigawatts (billions).

Power Administrations, Power Agencies, and Power Authorities. In general, a *power administration* is a government agency that does not own generation, but sells or manages it when produced by other governmental resources. For example, the Bonneville Power Administration (BPA), a part of the U.S. Department of Energy, sells power produced by generating plants owned and operated in the northwestern US by the US Army Corps of Engineers and the US Department of the Interior. Similarly, the Southeastern Power Administration administers power produced by the U.S. Army Corps of Engineers, and operates nine hydropower plants in the Appalachians. A power administration often owns transmission and operation facilities, but not the generation itself.

By contrast, a *power authority* denotes a government-owned generation and transmission utility. For example, the Tennessee Valley Authority (TVA) owns and operates generation and transmission facilities throughout the Appalachian area of the United States. It is the largest among several power authorities in the United States.

Often, a group of municipal utilities will form a *power agency* or *generating district* of their own so that they can jointly own and operate generation plants to provide power to themselves. For example, the Florida Municipal Power Agency, in Orlando, is owned by 19 municipal utilities, and operates five generating plants whose power output is shared by its owners. Agencies are government-owned in the sense that they are a shared resource of several municipal governments.

Power broker. An individual or company that arranges power sales between other parties, but never actually takes legal "title" to the power. Power brokers do not have to register with the Federal Energy Regulatory Commission (FERC). Thus, anyone can be a power broker. See also *Power marketer*.

Power exchange (PX). An organization that operates a marketplace for wholesale power, allowing buyers and sellers to sell power anonymously at the prevailing market price, determined through some sort of auction process.

Power marketer. Technically, an individual or company that sells electric power at the wholesale level, either power generated by its own facilities or power purchased from others. Generally, the term is used to indicate those who buy and then resell power, since the term "Genco" covers those who generate the power they sell. Unlike a power broker, a power marketer takes possession of the power, buying from A and reselling to B. By contrast a power broker introduces A to B, for a fee.

Power pool. An arrangement between two or more interconnected electrical systems, planned and operated to supply power in the most reliable cost-effective way for their combined load requirements. Among the benefits are: improved reliability of power supply, shared reserve generation capability, diversity of resources, coordinated maintenance programs, and joint ownership of new power plants. At the end of the regulated vertical utility era (circa 1995) there are more than 100 power pools in the United States.

Primary voltage. The voltage at which the primary portion of the power distribution system – the feeder system – operates. The feeder system carries power from substations to the vicinity of all homes and businesses within a one to five mile radius of the substation, and usually operates at steady primary voltage of between 34.5 kV and 2 kV, as determined by the electric distribution utility.

Privatization. The process of a government selling state-owned companies (in this case, electric utilities) to private, investor-owned firms.

Public Utility District (PUD). Essentially a county-owned, rather than a city-owned utility, which otherwise resembles a municipal utility.

Public Utility Regulatory Policies Act (PURPA). One of the five parts of the National Energy Act, passed by the U.S. Congress in 1978. PURPA created a new class of power producers called Qualifying Facilities (QFs), and required utilities to buy power from them at their avoided cost.

Qualifying Facility (QF). A power producer or co-generator that qualifies under the Public Utility Regulatory Policies Act of 1978 (PURPA) to supply electric energy and generating capacity to regulated electric utilities.

Ramp up. To increase the output of a generator. The terms "ramp up" and "ramp down" refer to the gradual, linear increase or decrease of a generator's power output. Except for a few special types of electrical generators, most cannot be "floored" and suddenly run up to full output, as can automobile engines, for example. In fact, some nuclear units can take as long as two days to run up to their output from zero to full capability.

RCM. See *Reliability-Centered Maintenance.*

Reactance. The quality of a material or device to oppose a change in the current passing through it. The imaginary part of the impedance.

Real-Time Pricing (RTP). Through the use of automation the utility communicates the current cost of power to its customers (i.e., in real time). Thus at peak periods the customers know that power is costing much more than at other times and can decide if they wish to cut back on consumption, etc.

Recloser. A recloser is an electrical device consisting of a circuit breaker and control logic that will interrupt a circuit if trouble is detected, but attempt to "reclose" (reconnect to turn the power back on) a moment later, often programmable so that utility operators can set it to try several times over a period of up to several minutes before giving up (see *Lock Out*).

Regional Transmission Group (RTG). An organization of transmission owners, transmission users, and other interested parties, voluntarily created for the purpose of coordinating transmission, planning, expansion, operation, and use within a region. At one time, RTGs were expected to be a way for utilities in a region to deal with FERC-mandated open access requirements. Independent System Operators (ISOs) developed, instead.

Reliability-Centered Maintenance (RCM). A method of allocating money and resources to the maintenance of equipment. The more important a unit of equipment is to customer reliability, the higher the priority given to its maintenance.

Resco. See *Retail energy services company*.

Resistance. The quality of a material to resist the flow of electric current through it. The real component of impedance.

Re-structuring. The process of setting up a new framework for how a company or industry will work, including perhaps creating new organizations and agencies. Most governments re-structure their electric industry after de-regulating it. Forming ISOs and dis-aggregating vertically integrated utilities is definitely re-structuring.

Retail energy services company (Resco). Involves selling the power along with various energy-related services, to homeowners and business persons.

Retail wheeling. The service of moving power at the distribution level. Retail wheeling implies open distribution system access: The portion of the electric system leading to individual homes and businesses is open for any competitive retail energy service company to use.

Rhinoceros. A splendid, anachronistic horned herbivore indigenous to Africa and parts of Asia and greatly endangered by man's encroachment on its (now) limited remaining territory: relatively peaceful if very territorial, immensely strong, but surprisingly fast when necessary. The rhinoceros has absolutely nothing to do with electric power, but the authors have worked the word into each power-engineering book they have written, anyway. The electric industry can learn a lot from this animal: keep your head down and your eyes focused on what's in front of you. Mind your own business. Have a thick skin. Share your turf with peaceful neighbors like the gazelles and never overreact to minor nuisances. Fight only when provoked, but then make certain your foe does not survive to attack another day. A portion of the proceeds from this book will go to support rhinoceros preservation programs around the world.

Right-of-way (ROW). Land that a utility buys and owns in order to have the required route for a transmission or distribution line (see also easement).

RTP. See *Real-Time Pricing.*

SAIDI (System Average Interruption Duration Index). The average amount of time in a period (usually a year) that a customer was without power.

SAIFI (System Average Interruption Frequency Index). The average amount of times in a period (usually a year) that a customer experienced a sudden cessation of power availability that lasted more than a few moments. (See *MAIFI*).

SCADA. See *Supervisory control and data acquisition* and *D-SCADA.*

Sectionalizer. A type of automatic switch designed to be placed on medium voltage (e.g., 15 kV) lines, that opens (i.e., prevents any more power flow) when the power going through it is momentarily disconnected. It can be programmed to coordinate with a recloser in such a way that a programmed set of one recloser and one or more sectionalizers, located on a distribution feeder, will operate to isolate a fault (downed line) but automatically reconnect as many customers as can be served with the faulty section still out of service. This reduces the number of energy consumers who must be out of service when a line is down to the minimum possible.

Security. (1) a condition where a power grid is operating so that it will continue to do so even if any element of the system fails unexpected or a large shock in load occurs, (2) immunity from physical or cyber damage or interference in the power system and its control, from terrorists, etc. See also *Interconnected Security.*

Shield wire. A wire running along the top of a transmission or distribution line (from pole top to pole top, from tower top to tower top) used to protect the power line conductors from lightning strikes (essentially, a lightning rod wire running above the power line). The shield wire is actually part of the structures, not the circuit (i.e., it has no role in the delivery or flow of the power).

SMD. See *Standard Market Design*

Spatial Frequency Analysis (SFA). A method of power engineering analysis developed in the early 1980s, that simultaneously evaluates the need for equipment capacity and the required locational needs, permitting accurate analysis of power right-of-way and substation siting needs, environmental and esthetic restrictions, and cost, all in a quantitative, coordinated manner.

Standard. A standard is a documented internal guideline used by a utility's planning, engineering, design, operation, diagnosis, or repair groups. Some standards are in fact legal "standards" in the sense that they are required by law or regulation. Most, however, are merely sets of documented guidelines.

Standard Market Design (SMD). SMD is a uniform, i.e., standardized, way of structuring and managing the transmission grid/wholesale energy marketplace in the United States.

Star circuit or configuration. A term used in the UK and other Commonwealth countries for *wye* circuit configuration.

Supervisory Control and Data Acquisition (SCADA) systems. The computer and monitoring systems used by a utility, power pool, Poolco, or independent system operator (ISO) to control a regional power transmission grid.

Sustainable Point. The stable long-term age/failure rate/maintenance cost point toward which a set of equipment put into service and repaired or replaced as it fails will tend over time.

Tesla. Arguably the smartest of the late 19th century electric experimenters and inventors, this Italian immigrant to the US, Nicola Tesla was a brilliant, clever innovator and meticulous experimenter.

Transformer. One of the basic building blocks of power systems, transformers raise or lower the voltage of electric power, basically changing its economy of scale with respect to transportation or usage. See also *Autotransformer*.

Transmission. Moving large amounts of electric power long distances, as bulk quantities. That portion of a power system that moves bulk and intermediate amounts of power over long distances, generally operating at voltages somewhere between 34.5 kV and 765kV (phase to phase).

Transmission companies (Transcos). De-regulated companies that own and operate transmission facilities and provide them to open-access users. Often integrated with a Disco into a "wires company."

Transmission system. The entire set of all interconnected transmission lines, transformers, and other equipment (circuit breakers, control computers) in a power system, used for the shipment of bulk power among many locations.

Transmission utilities. See *Transmission companies*.

Unbundling. To offer and price separately a number of items previously grouped into one and sold at one group price. See *Ancillary services.*

Uninterruptible Power Supply (UPS). Any of several types of devices that can supply power on a backup, or emergency, basis, so that even if power from the local power utility system is interrupted due to a storm or equipment failure, electric power is available. UPSs are available in small units (a cubic foot in size, perhaps) that can power only a single computer or health-monitoring system for perhaps an hour, up to house-sized units that can power an entire factory for a day or more.

Utilization voltage. The voltage at which power is used. Typically, this is a voltage sufficiently low that direct contact with open wires, while painful, can cause injury or death only under very extreme or unusual circumstances. Utilization voltage is 230-250 volts in most of Europe, 110 to 125 volts in the United States, and 100-110 volts in Japan.

Vertically integrated electric utility. The "traditional" electric utility that owned facilities and managed all the "vertical functions" from power production to customer billing in one operation (i.e., generation, transmission, distribution, and retail services) for producing, delivering, and selling electric power to the end-users.

Voltage. The electrical pressure that pushes electric current (and hence power) through wires, transformers, and other types of electrical equipment (and, for that matter, if they put themselves where they shouldn't be, people, too). The effect of voltage is usually proportional to its square – 240 volts can provide four times as much power as 120 volts, etc. Typical voltages are 120 and 240 volts in the home, 600 volts in industry, 12,000 volts for power distribution, and 138,000 to 765,000 volts for bulk power transmission.

Watt. A measure of electric power, equal to one amp of current at one volt. See *Power.*

Watt, James. An 18th century Scottish engineer who developed many of the concepts fundamental to practical steam power (which was used to drive nearly all early electric generators) and who was honored by the use of his name as the fundamental unit of electric power.

Westinghouse, George. An engineer-inventor-businessman and the commercial developer, but not the inventor, of the transformer and alternating current (AC) power systems. An early and vigorous proponent of alternating power, his AC power systems eventually triumphed in the marketplace over DC systems of his rival, Thomas Edison. A better engineer than businessman, he lost control of his Westinghouse Electric Company, which went on to become a giant in the electric equipment industry, only a few years after it was founded.

Wheeling. The service of transporting power from one location to another for a fee. The term was coined in the days of regulated electric utilities, when one utility would "wheel" power through another's power system, i.e., in one side of the system and out the other, and refers to an analogous railroad term for wheeling one company's cargo cars through another railroad's rail system without unloading them.

"Wire companies." Monopoly franchise "wire companies," private or governmental, are given responsibility to operate the sole transmission and distribution system in an area – the franchise territory – for the benefit of everyone who wants to move electricity. They charge a fee for use, negotiated with, and set by, local government regulation.

Wye circuit or configuration. A basic type of electric line configuration in which there are four conductors, one for each of the three phases, and a "neutral" or a ground wire (a wye circuit might also have an additional one or two wires which are *shield wires*). The predominant approach to distribution line design, but not as widely used in transmission. See also delta circuit or configuration.

Zonal pricing. Pricing methods for power or transmission services that apply different prices in each of several different areas. Often, a region is divided into several zones, each with different postage stamp rates.

Index

T - #0315 - 071024 - C516 - 229/152/23 - PB - 9780367392048 - Gloss Lamination